Diffusions and Waves

Mathematics and Its Applications

Managing Editor:

M. HAZEWINKEL

Centre for Mathematics and Computer Science, Amsterdam, The Netherlands

Volume 552

Diffusions and Waves

by

Henryk Gzyl

SPRINGER-SCIENCE+BUSINESS MEDIA, B.V.

A C.I.P. Catalogue record for this book is available from the Library of Congress.

ISBN 978-94-010-3949-9 ISBN 978-94-010-0293-6 (eBook)
DOI 10.1007/978-94-010-0293-6

Originally published by Kluwer Academic Publishers in 2002
Softcover reprint of the hardcover 1st edition 2002

CONTENTS

INTRODUCTION

Probability theory, and more specifically, probabilistic diffusion theory or functional integration has played a significant role in the theory of elliptic and parabolic differential equations. Take a look at Sections 1.1 and 2.1 to get some of the flavor of the field.

The basic link is provided by the probabilistic representation of the time-dependent Green functions (or propagators in the physicists' jargon) and the time-independent Green functions as path integrals. This connection between partial differential equations and diffusions does two things for us. On the more theoretical side, it provides us with elegant proofs of very many results, and, on the practical side, it provides us with explicit formulae which can be used (and have been used) as starting points for numerical work. As a matter of fact, the computations of Green functions from their probabilistic representation constitute outstanding examples of trivially parallelizable computations: at each CPU one averages over all trajectories issued from a given discretization point.

Actually, the material in this book was developed as a response to a challenge by some geophysicists friends of mine: How about putting all that mathematics to use to show us what a wave in a heterogeneous medium looks like!

So the first question was: How to associate a diffusion process with a wave in a heterogeneous medium? Or even simpler: How to associate a diffusion process with a wave propagating in a constant velocity medium? This problem seems to be open despite the fact that a negative answer provided by Duddley in [12–13] of Chapter 8 exists. If the wave equation in a three-dimensional space is Lorentz invariant, and there are no Lorentz invariant diffusions, we seem to be in a dead-end street. On the other hand, if the wave equation is factored into Dirac equations (or some other procedure of extracting square roots) and a way of associating processes with these equations exists, there may exist a way of going around Duddley's negative result. This is the spirit in which the reader should look at the results in Chapters 7 and 8: A quest to directly associate stochastic processes with wave equations.

Chapters 1, 2 and 3 are intended to make the volume self-contained, and there is a long list of commented definitions and the very basic stuff. Chapters 4 and 5 contain the bulk of the material on the best I can do so far in connecting waves to diffusions, first for the constant velocity case and then for scalar waves in heterogeneous media. The connection comes about by reduction to the elliptic case by means of Laplace transformation in the time variable.

The Laplace transform is used for two reasons: first because it is the natural way to do away with the time variable when dealing with problems with initial data in time. But this is not the important reason, for initial data can be made part of the source terms driving the wave. The important reason is that by taking Laplace transforms, instead of the traditional Fourier transforms used in scattering theory, the resulting elliptic problem is amenable to the direct and traditional probabilistic analysis. Perhaps our colleagues working with path integration in quantum mechanics could provide analogous results for the elliptic problems obtained by taking Fourier transforms.

As we shall see in Chapters 4, 5, 6, and 9, the probabilistic representation formulae will be important, not only in obtaining the solution (explicit or implicit) of initial and boundary value problems. It will be useful as well as a starting point for the solution of inverse problems.

This, as will be clearly seen below, is due to the fact that in the path integral or probabilistic, representations of the propagators and static Green functions, the velocity of propagation appears in a very explicit way.

The results collected in this volume are a first shot at very interesting inverse problems related to wave propagation. Many interesting open problems exist and my hope is that this volume may be of help and motivate others to look at them (from the present perspective, of course).

CHAPTER

ONE

BASIC PROBABILISTIC NOTIONS

This chapter is devoted to recalling the basic concepts used throughout. It really is a long list of commented definitions. It can be skipped without loss of continuity by those familiar with its contents.

1.1. Probability spaces

There are two basic sets in this game: the set of "elementary" events and the space of states accessible to observation. As far as structure goes, these have a lot in common: these are measurable spaces. A measurable space (X, Σ) is a pair in which X is a set and $\Sigma \subset P(X)$ is a collection of subsets which is a σ-algebra.

DEFINITION 1.1. A σ-algebra Σ of subsets of X is any collection of subsets of Ω such that

(i) $X \in \Sigma$.
(ii) Σ is closed under complementation.
(iii) Σ is closed under countable unions.

This is chosen because the algebra of set operations is coincidental with the algebra of logical connectives, and we are going to ask questions about measurements naturally associated with sets.

Comment: When X has some topology, the minimal σ-algebra containing all open sets is called the Borel σ-algebra on X and denoted by $\mathcal{B}(X)$.

DEFINITION 1.2. Let (X_1, Σ_1) and (X_2, Σ_2) be two measurable spaces and let $f : X_1 \to X_2$ be a function. It is said to be measurable whenever

$$f^{-1}(B) = \{f \in B\} \in \Sigma_1 \quad \forall B \in \Sigma_2.$$

Comments:

(i) One usually abbreviates by writing $f \in \Sigma_1 / \Sigma_2$.

(ii) When (X_2, Σ_2) is understood or fixed, one writes $f \in \Sigma_1$. Usually, $X_2 = \mathbb{R}$ and $\Sigma_2 = \mathcal{B}(\mathbb{R})$.

(iii) The notation $f \in b\Sigma_1$, $f \in \Sigma_{1+}$ is used to mean f is a bounded (resp., positive) Σ_1/Σ_2-measurable function.

The next proposition is useful in many instances, specially when computing conditional expectations. We do not state it in full generality.

PROPOSITION 1.1. *Let $\{f_i : i = 1, \ldots, n\}$ be a collection of measurable functions $f_i : (X, \Sigma) \to (E_i, \mathcal{E}_i)$. Let $f_i^{-1}(\mathcal{E}_i)$ be the sub-σ-algebras of Σ determined by f_i, and $\mathcal{G}$ be the smallest sub-σ-algebra containing all of them. Next, if*

$$H : (X, \mathcal{G}) \to (E, \mathcal{E})$$

is a measurable function, then there exists a measurable

$$\Phi : (E_1 \times \cdots \times E_n, \mathcal{E}_1 \otimes \cdots \otimes \mathcal{E}_n) \to (E, \mathcal{E})$$

such that

$$H = \phi(f_1, \ldots, f_n).$$

Comment: $E_1 \times \cdots \times E_n$ denotes the Cartesian product and $\mathcal{E}_1 \otimes \cdots \otimes \mathcal{E}_n$ the σ-algebra generated by products of sets in $\mathcal{E}_1, \ldots, \mathcal{E}_n$ (what else?).

Measures and probabilities describe our intuitive idea of volume.

DEFINITION 1.3. (a) A measure m on (X, Σ) is a mapping $m : \Sigma \to [0, \infty)$ such that

(i) $m(\phi) = 0$,

(ii) $m(U A_i) = \Sigma m(A_i)$ for any countable, disjoint, collection $\{A_i : i = 1, 2, \ldots\}$ of elements $A_i \in \Sigma$.

(b) A probability P on (X, Σ) is a measure on (X, Σ) such that $P(X) = 1$.

(c) A kernel on (X_1, Σ_1), (X_2, Σ_2) is a mapping $K(x, B) : X_1 \times \Sigma_2 \to [0, \infty)$ such that

(i) $\forall B \in \Sigma_2$, $x \to K(x, B)$ is X_1-measurable.

(ii) $\forall x \in X_1$, $B \to K(x, B)$ is a measure on Σ_2.

The integration with respect to a measure m is defined stepwise: first, on the class of simple functions, i.e. $f = \sum a_i I_{A_i}$ where finitely many of the a's A_i are nonzero (and $I_A(x) = 1$ or depending on whether $x \in A$, or not), then, on functions f which are increasing limits of increasing sequences f_n of simple functions, and then for arbitrary functions $f = f_1 - f_2$ written as differences of positive functions:

$$\int f \,\mathrm{d}m = \sum a_i m(A),$$
$$\int f \,\mathrm{d}m = \lim_n \int f_n \,\mathrm{d}m = \lim \sum a_i^{(n)} m(A_i^{(n)}),$$
$$\int f \,\mathrm{d}m = \int f_1 \,\mathrm{d}m - \int f_2 \,\mathrm{d}m$$

with the proviso that at least one of the two last integrals is finite.

We are not proving it, but countable arithmetic or lattice operations with measurable functions yield measurable functions (whenever defined, of course). For example,

$$\liminf f_n(x) := \underline{\lim} f_n(x) := \sup_n \inf_{k \geqslant n} f_k(x),$$
$$\limsup f_n(x) := \overline{\lim} f_n(x) = \inf_n \sup_{k \geqslant n} f_k(x)$$

are measurable functions. By the way, when these numbers are identical, they equal $\lim f_n(x)$, of course.

Also, composition of measurable functions yields measurable functions.

The operation of integration is linear (as a function of the f's) and it commutes with some limiting operations.

We also have results like

PROPOSITION 1.2. (i) *Let $N(x, B)$ be a kernel on $X_1 \times \Sigma_2$ such that $N(x, X_2) \in b\Sigma_1$. Let $f \in \Sigma_{2+}$ or $f \in b\Sigma_2$, then $g(x) = \int f(y) N(x, \mathrm{d}y) \in \Sigma_1$.*

(ii) *Let m be a finite measure on (X_1, Σ_1). Under the same assumptions as above,*

$$n(B) = \int m(\mathrm{d}x) N(x, B)$$

is a measure on (X_2, Σ_2).

Comments: More standard notation: $f \in L_p(X, \Sigma, m)$ or $f \in L_p(m)$ when other arguments are clear from the context means $\int |f|^p \,\mathrm{d}m < \infty$. Similarly $f \in bL_p$ (resp., $f \in L_{p+}$) denote bounded (resp., positive) functions with m-integrable p-th power.

Similarly, when there's a topology on X, $f \in C(X)$ (resp., $C_0(X), C_c(X), bC(X)$) means f is continuous (resp., continuous and vanishing at infinity, with compact support, or bounded continuous).

In setting up a probability space $(\Omega, \mathcal{F}, P)$, where the points ω of Ω are to be thought of as experiments or elementary events, the σ-algebra $\mathcal{F}$ contains the allowed questions we can ask about experiments, the probability P you interpret as you please.

The results of a measurement in an experiment can be a symbol (head or tails, the label in the face of a die) or a point in a manifold. To describe them we introduce.

DEFINITION 1.4. An $(E, \mathcal{E})$-valued random variable X is just an $\mathcal{F}/\mathcal{E}$-measurable function $X : \Omega \to E$.

Thus $\{X \in A\} = X^{-1}(A)$ contains all experiments ω which result in A and $P(X \in A)$ describes how likely a measurement of X is to fall in A.

Comments: Usually E is some metric space and $\mathcal{E}$ denotes the Borel subsets if E. When $(E, \mathcal{E})$ stays fixed we just say X is $\mathcal{F}$-measurable. When both $\mathcal{F}$ and $\mathcal{E}$ stay fixed in an argument, we say X is measurable.

Usually, when $E = \mathbb{R}, \mathcal{E} = B(\mathbb{R})$ and $X \in \mathcal{F}/B(\mathbb{R})$, we shall only say X is a random variable without further ado.

DEFINITION 1.5. The expected value (or mean value) of a random variable is defined to be

$$EX = \int X \, \mathrm{d}P = \int_\Omega X(\omega) \, \mathrm{d}P(\omega)$$

whenever $X \in L_1(\Omega, P)$.

Thus, taking expectations coincide with computing integrals and have the properties described above.

The verification of these is in any book on measure theory. Let us close this section with an important inequality.

PROPOSITION 1.3 (Jensen's inequality). *Let X be a random variable with finite expectation and $\Phi : \mathbb{R} \to \mathbb{R}$ be a convex function. Then*

$$\Phi(EX) \leqslant E\Phi(X).$$

1.2. Conditional expectations

The notions of conditional expectation and that of independence are what separates probability theory from measure theory.

DEFINITION 1.6. Let $(\Omega, \mathcal{F}, P)$ be a probability space and $\mathcal{G}$ be a sub-σ-algebra of $\mathcal{F}$. Let X be an integrable random variable. The conditional expectation of X given $\mathcal{G}$, denoted by $E[X|\mathcal{G}]$, is a random variable such that

(i) $E[X|\mathcal{G}]$ is $\mathcal{G}$-measurable.

(ii) $\forall H \in b\mathcal{G},\ E[XH] = E[H\,E[X|\mathcal{G}]]$.

Since conditional expectations are expectations, they must be linear, continuous functionals as well. These and additional properties are contained in

PROPOSITION 1.4. *All random variables involved are assumed integrable. All σ-algebras involved are sub-σ-algebras of $\mathcal{F}$.*

(a) *If* $X \geqslant 0$, *then* $E[X|\mathcal{G}] \geqslant 0$.

(b) $E[aX_1 + bX_2|\mathcal{G}] = aE[X_1|\mathcal{G}] + bE[X_2|\mathcal{G}]$.

(c) *If* $Y \in b\mathcal{G}$, $E[XY|\mathcal{G}] = YE[X|\mathcal{G}]$.

(d) $E[E[X|\mathcal{G}]] = E[X]$.

(e) *If* $\mathcal{G}_1 \subset \mathcal{G}_2$, $E[E[X|\mathcal{G}_2]|\mathcal{G}_1] = E[X|\mathcal{G}_1]$.

(f) *If* X_n *is a monotone sequence or it is uniformly bounded by an integrable random variable, and* $X_n \to X$, *then*

$$E[\lim X_n|\mathcal{G}] = \lim E[X_n|\mathcal{G}].$$

(g) Φ *is a convex function defined on the range of* X, *then*

$$\Phi(E[X|g]) \leqslant E[\Phi(X)|g].$$

(h) *Let* $g(x, y)$ *be a bounded function on* $\mathbb{R}^2$ *and* $Y \in \mathcal{G}$. *Then*

$$E[\mathcal{G}(X, Y)|\mathcal{G}] = E[\mathcal{G}(X, y)|\mathcal{G}]_{y=Y}.$$

1.3. Examples of conditional expectations

Consider first the σ-algebra $\mathcal{G}$ determined by a partition $\{A_1, \ldots, A_n\}$ into sets in $\mathcal{F}$. For example, if Y is a random variable taking finitely many values $y_1, \ldots, y_n$, then $A_i = \{Y = y_i\}$.

Any $\mathcal{G}$-measurable function is constant on the sets A_i, thus $E[X|\mathcal{G}] = \sum \zeta_i I_{A_i}$. Since any $\mathcal{G}$-measurable H is a linear combination of I_{A_j}, to find this we compute the defining identity

$$E[I_{A_j} E[X|\mathcal{G}]] = E[I_{A_j} X]$$

which yields

$$\zeta_j P(A_j) = E[X; A_j] = \int_{A_j} X_i \, \mathrm{d}P$$

or (when $P(A_j) \neq 0$)

$$E[X|\mathcal{G}] = \sum \frac{1}{P(A_j)} \int_{A_j} X \, \mathrm{d}P. \tag{1.1}$$

Thus, $E[X|\mathcal{G}]$ is a function having a constant value $\int_{A_j} X \,\mathrm{d}P/P(A_j)$ over each A_j. By the way, if Z is a random variable that also takes discrete values $z_y, \ldots, z_k, \ldots$ and X above is $I_{\{Z=z_k\}}$, then $E[I_{\{Z=z_k\}}|\mathcal{G}]$ has the value $P[Z = z_k|Y = y_i]/P(Y = y_i)$ over the set $A_j = \{Y = y_i\}$.

The other standard example consists in assuming that the two random variables X and Y have a known joint distribution function

$$P(X \in A, Y \in B) = \int_{A\times B} P(x, y)(\mathrm{d}x, \mathrm{d}y).$$

We want to find $E[X|Y]$, which, according to Proposition 1.3, equals $h(Y)$ for some $h \in B(\mathbb{R})$. Using the definitions, we have

$$E[Xg(Y)] = \int xg(y)p(x, y)\,\mathrm{d}x\,\mathrm{d}y = \int g(y)\left(\int xp(x, y)\,\mathrm{d}x\right)\mathrm{d}y$$

as well as

$$E[E[X|Y]g(Y)] = \int h(y)g(y)\left(\int p(x, y)\,\mathrm{d}x\right)\mathrm{d}y.$$

The two identities being equal for all bounded $g(x)$, it follows that

$$E[X|Y] = \frac{\int xp(x, Y)\,\mathrm{d}x}{\int p(x, Y)\,\mathrm{d}x}. \tag{1.2}$$

Actually, a variation on this theme yields (1.1) as well. Try it!

Akin to the notion of conditional expectation is the notion of independence.

Definition 1.7. Let $\mathcal{G}_1$ and $\mathcal{G}_2$ be sub-σ-algebras. We say that they are independent whenever

$$EX_1X_2 = EX_1EX_2$$

for any $X_1 \in b\mathcal{G}_1$ and $X_2 \in b\mathcal{G}_2$. Equivalently, whenever (verify)

$$E[X_1|\mathcal{G}_2] = E[X_1] \tag{1.3}$$

for any integrable $X_1 \in \mathcal{G}_1$.

Comment: If X is an E-valued random variable, $\sigma(X) = \{\{X \in A\}; A \in B(E)\}$ is the σ-algebra generated by X, and any random variable $X_1 \in b\sigma(X)$ is of the form $f(X)$ with $f :\in B(E)$. To say that two random variables (taking values where it pleases you) are independent, amounts to say that for bounded f and g

$$E[f(X)g(Y)] = Ef(X)Eg(Y).$$

Taking $f = I_A$ and $g = I_B$, we get the usual

$$P(X \in A, Y \in B) = P(X \in A)P(Y \in B)$$

but Definition 1.3 puts the emphasis where it belongs.

1.4. Processes

A random process X with a state space $(E, \mathcal{E})$ is just a collection $\{X_i : i \in I\}$ of E-valued random variables.

Usually the word process is reserved for processes whose index set is linearly ordered and can be thought of as "time" parameters. When the points of I label "spatial" points, one refers to X as a random field on I.

From now on we consider $I = [0, \infty)$, and to add structure to our set up, we assume we are given a filtration $(\mathcal{F}_t)_{t \geqslant 0}$ that is an increasing collection of sub-σ-algebras $\mathcal{F}_t$ of $\mathcal{F}$, and we assume that X is adapted to $(\mathcal{F}_t)_{t \geqslant 0}$, i.e., $X_t \in \mathcal{F}_t, \forall t \geqslant 0$.

To consider a few examples of processes we also assume we have been with a measure P on $(\Omega, \mathcal{F})$.

DEFINITION 1.8. A Markov process $\{X_t; t \geqslant 0\}$ with space $(E, \mathcal{E})$, is a stochastic process satisfying

(i) $X_t : \Omega \to E$, X_t is $\mathcal{F}_t$ measurable.

(ii) For any $f \in b\mathcal{E}$, and any $t > 0$, $s \geqslant 0$

$$E[f(X_{t+s})|\mathcal{F}_s] = E[f(X_{t+s})|X_s].$$

Comments:

(a) Condition (ii) amounts to asserting that in order to predict the future after time s, the knowledge of the state of the system at times s is sufficient.

(b) Note that for fixed f, t and s, $E[f(X_{t+s})|X_s]$ is measurable with respect to X_s, so it must be a function of X_s, i.e., $E[f(X_{t+s})|X_s] = H_t(f, X_s)$.

(c) Note that for fixed t, s, $H_t(f, X_s)$ is a bounded, linear, positive functional of f, which can be restricted to $C_0(E)$. By the Riesz representation theorem (see also Getoor's [11]), there exists a kernel $P_{s,t}(x, \mathrm{d}y)$ such that

$$E[f(X_{t+s})|X_s] = \int P_{s,t+s}(X_s, \mathrm{d}y) f(y).$$

(d) When $P_{s,t}(x, \mathrm{d}y)$ depends on $t - s$ only, we say that the process X_t is time homogeneous and write for $t > 0$, $s \geqslant 0$

$$E[f(X_{t+s})|X_s] = \int P_t(X_s, \mathrm{d}y) f(y) \tag{1.4}$$

where the right-hand side is an "object" associated with $(E, \mathcal{E})$. Such a right-hand side provides us with explicit means of computing the left-hand side in terms of kernels $P_t(x, \mathrm{d}y)$.

Another important consequence of (1.4) is the Chapman–Kolmogorov equation for the family of kernels $P_t(x, \mathrm{d}y)$.

To begin with, note that we would like to think of $P(X_0 \in A) = \mu(A)$ as an initially known distribution of the process X at $t = 0$. For consistency, one imposes the condition $P_0(x, \mathrm{d}y) = \delta_x(\mathrm{d}y) = \delta(y - x)\,\mathrm{d}y$ in the usual notation or $P_0(x, \mathrm{d}y) = \delta_x(\mathrm{d}y)$ in a more "formal" notation. (But the theory allows for more general conditions.) Now we can have both $s = t = 0$ in (1.4).

Let now f and g be two bounded real-valued functions defined on E and consider $E[f(X_{t+s})g(X_0)]$. We shall compute this in two different ways

$$
\begin{aligned}
E[f(X_{t+s})g(X_0)] &= E[E[f(X_{t+s})|X_s]g(X_0)] \\
&= E\left[g(X_0)\int P_t(X_s, \mathrm{d}y)f(y)\right] \\
&= E\left[g(X_0)E\left\{\int P_t(X_s, \mathrm{d}y)f(y)/X_0\right\}\right] \\
&= E\left[g(X_0)\int P_s(X_0, \mathrm{d}z)\int P_t(z, \mathrm{d}y)f(y)\right] \\
&= \int \mu(\mathrm{d}x)\left(\int P_s(x, \mathrm{d}z)\int P_t(x, \mathrm{d}y)f(y)\right)g(x).
\end{aligned}
$$

Here the first identity comes from (1.4) and the second one from Definition 1.6. To get the third one note that $\int P_t(X_s, \mathrm{d}y)f(y)$ is a function of X_s written in a particular way. For the third and fourth ones, repeat the procedure, this time conditioning with respect to $\mathcal{F}_0$. For the last step, just use the definition of $\mu(\mathrm{d}x)$.

A different approach would be to the condition with respect to $\mathcal{F}_0$ from the very beginning yielding

$$
\begin{aligned}
E[f(X_{t+s})g(X_0)] &= E[g[f(X_{t+s})|X_0]g(X_0)] \\
&= E\left[g(X_0)\int P_{t+s}(X_0, \mathrm{d}y)f(y)\right] \\
&= \int \mu(\mathrm{d}x)\left(\int P_{t+s}(x, \mathrm{d}y)f(y)\right)g(x).
\end{aligned}
$$

Since the ends of both chains are the same for any $g(x)$, we conclude that (a.s.-μ)

$$
P_{t+s}f(x) = P_s P_t f(x),
$$

which is just another way of writing

$$\int P_{t+s}(x, \mathrm{d}y) f(y) = \int P_s(x, \mathrm{d}z) \int P_t(z, \mathrm{d}y) f(y). \tag{1.5}$$

Note that this is a precise way of writing the operator identity $P_{t+s} = P_t P_s$ $(= P_s P_t)$, exercise!

Comment: Similar computations would yield analogous

$$P_{s,t} f(x) = \int P_{s,t}(x, \mathrm{d}y) f(y) = \int P_{s,u}(x, \mathrm{d}z) \int P_{u,t}(z, \mathrm{d}y) f(y) \tag{1.6}$$

for any $0 \leqslant s < u < t$.

Identity (1.6) is usually called the Chapman–Kolmogorov equation for the family of kernels $\{P_t(x, \mathrm{d}y)\}$.

The other important class of stochastic processes, which plays a fundamental role in the theory of Markov processes (but we shall not exploit their full potential) is contained in

DEFINITION 1.9. A (continuous time) martingale $\{X_t : t \geqslant 0\}$ is a process satisfying

(i) $X_t \in \mathcal{F}_z$, $t \geqslant 0$.
(ii) $E|X_t| < \infty \; \forall t \geqslant 0$.
(iii) For any $t \geqslant s \geqslant 0$, $E[X_t|\mathcal{F}_s] = X_t$.

Below we provide examples of some standard martingales arising in the study of Markov processes.

Even though below we shall come to these concepts, we bring them up here for this is the natural place to state the Optional Sampling Theorem.

DEFINITION 1.10. A positive random variable T is called an optional time (relative to $\mathcal{F}_t$) whenever $\{T \leqslant t\}$ is in $\mathcal{F}_t$ for all t. We put

$$\mathcal{F}_T = \{\Lambda \in \mathcal{F} : \Lambda \cap \{T \leqslant t\} \in \mathcal{F}_t \; \forall t \geqslant 0\}.$$

THEOREM 1.1 (Optional Sampling Theorem). *Let $\{X_t : t \geqslant 0\}$ be a martingale and $0 \leqslant S \leqslant T < \infty$ be two optional stopping times such that X_T and X_S are integrable. Then $E[X_T|\mathcal{F}_T] = X_S$.*

Comment: This is quite a diluted version of a collection of theorems. But it will suit our applications.

To finish, we should mention that as the time index consists of the natural integers $\{0, 1, 2, \ldots\}$, we speak of a (discrete time) Markov chain and of a discrete time martingale.

1.5. Semigroups, generators and resolvents

1.5.1. Semigroups

This section is designated to some standard analytical properties of semigroups and allied notions.

We saw above how semigroups appeared as a consequence of the Markov property. In this section, we examine semigroups in their natural historical settings as propagators of initial data for evolution equation.

Let $(E, \mathcal{E})$ be our usual (state) space and $\mathbf{B}$ be a Banach space of functions on E. Two standard examples are $C_b(E)$ and $C_0(E)$ which denote, respectively, the continuous bounded functions on E and the continuous functions vanishing at infinity, both having the norm $\|f\| = \sup\{|f(x)| : x \in E\}$. Sometimes one considers

$$L_p(E, m) = \left\{ f : E \to \mathbb{R}\left[\int |f(x)|^p \, \mathrm{d}m(x)\right]^{1/p} = \|f\|_p < \infty \right\}$$

for some appropriate measure m.

DEFINITION 1.11. A semigroup on a Banach space $\mathbf{B}$ is a family of operators $\{T_t :\geqslant 0\}$, $T_t : \mathbf{B} \to \mathbf{B}$ satisfying

$$T_{t+s} = T_t T_s \quad \text{for } t, s > 0.$$

DEFINITION 1.12. If the semigroup $\{T_t : t \geqslant 0\}$ satisfies

$$\begin{aligned} &T_{t+s} = T_t T_s, \quad t, s \geqslant 0, \\ &T_0 = 0, \\ &\lim_{t \downarrow 0} \|T_t f - f\| = 0, \end{aligned}$$

it is called a semigroup of class C_0 (or a C_0-semigroup).

For a semigroup to be a candidate for the transition semigroup of a Markov process, the operators must have additional properties.

DEFINITION 1.13. A Feller–Dynkin semigroup is a (sub-Markov) semigroup $\{P_t : t \geqslant 0\}$ of class C_0 on $C_0(E)$ such that if $0 \leqslant f \leqslant 1$, then $0 \leqslant P_t f \leqslant 1$.

Whenever $P_t : b\mathcal{E} \to C_b(E)$ $\forall t > 0$, we say that P_t is a strong-Feller semigroup.

Comments: When $P_t 1(x) = 1(x)$ (where $1(x)$ is obviously the function everywhere equal to 1), the semigroup is said to be Markov. When $P_t 1(x) < 1$ for some x, P_t is only sub-Markov. In this case, we say that mass is being lost or cooling is occurring.

To capture the missing mass, one adds a cemetery point Δ to E, forming $E_\Delta = E \cup \{\Delta\}$, extending the σ-algebra $\mathcal{E}$ to $\mathcal{E}_\Delta$ in the obvious way and setting for any $x \in E$.

$$P_t(x, \Delta) = 1 - P_t(x, E),$$
$$P_t(\Delta, E) = 0, \qquad P_t(\Delta, \Delta) = 1, \quad t \geqslant 0.$$

We leave for the reader to verify that this yields a Markov semigroup.

Examples: For the following two examples, let $E = \mathbb{R}$ and $\mathcal{E} = B(\mathbb{R})$ and $f \in b\mathcal{E}$.

Example 1: For $v \in \mathbb{R}$ and $\lambda \geqslant 0$,

$$P_t f(x) = e^{-\lambda t} f(x + vt). \tag{1.7}$$

It is clear that P_t is a sub-Markov semigroup and that $P_t f$ has the same regularity properties that f does. Also, and very important, this example hints that deterministic flows, associated with differential equations, determine Markov processes. (A fact that confuses some people for they mix up randomness and the Markov property.)

Example 2: As $\lambda > 0$, $h > 0$, put, for $f \in b\mathcal{E}$,

$$P_t f(x) = \sum_{n=0}^{\infty} \frac{(\lambda t)^n}{n!} e^{-\lambda t} f(x + nh), \tag{1.8}$$

which describes a particle that moves to the right by jumps of size h, jumping n times with probability $\frac{(\lambda t)^n}{n!} e^{-\lambda t}$.

For the next two, let $E = \mathbb{R}^d$ and $\mathcal{E} = \mathcal{B}(\mathbb{R}^d)$ and let again $f \in \mathcal{E}$. Consider

Example 3: For $x \in \mathbb{R}^d$, $t > 0$

$$P_t f(x) = \int f(x + y) e^{-|y|^2/2t\sigma} \frac{dy}{(2\pi t\sigma)^{d/2}},$$
$$P_0 f(x) = f(x). \tag{1.9}$$

This is the usual semigroup describing the motion of a diffusing (Brownian) particle proposed by Einstein. It is the canonical example of the strong-Feller semigroup. (It regularizes anything it acts on.)

Example 4: To finish, consider

$$P_t f(x) = \frac{1}{\pi^{d/2}} \int \frac{t f(y)\, dy}{\|(x - y)^2 + t^2\|} \frac{d}{2} \tag{1.10}$$

which, when $d = 2$, is the standard Cauchy semigroup. It will play a role in some computations below, and we shall attach a probabilistic meaning to it.

1.5.2. Resolvents

For any $\alpha > 0$ (or for any complex α with $\operatorname{Re}\alpha > 0$), the *resolvent* family $\{U^\alpha; \alpha > 0\}$ associated with the semigroup P_t is defined by

$$U^\alpha f(x) = \int_0^\infty \mathrm{e}^{-\alpha t} P_t f(x)\,\mathrm{d}t. \tag{1.11}$$

Comment: It is a standard result that follows from the semigroup property that $\|P_t\| \leqslant \mathrm{e}^{\beta t}$ for some real β. Thus, properly speaking, U^α is defined for x (or $\operatorname{Re}\alpha > \beta$). When the semigroup is (sub)Markov, $\beta < 0$ and everything works fine.

In the probabilistic literature $U^\alpha f$ is usually called the α-potential of f, and when $U^0 f$ is defined, it is called the potential of f.

PROPOSITION 1.5. *The family $\{U^\alpha : \alpha > 0\}$ associated with a Feller–Dynkin semigroup satisfies*

(i) $\|U^\alpha\| \leqslant 1/\alpha$.
(ii) *The resolvent equation for $\alpha, \beta > 0$, $U^\alpha - U^\beta = (\beta - \alpha)U^\alpha U^\beta$.*
(iii) *For $f \in C_0(E)$, $\|\alpha U^\alpha f - f\| \to 0$ as $\alpha \to \infty$.*

These follow rapidly from the definitions. Actually we can go a bit further.

COROLLARY 1.1. *The range $\mathbf{R} = \operatorname{Ran}(U^\alpha)$ is independent of α. Define $\mathbf{B}_0 = \{f \in \mathbf{B} : \|\alpha U^\alpha f - f\| \to 0$ as $\alpha \to \infty\}$. Then U^α is injective $(1:1)$ on $\mathbf{B}_0$ and is a bijection on $\mathcal{D} = U^\alpha(\mathbf{B}_0)$ (which is independent of α).*

PROOF. Let $g = U^\alpha f$. Then for any β, $g = U^\beta(f - (\beta - \alpha)U^\alpha f)$, i.e., $g \in \operatorname{Ran} U^\beta$. To see that for $g \in U^\beta f, \alpha U^\alpha g \to g$, it suffices to notice from the definition that $U^\alpha f \to 0$, as $\alpha \to \infty$ and that

$$\alpha U^\alpha U^\beta f = \alpha(U^\beta - U^\alpha)f/(\alpha - \beta) \to U^\beta f,$$

as $\alpha \to \infty$. To see that U^α is bijective on $\mathcal{D}$, it suffices to note that, if $U^\alpha f = 0$ for $f \in \mathbf{B}_0$, then $0 = \alpha U^\alpha f \to f$ as $\alpha \uparrow \infty$. □

DEFINITION 1.14. The infinitesimal generator of P_t is defined on $\mathcal{D}$ by

$$Gf = (U^\alpha)^{-1} f - \alpha f. \tag{1.12}$$

Comment: Observe that, since $U^\alpha(\mathcal{B}_0)$ is independent of α, the definition of G does not depend on the chosen α, and that from (1.10) if follows that on $\mathcal{D}$

$$U^\alpha(\alpha - G)f = f. \tag{1.13}$$

And replacing, in (1.12), f by $U^\alpha f$, we obtain

$$(\alpha - G)U^\alpha f = f \tag{1.14}$$

for all f in $\mathcal{B}_0$.

Again, let $g \in \mathcal{D}$, $g = U^\alpha f$ with f in $\mathbf{B}_0$. Since $P_t f \to f$ and $P_t U^\alpha f \to U^\alpha f$ as $t \downarrow 0$, and since an elementary computation shows that $P_t U^\alpha f = e^{\alpha t} \int_t^\infty e^{-\alpha s} P_s f \, ds$, from this it is clear that setting

$$\widetilde{G}(U^\alpha f) := \frac{d}{dt} P_t U^\alpha f|_{t=0} = \alpha U^\alpha f - f$$

or, equivalently,

$$(\alpha - \widetilde{G})U^\alpha f = f$$

from which, combining with (1.14), it follows that on $\mathcal{D}$

$$\left.\frac{dP_t}{dt} g\right|_{t=0} = Gg. \tag{1.15}$$

To conclude, we list the obvious consequences of this definition result, which we shall make use of below

$$P_t g = g + \int_0^t P_s G g \, ds = g + \int_0^t G P_s g \, ds. \tag{1.16}$$

1.6. Examples

Let us list the infinitesimal generators of the semigroups, described above, as some typical martingales associated with Markov processes.

Example 1: For $P_t f(x)$, given by (1.7), and f having one continuous derivative,

$$Gf(x) = vf'(x) - \lambda f. \tag{1.17}$$

Example 2: For $P_t f(x)$, given by (1.8),

$$Gf(x) = \lambda(f(x+h) - f(x)). \tag{1.18}$$

Example 3: For $P_t f(x)$, given by (1.9), and f having at least two continuous derivatives,

$$Gf(x) = \frac{1}{2}\Delta f(x) = \frac{1}{2}\left(\sum_{i=1}^n \partial^2 f/\partial x_i^2\right). \tag{1.19}$$

Example 4: For $P_t f(x)$, given by (1.9), and f such that the integral described is convergent,

$$Gf(x) = \frac{1}{\pi^{d/2}} \int \frac{y \cdot \Delta f(x+y)}{\|y\|^d} \mathrm{d}y, \tag{1.20}$$

where Δf denotes the usual gradient of f.

We have a difficult problem of describing $\mathcal{D}$ for each case.

Let us now assume that X is a Markov process with semigroup P_t and generator G. There are a few martingales one can readily construct which we shall use below.

Let $f \in \mathcal{D}$, then

Example 5: The following stochastic process

$$Y_t = f(X_t) - f(X_0) - \int_0^t Gf(X_u)\,\mathrm{d}u \tag{1.21}$$

is a martingale. Let us verify it. Let $t \geqslant 0,\ s > 0$.

$$\begin{aligned} E[Y_{t+s}|\mathcal{F}_t] &= Y_t + E[Y_{t+s} - Y_t|\mathcal{F}_t] \\ &= Y_t + E\left[f(X_{t+s}) - f(X_t) - \int_t^{t+s} Gf(X_u)\,\mathrm{d}u|\mathcal{F}_t \right]. \end{aligned}$$

To see that the conditional expectation vanishes, we use the Markov property and (1.16). To begin with, note that

$$\begin{aligned} E[f(X_{t+s}) - f(X_t)|\mathcal{F}_t] &= E[f(X_{t+s})|X_t] - f(X_t) \\ &= P_s f(X_t) - f(X_t). \end{aligned}$$

Also note that

$$\begin{aligned} E\left\{ \int_t^{t+s} (Gf)(X_u)|\mathcal{F}_t \right\} &= E\left\{ \int_0^s (Gf)(X_{u+t})\,\mathrm{d}u|\mathcal{F}_t \right\} \\ &= \int_0^s E\{(Gf)(X_{t+u})|\mathcal{F}_t\}\,\mathrm{d}u = \int_0^t P_u(Gf)(X_t)\,\mathrm{d}u. \end{aligned}$$

Now invoke (1.16) to finish. We leave the justification of the steps for the interested reader. If we rewrite (1.21) as

$$\mathrm{d}f(X_t) = Gf(X_t)\,\mathrm{d}t + \mathrm{d}M_t \tag{1.22}$$

and make use of an obvious symbolic (mechanical) meaning of integrals with respect to $\mathrm{d}M_t$ to obtain martingales, we can prove that

LEMMA 1.1. *Let $f > 0$ be in $\mathcal{D}$. Then*

$$\mathcal{E}(t) = \frac{f(X_t)}{f(X_0)} \exp\left(- \int_0^t \left(\frac{Gf}{f} \right)(X_u)\, du \right)$$

is a martingale.

PROOF. Use (1.22) and "differentiate" it to obtain

$$\begin{aligned} d\mathcal{E}(t) &= \frac{df(X_t)}{f(X_0)} \exp - \left\{ \int_0^t \left(\frac{Gf}{f} \right)(X_u)\, du \right\} \\ &\quad - \frac{f(X_t)}{f(X_0)} \left(\frac{Gf}{f} \right)(X_t) \left(\exp - \left\{ \int_0^t \left(\frac{Gf}{f} \right)(X_u)\, du \right\} \right) dt \\ &= \frac{1}{f(X_0)} \left(\exp - \int_0^t \left(\frac{Gf}{f} \right)(X_u)\, du \right) dM_t. \end{aligned}$$

Or, equivalently,

$$\mathcal{E}(t) = 1 + \int_0^t \frac{\mathcal{E}(s)}{f(X_s)} dM(s)$$

and admitting, as we said above, that integration with respect to M_t yields a martingale, we are finished. □

1.7. Construction of Markov processes

1.7.1. Basics

The purpose of this section is threefold: to introduce more terminology, to make definitions a bit more rigorous, and to recall what is involved in "constructing" such objects.

DEFINITION 1.15. A time homogeneous (Hunt) Markov process with state space $(E, \mathcal{E})$ is specified by the collection

$$X = \big(\Omega, \mathcal{F}, (\mathcal{F}_t)_{t \geqslant 0}, (X_t)_{t \geqslant 0}, (\theta_t)_{t \geqslant 0}, (P^x)_{x \in E}\big).$$

In that list, $(\Omega, \mathcal{F})$ is our old friend, the basic space of events. The filtration $(\mathcal{F}_t)_{t \geqslant 0}$, is increasing, right continuous ($\mathcal{F}_z = \bigcap_{\varepsilon > 0} \mathcal{F}_{t+\varepsilon}$) and complete (containing the sets of measure zero with respect to $(P^x, x \in E)$.

The coordinate maps $X_t : \Omega \to E$ are $\mathcal{F}_t$-measurable and right continuous (i.e., $X_{t+\varepsilon}(\omega) \to X_t(\omega)$ as $\varepsilon \downarrow 0$) for all $t \geqslant 0$, and having left limits (i.e., $X_t(\omega) = \lim X_{t-\varepsilon}(\omega)$ as $\varepsilon \downarrow 0$) for all $t > 0$.

The shift operators $\theta_t : \Omega \to \Omega$ are defined by

$$X_t(\theta_t(\omega)) = X_{t+s}(\omega),$$

for all $s, t \geqslant 0, \omega \in \Omega$.

The family of measures $\{P^x, x \in E\}$ defined on $(\Omega, \mathcal{F})$ is such that for every $\Lambda \in \mathcal{F}$, the mapping $x \to P^x(\Lambda)$ is $\mathcal{E}$-measurable. Intuitively $P^x(\cdot)$ serve to define regular versions for the conditional expectations. $P[\cdot\, |X_0] = P^{X_0}[\,]$ and $P^x(X_0 \in A) = I_A(x)$.

Now, we can further restate the Markov property as

DEFINITION 1.16. The process X is Markov if and only if, for every bounded, $\mathcal{F}$-measurable $H : \Omega \to \mathbb{R}$ and every $t > 0$,

$$E^x[H \circ \theta_t | \mathcal{F}_t] = E^x[H \circ \theta_t | X_t] = E^{X(t)}[H], \tag{1.23}$$

for any x in E. (E^x denotes integration with respect to $\mathrm{d}P^x$.)

Here, the right-most side of the equation tells us how to compute any of the two conditional expectations.

If $H = f(X_s), s \geqslant 0$ and f is in $\mathcal{E}$, then

$$E^x[f(X_{t+s}) | \mathcal{F}_t] = E^{X(t)}[f(X_s)]$$

which, combined with

$$E^x[f(X_s)] := P_s f(x), \tag{1.24}$$

yields

$$E^x[f(X_{t+s}) | \mathcal{F}_t] = P_s f(X_t).$$

Note that in (1.24) the symbol on the right-hand side is defined by the left-hand side. We leave for the reader to verify that, as operators acting on bounded functions $f : E \to \mathbb{R}$, we have

$$P_{t+s} = P_t P_s$$

as a consequence of the Markov property (1.23) and also

$$P_0 f(x) = E^x[f(X_0)] = f(x).$$

When P_t happens to be only a sub-Markov semigroup, it can be made Markov (see Section 4) by extending if to the augmented state space $(E_\Delta, \mathcal{E}_\Delta)$. Trajectorywise this amounts to extend X_t to E_Δ by putting $X_t(\omega) = \Delta$ if $X_s(\omega) = \Delta$ for $s \leqslant t$.

1.7.2. Sketch of constructions

It is the purpose of this subsection to make the notation introduced under basics to be natural.

In a variety of applications one starts from an equation like

$$\frac{\partial u}{\partial t}(t,x) = GU(t,x), \quad t > 0, \qquad u(0,x) = f(x), \tag{1.25}$$

where G is a linear operator defined on some $D \subset \mathbf{B}$, and f is in $\mathbf{B}$.

In solving such equations one ends up with

$$u(t,x) = P_t f(x) = \int P_t(X, \mathrm{d}y) f(y), \tag{1.26}$$

$\{P_t : t \geqslant 0\}$ being a C_0 semigroup on the space $\mathbf{B}$ of initial data.

Whenever (1.25) has some probalibistic interpretation and P_t is a Feller–Dynkin semigroup, one is confronted with the issue of producing a setup to describe the "microscopic" behavior underlying (1.25), which would allow for more refined questions than those provided by the averaging out implicit in (1.26).

What follows is only sketchy and approximate (from the point of view of mathematical rigor).

The first step is to construct $(\Omega, \mathcal{F}^0, \mathcal{F}_t^0, X_t, \theta_t)$, where either

$$\Omega = \{\omega : [0,\infty] \to E, \omega \text{ being continuous}\},$$

which will describe microscopic motions of diffusing particles, or

$$\Omega = \{\omega : [0,\infty] \to E, \omega \text{ right continuous with left limits}\},$$

which describes particles that move continuously between jumps.

Now, in each case, X_t is defined as the obvious coordinate mapping

$$X_t : \Omega \to E, \quad X_t(\omega) = \omega(t)$$

whose intuitive meaning is clear: $X_t(\omega)$ is the actual position of the "particle" whose motion in E is described by the random trajectory ω.

The basic σ-algebras $\mathcal{F}^0$and $\mathcal{F}_t^0$ are the minimal σ-algebras making $X(t)$ measurable

$$\begin{aligned} \mathcal{F}^0 &= (X_t : t \geqslant 0), \\ \mathcal{F}_t^0 &= \sigma(X_s; s \leqslant t) \end{aligned}$$

and which are generated by the cylinder sets

$$\{X_{t_1} \in A_1, X_{t_2} \in A_2, \ldots, X_{t_n} \in A_n; 0 \leqslant t_1 < t_2 < \cdots < t_n\},$$

where $A_i \in \mathcal{E}$ and $t_n < \infty$, or $t_n \leqslant t$ for $\mathcal{F}^0$ or $\mathcal{F}_t^0$. These σ-algebras contain the basic questions we can ask about the microscopic behavior of the particles.

Also, the shift operators $\theta_t : \Omega \to \Omega$ are defined as above $X_t(\theta_s \omega) = X_{t+s}(\omega) = \omega(t+s)$, and are $\mathcal{F}^0$-measurable.

On the cylinder sets P^x are easy to define, using the semigroup P_t introduced in (1.26)

$$P^x\left(\bigcap_{i=1}^{n}\{X_{t_i} \in A_i\}\right) = \int_{A_1} \cdots \int_{A_n} p_{t_1}(x, \mathrm{d}x_1) \cdots p_{t_n - t_{n-1}}(x_{n-1}, \mathrm{d}x_n) \tag{1.27}$$

and since any set in $\mathcal{F}^0$ can be approximated by cylinder sets, P^x as obtained from (6.7). An equivalent description would be to define the expected value with respect to P^x of a function $F(X_{t_i}, \ldots, X_{t_n})$. Write

$$\begin{aligned} &E^x[F(X_{t_1}, \ldots, X_{t_n})] \\ &= \int_E \cdots \int_E F(x_1, \ldots, x_n) P_{t_1}(x, \mathrm{d}x_1) \cdots P_{t_n - t_{n-1}}(x_{n-1}, \mathrm{d}x_n). \end{aligned} \tag{1.28}$$

The completion process is achieved by throwing into $\mathcal{F}_0^0$ all sets which are of zero measure with respect to all $\{(P^x) : x \in E\}$ and throwing into E the sets of zero measure with respect to all probability measures μ son $(E, \mathcal{E})$.

Comment: We set $P^\mu(X_0 \in A) = \int \mu(\mathrm{d}x) P^x(\cdot)$. This leads to

$$P^\mu(\in A) = \int \mu(\mathrm{d}x) P^x(\in A) = \int \mu(\mathrm{d}x) I_A(x) = \mu(A)$$

making the notation P^μ clear: it is the probability measure on the space of all trajectories such that at time $t = 0$ the particles are released with distribution μ. This makes, for example, the notation

$$\begin{aligned} \mu_t(A) &= P^\mu(x_t \in A) = \int \mu(\mathrm{d}x) P_t(x, \mathrm{d}x_1) I_A(x_1) \\ &= \int \mu(\mathrm{d}x) P_t(x, A) \end{aligned}$$

clear.

1.7.3. Stopping times and the strong Markov property

What follows is a key ingredient in the probabilistic representations of solutions of boundary problems associated with the operator G in (1.25).

We know that a positive random variable $T : \Omega \to [0, \infty)$ is called a stopping time whenever $\forall t \geqslant 0$

$$\{T \leqslant t\} \in \mathcal{F}_t. \tag{1.29}$$

Whenever the filtration $\mathcal{F}_t$ is right continuous (as in our case), we can replace (1.29) by

$$\{T < t\} \in \mathcal{F}_t.$$

The σ-algebra of events up to (random time) T is defined by

$$\mathcal{F}_T = \{\Lambda \in \mathcal{F} : \Lambda \cap \{T \leqslant t\} \in \mathcal{F}_t\}.$$

Example: To verify that the following are stopping times will depend on many technical facts that we have omitted. Let $f \in \mathcal{E}$, then

$$T = \inf\{t > 0 : f(X_t) > a\}$$

is a stopping time for any real a. In particular, when $A \in E$, $f = I_A$ and $a = 0$

$$T_A = \inf\{t > 0 : X_t \in A\} = \inf\{t > 0 : I_A(X_t) > 0\}.$$

When A is either an open set or a closed set, which is the particular case in most applications, one can easily verify that T_A is a stopping time.

When A is an open set and Q denotes the rational numbers

$$\{T_A \leqslant t\} \quad \bigcup_{q \in Q, q \leqslant t} \{X_q \in A\} \in \mathcal{F}_t,$$

and when A is closed, A^c is open and clearly

$$\{T_A > t\} = \bigcap_{q \in Q, q \leqslant t} \{X_q \in A^c\}.$$

In both cases we have to make use of the right continuity of X_t and the fact that A (or resp., A^c) is open to verify the identities written down.

DEFINITION 1.17. For any A, the stopping times

$$\begin{aligned} T_A &= \inf\{t > 0 : X_t \in A\}, \\ D_A &= \inf\{t \geqslant 0 : X_t \in A\} \end{aligned}$$

are called, respectively, the (first) hitting time and the (first) entry time of A by X.

DEFINITION 1.18. The random variables X_T, θ_T are defined in the obvious way

$$\begin{aligned} (X_T)(\omega) &= X_{T(\omega)}(\omega) = \omega(T(\omega)), \\ (\theta_T)(\omega) &= (\theta_{T(\omega)})(\omega). \end{aligned}$$

When the semigroup P_t is not Markov on E, i.e., $P_t(x, E) < 1$ for some $x \in E$, and we recur to the extended space E_Δ, it is necessary to introduce the "lifetime" ζ of the process X

$$\zeta = \inf\{t > 0 : X_t(\omega) = \Delta\} = \inf\{t > 0 : X_t(\omega) \notin E\}.$$

Comment: Clearly, whenever $T \geqslant \zeta$, $X_T = \Delta$.

LEMMA 1.2. *Let S, T be stopping times. Then $S + T_0\theta_S$ is a stopping time, and*

$$X_T \circ \theta = X_{S+T\circ\theta_S}.$$

LEMMA 1.3. *Let S, T be stopping times. Then*

(a) $S \wedge T$, $S \vee T$, $S + T$ *are stopping times.*
(b) *If* $S \leqslant T$, $\mathcal{F}_S \subset \mathcal{F}_T$.
(c) $\{S < T\}$ *and* $\{S \leqslant T\}$ *belong to* $\mathcal{F}_S \cap \mathcal{F}_T$.

If T_n, T *are stopping times, then*

(d) $\sup T_n, \inf T_n$ *are stopping times.*
(e) *If* $T_n \downarrow T$, *then* $\mathcal{F}_T = \bigcap^n \mathcal{F}_{T_n}$.

DEFINITION 1.19. The process X is strong Markov, whenever for any $H \in b\mathcal{F}$ and any stopping time T

$$E^x[H \circ \theta_T | \mathcal{F}_T] = E^{X(T)}[H] \quad \text{on } \{T < \zeta\} \text{ a.s. } P^x \text{ for any } x \in E. \tag{1.30}$$

DEFINITION 1.20. The process X_t is quasi-left-continuous, whenever for a sequence of stopping $T_n \uparrow T$,

$$\lim X_{T_n} = X_T \quad \text{on } \{T < \infty\}.$$

To finish this long list, we give a proper name to the processes we are going to be working with, which includes diffusions and processes that move by jumps.

DEFINITION 1.21. A Hunt process is a Markov process on a separable metric space $(E, \mathcal{E})$ as above, having right continuous trajectories with left limits at every positive time $t < \zeta$ such that

(i) X is strong Markov,
(ii) X is quasi-left-continuous on $[0, \infty)$.

1.8. Transformations of Markov processes

1.8.1. Additive and multiplicative functionals

From the constructions sketched in Section 6 it must be clear that, in the description of a Markov process by the collection $X = (\Omega, \mathcal{F}, \mathcal{F}_t, X_t, \theta_t, P^x)$, which really determines the nature of the process, there is the collection $\{P^x : x \in E\}$.

Thus, starting from a family $\{P^x : x \in E\}$, a way of modifying a process is modifying the laws P^x. Another way is to modify the trajectories or the speed at which they are traversed.

DEFINITION 1.22. A continuous additive functional (CAF) is a process $\{A_t : t \geqslant 0\}$ such that

(i) $A_t \in \mathcal{F}_t$ and $A_t \geqslant 0$ for all $t \geqslant 0$,

(ii) $A_t(\omega)$ is continuous in ω,

(iii) $A_{t+s}(\omega) = A_t(\omega) + A_s \circ \theta_t(\omega)$ for all $t, s \geqslant 0$, all ω.

Comment: Note that the third condition implies $A_0 = 0$, and also, as mentioned, we are skipping the a.s. P^x all x qualifier.

Example: Let $f \in b\mathcal{E}$. Then

$$A_t = \int_0^t f(X_s)\,\mathrm{d}s$$

is a *CAF*. Verify it! An important example occurs in the case, where $f = I_S$ for $S \in \mathcal{E}$. Then

$$A_t = \int_0^t I_S(X_u)\,\mathrm{d}u$$

will grow only when the trajectory of X_t passes through S. In many applications (when we consider diffusions, for example) our semigroup $P_t(x, \mathrm{d}y)$ has a density with respect to some reference measure $m(\mathrm{d}x)$ on $(E, \mathcal{E})$, and then, in some cases, $A_t(S)$, regarded as a random measure for each fixed t and ω, has density L_t^x with respect to $m(\mathrm{d}x)$, i.e.,

$$A_t(\omega, S) = \int_0^t I_s(X_u(\omega))\,\mathrm{d}u = \int_S L_t^x(\omega)m(\mathrm{d}x) \tag{1.31}$$

or if you prefer, for every $f \in b\mathcal{E}_t$

$$A_t(\omega) = \int_0^t f(X_s)\,\mathrm{d}s = \int_E f(x)L_t^x(\omega)\,\mathrm{d}m(x) \tag{1.32}$$

which is obtained from (1.30) by the usual procedure: first try it on simple functions and then pass to the limit.

Since for each fixed f, A_t in (1.32) satisfies the additivity condition (iii), it is only reasonable to expect that

$$L_{t+s}^x = L_t^x + L_s^x \circ \theta_t \tag{1.33}$$

at least for almost all x, a.s. $m(\mathrm{d}x)$. We assume that it actually holds for all x in E. There are a variety of instances in which one can actually prove that

$$(x, t) \to L_t^x(\omega)$$

is continuous for all ω.

The *CAF* L_t^x is called the local time (of X) at x, and we assume that $L_t^x(\omega)$ is actually continuous in (x, t). We make good use of them when dealing with horizontally stratified media.

Actually, an important variation on the theme occurs when we consider the local time at the boundary of an open set D in E.

The first step consists of defining the boundary measure $\widetilde{m}(\mathrm{d}x)$ on the boundary ∂D of D. Let f be a continuous function vanishing off in a neighborhood $D(\varepsilon) = \{x \in E : d(x, \partial D) < \varepsilon\}$ of ∂D. Define the boundary (or surface) measure $\widetilde{m}(\mathrm{d}x)$ by

$$\int_{\partial D} f(x)\widetilde{m}(\mathrm{d}x) = \lim \frac{1}{\varepsilon} \int_{D(\varepsilon)} f(y)m(\mathrm{d}y) \tag{1.34}$$

when the limit exists and happens to be independent of f and its extensions to $D(\varepsilon)$.

Another local time we are interested in, is the local time L_t at the boundary ∂D of D. If can be defined by the limit

$$L_t = \lim \frac{1}{\varepsilon} \int_0^t I_{D(\varepsilon)}(X_s)\,\mathrm{d}s \tag{1.35}$$

and satisfies

$$L_t = \int_0^t I_{\partial D}(X_s)\,\mathrm{d}L_s. \tag{1.36}$$

Furthermore, when L_t^x exists at every x, combining (1.33) and (1.35) one obtains

$$L_t = \int_{\partial D} L_t^x \widetilde{m}(\mathrm{d}x). \tag{1.37}$$

These identities will be important when dealing with the probabilistic version of the Neumann problem.

DEFINITION 1.23. The (right) continuous inverse of a *CAF* A is the family of stopping times $\{\tau_t : t \geqslant 0\}$ defined by

$$\tau_t = \inf\{s > 0 : A_s > t\}. \tag{1.38}$$

DEFINITION 1.24. Note that τ_t jumps where A_t is flat. It is right continuous and has left limits. The size of the jumps equals the length of the flat plateaus. Also, introducing the σ-algebras $\widehat{\mathcal{F}}_t = \mathcal{F}_{\tau_t}$ and the shift operators $\hat{\theta}_t = \theta(\tau_t)$, we have

LEMMA 1.4. *The process τ_t satisfies:*

(i) $\tau_t \in \widehat{\mathcal{F}}_t$,

(ii) $\tau_{t+s} = \tau_t + \tau_s \circ \theta_{\tau_t}$.

If A_t is as in (1.31) *with $f \geqslant c > 0$, then A_t is strictly increasing and τ_t satisfies*

$$\tau_t = \int_0^t \mathrm{d}s / f(X(\tau_s)). \tag{1.39}$$

DEFINITION 1.25. A multiplicative functional is a process $\{M_t : t \geqslant 0\}$ such that

(i) $0 \leqslant M_t \leqslant 1$, $M_t \in \mathcal{F}_t$ for all $t \geqslant 0$.

(ii) $M_{t+s} = M_t M_s \circ \theta_t$.

Example: Three standard examples are as follows. Let f be strictly positive, and put

$$M_t = f(X_t)/f(x_0). \tag{1.40}$$

Or let A_t be an *CAF*, and put

$$M_t = \exp(-A_t) \tag{1.41}$$

and we leave for the reader to verify that both examples satisfy Definition 1.23.

The martingale

$$\mathcal{E}(t) = \frac{f(t)}{f(0)} \exp\left\{ - \int_0^t \left(\frac{Gf}{f}\right) x(s)\, \mathrm{d}s \right\}$$

introduced in Lemma 1.1 is clearly a multiplicative functional.

To finish the list of examples, let D be an open set

$$T = \inf\{t > 0 : X(t) \in D^c\}.$$

Define

$$M_t = I_{\{t<T\}}. \tag{1.42}$$

We leave for the reader to verify that it is a multiplicative functional. It will show up two sections below. The choice of D to be open is restrictive, but sufficient for our needs.

1.8.2. Time changing a Markov process

Let X be a Markov process and A a CAF of X with right continuous inverse τ. The time change of X by A is defined to be the process $\widehat{X}$ whose trajectories are specified by

$$\widehat{X}_t(\omega) = X_{\tau(t)}(\omega) \tag{1.43}$$

and (obviously) defining a Markov process with the transition semigroup

$$\widehat{P}_t f(x) = E^x\big[f(\widehat{X}_t)\big] = E^x\big[f(X_{\tau(t)})\big] \tag{1.44}$$

for $f \in b\mathcal{E}$ or $f \in \mathcal{E}_+$. Actually, it can be shown that the new state space for the process is $\widehat{E} = \{x \in E : P^x(\tau_0 = 0) = 1\}$ with inherited σ-algebra $\widehat{\mathcal{E}} = \mathcal{E} \cap \widehat{E}$.

For example, if we deal with diffusions, and $A_t = \int_0^t f(X_s)\,\mathrm{d}s$, where f has support $\overline{D}$, with $D = \{x : f(x) > 0 : X(t) \in D\}$, having a regular boundary ∂D, then $\widehat{E} = \overline{D}$.

We say that $x \in \partial D$ is regular (for D) when $P^x(T = 0) = 1$, where $T = \inf\{t > 0 : X(t) \in D\}$.

A different, but related characterization of the points of $\widehat{E}$, is as the set of points x such that A increases right away when $X_0 = x$.

Important for us is the following

PROPOSITION 1.6. *Let $a(x) \geqslant a_0 > 0$ be a continuous function of E and $A_t = \int_S^t a(X_s)\,\mathrm{d}s$.*

Let $f \in D(G)$, be the domain of the infinitesimal generator of the transition semigroup P of X. Then $f \in D(\widehat{G})$ and

$$\widehat{G} f(x) = \frac{1}{a(x)} G f(x). \tag{1.45}$$

PROOF. We present two proofs. The longer one uses basics. Both involve some martingale stuff.

Let $f \in D(G)$, and note that

$$\begin{aligned}
\widehat{U}^\alpha f(x) &= \int_0^\infty \mathrm{e}^{-\alpha t}\widehat{P}_t f(x)\,\mathrm{d}x = E^x \int_0^\infty \mathrm{e}^{-\alpha t} f(X_{\tau(t)})\,\mathrm{d}t \\
&= E^x \int_0^\infty \mathrm{e}^{-\alpha A(t)} f(X_t)\,\mathrm{d}A(t) = -\frac{1}{\alpha}E^x \int_0^\infty f(X_t)\,\mathrm{d}\big(\mathrm{e}^{-A(t)}\big) \\
&= -\frac{1}{\alpha}E^x \int_0^\infty \mathrm{d}\big(f(X_t)\mathrm{e}^{-\alpha A(t)}\big) + \frac{1}{\alpha}E^x \int_0^\infty \mathrm{e}^{-\alpha A(t)}\mathrm{d}f(X(t)).
\end{aligned}$$

So far we used a simple change of variables $t \to A(t)$ at the third step and then integration by parts. Since our f is bounded and $A(t) \geqslant a_0 t$, the next to the last integral is

$$-E^x \int_0^\infty \mathrm{d}\big(f(X_t)\mathrm{e}^{-\alpha A(t)}\big) = f(x).$$

We "saw" in Section 5 that $\mathrm{d}f(X_t) = Gf(X_t)\,\mathrm{d}f + \mathrm{d}M(t)$, which allows us to rewrite the last integral as

$$E^x \int_0^\infty \mathrm{e}^{-\alpha A(t)}\,\mathrm{d}f(X(t)) = E^z \int_0^\infty \mathrm{e}^{-\alpha A(t)} Gf(X_t)\,\mathrm{d}t + E^x \int_0^\infty \mathrm{e}^{-\alpha A(t)}\,\mathrm{d}M(t).$$

The last integral above is zero because of the martingale property of $\int_0^\infty \mathrm{e}^{-\alpha A(t)}\,\mathrm{d}M_t$. Writing $\mathrm{d}t = \frac{\mathrm{d}A(t)}{a(X(t))}$ and undoing the change of variables, the first integral above becomes

$$E^x \int_0^\infty \mathrm{e}^{-\alpha A(t)} \frac{Gf(X(t))}{a(X(t))}\,\mathrm{d}A(t) = E^x \int_0^\infty \mathrm{e}^{-\alpha t} \frac{Gf(\widehat{X}(t))}{a(\widehat{X}(t))}\,\mathrm{d}t;$$

and rearranging our first chain of identities, we obtain

$$\widehat{U}^\alpha\left(\alpha - \frac{1}{a(x)}G\right)f(x) = f(x)$$

which amounts to $\widehat{G}f(x) = Gf(x)/a(x)$ as desired.

A slicker proof starts from the identity

$$f(X(t)) = f(X_0) + \int_0^t (Gf)(X_s)\,\mathrm{d}s + M_t$$

and now one evaluates both sides at $\tau(t)$, changes variables in the integral, and appeals to the optional sampling theorem to insure that $M_{\tau(t)}$ is an $\mathcal{F}_{\tau(t)}$ martingale for each P^x. With all this

$$f(X(\tau_t)) = f(X(\tau_0)) + \int_0^t Gf(X(\tau(t)) \frac{\mathrm{d}s}{a(X(\tau_{(s)}))} + M(\tau_t),$$

where we made use of $P^x(\tau_0 = 0) = 1$ for all x since $a_0 > 0$.

From this one readily obtains

$$\widehat{G}f(x) = \lim^{t\downarrow 0} \frac{\widehat{P}_t f(x) - f(x)}{t} = \lim \frac{1}{t}\int_0^t E^x\left[\frac{Gf(X(\tau_s))}{a(x(\tau_{(s)}))}\right]\mathrm{d}s = Gf(x)/a(x). \qquad \square$$

1.8.3. The killing of a Markov process

Given a Markov process X and a multiplicative functional M_t, we obtain a new family of measures $\widehat{P}^x$ by defining

$$\widehat{E}^x[H] = E^x[HM_t] \tag{1.46}$$

whenever $H \in b\mathcal{F}_t$. These measures induce a semigroup defined by

$$\widehat{P}_t f(x) = E^x\{f(X_t)M_t\}. \tag{1.47}$$

It is simple to verify that this is a transition semigroup, and that the new state space on which the new process is to live is $\widehat{E} = \{x : P^x(M_0 = 1) = 1\}$; this is the function on which $\widehat{P}_t$ is defined on $\widehat{E}$ only.

The details of the construction of a process $\widehat{X}$ starting from the process X are natural but involved and we skip them.

When $M_t = \exp\{-A(t)\}$, with $A(t) = \int_0^t q(X_s)\,\mathrm{d}s$ with $q(x) \geqslant 0$ and continuous, the stochastic calculus trick allows us to conclude that $\widehat{P}_t$ has an infinitesimal generator $\widehat{G}$ with $D(G) = D(\widehat{G})$ and given by

$$\widehat{G}f(x) = Gf(x) - q(x)f(x). \tag{1.48}$$

This $q(x)$ is called the cooling rate or killing rate, meaning that one removes the randomly moving particles at the rate $\mathrm{d}M(t)/M(t)$, and sends them to a cemetery point Δ.

Another important example corresponds to the case $M_t = I_{\{t<T\}}$, and $T = \inf\{t > 0 : X(t) \in D^c\}$ for D open in E. In this case, the semigroup (1.46) is:

$$\widehat{P}_t f(x) = E^x\{f(X_t); t < T\} \tag{1.49}$$

and one sees that T acts as the lifetime of the killed process. In this example the trajectories of the killed process $\widehat{X}(t)$ coincide with those of $X(t)$ up to T, i.e., $\widehat{X}(t) = X(t)$ for $t < T$, and $\widehat{X}(t) = \Delta$ for $t \geqslant T$.

The resolvent of the new semigroup (2) is defined in the obvious way

$$\widehat{U}^\alpha f(x) = E^x \int_0^\infty \mathrm{e}^{-\alpha t}\widehat{P}_t f(x)\,\mathrm{d}t \tag{1.50}$$

which, for $M_t = I_{\{t<T\}}$ becomes (verify it)

$$\widehat{U}^\alpha f(x) = E^x \int_0^T \mathrm{e}^{-\alpha t} f(X_t)\,\mathrm{d}t. \tag{1.51}$$

When there is measure a $m(\mathrm{d}x)$ on $(E, \mathcal{E})$ with respect to which $P_t(x, \mathrm{d}y)$ has a density $p(t, x, y) = p_t(x, y)$, one an prove that the killed semigroup $\widehat{P}$ (or the killed process $\widehat{X}$) has the density $\widehat{p}_t(x, y)$ with respect to $m(\mathrm{d}x)$ such that

$$\widehat{P}_t f(x) = E^x[f(X_t); t < T] = \int \hat{p}_t(x, y) f(y) m(\mathrm{d}y).$$

In this case, the density $\hat{p}_t(x, y)$ satisfies $\hat{p}_t(x, y) = 0$ whenever $x \in D^c,\ y \in D$.

When $u(t) = \exp\{-A(t)\}$, with $A(t) = \int_0^t q(X_s)\,\mathrm{d}s$ for positive continuous $q(x)$, using the obvious identity

$$M(t) = 1 - \int_0^t a(x(s))M(s)\,\mathrm{d}s$$

and the simple Markov property, it is easy to see that

$$\widehat{P}_t f(x) = P_t f(x) - \int_0^t \widehat{P}_{t-s} q P_s f(x)\, ds. \tag{1.52}$$

Taking the Laplace transform on each side, we obtain the identity

$$\widehat{U}^\alpha f(x) = U^\alpha f(x) - \widehat{U}^\alpha q U^\alpha f \tag{1.53}$$

and differentiating both sides of (1.51) with respect to t, we conclude that $f \in D(G)$. Then $f \in D(\widehat{G})$ and

$$\widehat{G} f(x) = G f(x) - q(x) f(x).$$

References

Three classics on measure theory, containing an introduction to stochastic processes are

1. Chung, K. L. and Zhao, Z.: *From Brownian Motion to Schroedinger's Equation*, Springer-Verlag, Berlin, 1995.
2. Sznitman, A.-S.: *Brownian Motion and Random Media*, Springer-Verlag, Berlin, 1998.
3. Bauer, H.: *Probability Theory and Elements of Measure Theory*, Holt, Rinehart and Winston, New York, 1972.
4. Kingman, J. F. C. and Taylor, J. S.: *Introduction to Measure and Probability*, Cambridge Univ. Press, Cambridge, 1966.
5. Neveu, J.: *Bases Mathematiques du Calcul des Probabilites*, Mason, Paris, 1970.

All you wanted to know about semigroups, and their connection to stochastic processes can be checked in

6. Dynkin, E. B.: *Markov Processes*, Vols. I, II, Springer-Verlag, Berlin, 1965.
7. Ethier, S. N. and Kurtz, T. G.: *Markov Process: Characterization and Convergence*, Wiley-Interscience, New York, 1986.

And to learn a lot about probabilistic potential theory

8. Port, S. C. and Stone, C. J.: *Brownian Motion and Classical Potential Theory*, Academic Press, New York, 1978.
9. Chung, K. L.: *Lectures from Markov Processes to Brownian Motion*, Springer-Verlag, Berlin, 1982.
10. Blumenthal, R. M. and Getoor, R. K.: *Markov Processes and Potential Theory*, Academic Press, New York, 1968.
11. Getoor, R. K.: *On the Construction of Kernels*, Lecture Notes in Math. 714, Springer-Verlag, Berlin.

CHAPTER

TWO

FROM BROWNIAN MOTION TO DIFFUSIONS

The mathematical study of diffusions apparently began when A. Einstein published his papers on the study of small particles suspended in a fluid. In this chapter, we do two things: we present many of the results described in Chapter 1 in the context of Brownian motion and then, after an interlude devoted to stochastic integration, we show how to go from Brownian motion to more general diffusions.

2.1. Brownian motion

2.1.1. The classical approach

Using phenomenological arguments, Einstein established the following equation to describe the concentration $u(t,x)$, of particles at a point x in $\mathbb{R}^d$ at time t, when the initial concentration was $u_0(x)$:

$$\frac{\partial u}{\partial t} = \frac{\sigma}{2}\Delta u, \qquad u(0+,x) = u_0(x), \tag{2.1}$$

where σ is called the diffusion constant. The same equation was established much earlier by Fourier in his studies of propagation of heat in homogeneous media. There are many ways of solving (2.1); they all lead to

$$\begin{aligned} u(t,x) &= \int \frac{e^{-(x-y)^2/2\sigma t}}{(2\pi\sigma t)^{d/2}} u_0(y)\,dy \\ &= \int p_t(x-y)u_0(y)\,dy = \int u_0(x-y)p_t(y)\,dy, \end{aligned} \tag{2.2}$$

where, of course, $p_t(x-y) = \mathrm{e}^{-(x-y)^2/2\sigma t}/(2\pi\sigma t)^{d/2}$ is to be interpreted as the probability of finding a particle at y at time t, if the particle was released from x at $t=0$. Based on this interpretation, Einstein produced the result

$$\langle x^2\rangle = \int y^2 p_t(y)\,\mathrm{d}y = \sigma t$$

which allowed physicists to infer Avogadro's number, which enters in σ. Even though it is clear that (2.2) is defined for quite a large class of functions $u_0(x)$, we only assume $u_0(x)$ to be bounded. If, besides boundedness, we assume continuity of $u_0(x)$, then, after a change of variables, we see that

$$u(t,x) = \int u_0(x - y\sqrt{\sigma t})\mathrm{e}^{-y^2/2}\,\mathrm{d}y \to u_0(x) \tag{2.3}$$

by invoking standard convergence results (the dominated convergence theorem). Furthermore, we have

LEMMA 2.1. *If $u_0(x)$ is continuous and has a compact support, then*

$$\sup_x |u(t,x) - u_0(x)| \to 0 \quad \text{as } t \downarrow 0. \tag{2.4}$$

Comment: Using the fact that the continuous functions of compact support are a dense subset of $L_p(\mathbb{R}, \mathrm{d}x)$, we can remove the continuity assumption in part (b). As in Chapter 1, define the operator $P_t : B(\mathbb{R}^d) \to B(\mathbb{R}^d)$ by

$$P_t f(x) = \int p_t(x-y) f(y)\,\mathrm{d}y. \tag{2.5}$$

We have

LEMMA 2.2. *With the notation introduced above*

(i) $P_t 1(x) = 1$,
(ii) $P_t f(x) \leqslant \|f\|$,
(iii) *If* $f(x) \geqslant 0$, *then* $P_t f(x) \geqslant 0$,
(iv) $P_t(B) \subset C_0^\infty(\mathbb{R}^n)$,
(v) $P_t(L_p) \subset L_p$,
(vi) $\forall t > 0$, $\forall s \geqslant 0$, $\forall f \in B$, $P_{t+s} f(x) = P_s P_t f(x)$.

The proof of these assertions is simple and left for the reader. A comment is due: we set $P_0 f(x) = f(x)$ without forgetting that $\|P_t f - f\| \to 0$ only for some f's and in some norms. Using the interpretation provided above, all simple questions about the statistical nature of Brownian motion can be answered. In particular, if we denote the average of a function $f(X_{t}, \ldots, X_{t_n})$ depending on the values of the position of the diffusing particle at times $t_1 < t_2 < \cdots < t_n$ by

$$\langle f(t_1, \ldots, t_n)\rangle_\rho$$

the initial distribution density being ρ, we have

$$\begin{aligned}\langle f(t_1, \ldots, t_n)\rangle_\rho \\ = \int \cdots \int \rho(\mathrm{d}x_0) p_{t_1}(x_1 - x_0) \cdots p_{t_n - t_{n-1}}(x_n - x_{n-1}) f(x_1, \ldots x_n)\, \mathrm{d}x_1 \cdots \mathrm{d}x_n.\end{aligned} \tag{2.6}$$

It is very easy to see that for any function f with two continuous bounded derivatives and vanishing off some bounded set we can compute the infinitesimal generator of P_t to be

$$Gf(x) = \frac{\sigma}{2}\Delta f. \tag{2.7}$$

But actually, the class on which G acts is much larger. All functions such that $P_t f(x) \to f(x)$ at each x or for all x in compact regions include functions like x^n and e^{ax}, and G is defined on these as well. These will be useful to construct some martingales below. Set $\sigma = 1$ from now on. The resolvent of P_t is computed to be

$$U^\alpha f(x) = \int \mathrm{e}^{-\alpha t} P_t f(x)\, \mathrm{d}t = \int u_\alpha(x - y) f(y)\, \mathrm{d}y,$$

where

$$u_\alpha(x - y) = \int \mathrm{e}^{-\alpha t} p_t(x - y)\, \mathrm{d}t = \frac{2}{(2\pi)^{d/2}} \left(\frac{\sqrt{2\alpha}}{|x - y|}\right)^h K_n\big(\sqrt{2\alpha}|x - y|\big) \tag{2.8}$$

with $h = (d - 2)/2$. $K_\nu(2)$ denoting the standard Bessel function of purely imaginary arguments. When $d = 1$

$$u_\alpha(|x - y|) = \frac{\mathrm{e}^{-\sqrt{2\alpha}|x-y|}}{\sqrt{2\alpha}}$$

which becomes singular as $\alpha \downarrow 0$. This singularity also occurs for $d = 2$, whereas for $d \geqslant 3$ we have

$$u_0(|x - y|) = \frac{\Gamma(d/2 - 1)}{2\pi^{d/2}} \frac{1}{\|x - y\|^{d-2}}. \tag{2.9}$$

For $d = 3$, $u_\alpha(|x - y|)$ happens to be

$$u_\alpha(|x - y|) = \frac{1}{2\pi} \frac{\mathrm{e}^{\sqrt{2\alpha}|x-y|}}{\|x - y\|}.$$

2.1.2. The modern approach

This approach was outlined above. It was originally called functional integration or path integration. Throughout the remainder of this chapter we shall deal with the following sample space

$$\Omega = \{w : [0,\infty) \to \mathbb{R}^d,\ w \text{ continuous}\}$$

and define the usual coordinate maps $X_t : \Omega \to \mathbb{R}^d$, $X_t(w) = w(t)$, the usual filtrations $\mathcal{F}^0 = \sigma(X_t : t \geqslant 0)$ and $\mathcal{F}_t^\infty = \sigma(X_s : s \leqslant t)$ containing the basic questions about the behavior of the process. The shift operators θ_t are defined as in Section 6 of Chapter 1. For each measure μ on $(\mathbb{R}^d, B(\mathbb{R}^d))$, the measures P^μ are constructed so that:

(a) For bounded $f(x_1, \ldots, x_n)$ and $0 \leqslant t_1 < \cdots < t_n$

$$E^\mu[f(X_{t_1}, \ldots, X_{t_n})] = \int \cdots \int \mu(\mathrm{d}x_0) p_{t_1}(x_1 - x_0) \cdots p_{t_n - t_{n-1}}(x_n - x_{n-1}) f(x_1, \ldots, x_n)\, \mathrm{d}x_1 \cdots \mathrm{d}x_n \tag{2.10}$$

as proposed in (2.6).

(b) For any $H \in b\mathcal{F}^0$ or $H \in \mathcal{F}_+^0$, approximate it by functions as in (a), and pass to the limit. Thus

$$E^\mu[H] = \int H(w) P^\mu(\mathrm{d}w) \tag{2.11}$$

is well defined.

(c) When $\mu = \delta_x$, the unit point (or Dirac) mass at x, (2.1) yields

$$E^x[f(X_0)] = f(x) \tag{2.12}$$

as consistency requires.

At this stage, the family $\mathcal{F}_t^0, \mathcal{F}^0$ is completed and the filtration is made right continuous. The resulting objects are denoted by $\mathcal{F}_t$ and $\mathcal{F}$. From (2.1) it also follows that $x \to E^x[H]$ is Borel measurable for $H \in b\mathcal{F}^0$ and that

$$E^\mu[H] = \int \mu(\mathrm{d}x) E^x[H]$$

as consistency would require. This completes the basics about the path integration.

2.1.3. Some important hitting times and hitting distributions

Consider a one-dimensional Brownian motion $X(t)$, and let x, y be distinct points in $\mathbb{R}$. We are interested in the distribution $P^x(T_y \leqslant t)$ of the hitting time T_y of y by X started at x. For that it suffices to compute the Laplace transform $E^x[\mathrm{e}^{-\alpha T_y}]$ of the distribution of T_y. Note that the invariance of the distributions of one-dimensional Brownian motion under translations and reflections about the origin imply that $E^x[\exp(-\alpha T_y)] = E^0[\exp(-\alpha T_{|x-y|})]$, thus it suffices to consider say $y > x$. Let $f(x) = \exp \beta x$; thus $Gf = \beta^2 f/2$ and, according to Lemma 1 in Section 1.5,

$$M(t) = \exp\left\{\beta(X(t) - X(0)) - \frac{\beta^2}{2}t\right\}$$

is a martingale. Since now $X(T_y) = y$ and $E^x[e^{-\alpha T_y}] < \infty$ we have by the optional sampling theorem

$$E^x[M(T)] = 1 = e^{\beta(y-x)} E^x\left[e^{-\frac{\beta^2}{2}T_y}\right]$$

and, setting $\beta = \sqrt{2\alpha}$, we obtain

$$E^x\left[e^{-\alpha T_y}\right] = e^{-\sqrt{2\alpha}|x-y|} = \int e^{-\alpha t} \frac{|x-y| e^{-(x-y)^2/2t}}{(2\pi^3)^{1/2}} dt$$

according to any table of Laplace transforms. Thus, for $t > 0$

$$dP^x(T_y \leqslant t) = \frac{|x-y| e^{-|x-y|^2/t}}{(2\pi t^3)^{1/2}}. \tag{2.13}$$

A different, more hands on, derivation of the same result is as follows. Note that $P^x(X(t) > y) = P^x(T_y \leqslant t, X(t) > y)$, because of the continuity of the trajectories of X. Invoke the strong Markov property to write the right-hand side as

$$P^x\left(T_y \leqslant t, P^y(X_{t-T_y(\cdot)} > y)\right),$$

where the dot in $T_y(\cdot)$ is to mean that its argument depends on the trajectories under the outer integration. Since for any $s \geqslant 0$, and any z, $P^x(X_s > x) = 1/2$, we have

$$P^x(T_y \leqslant t) = 2P^x(X(t) > y).$$

Now, introduce the change of variables $s = (y-x)^2 t/z^2$ to obtain

$$\begin{aligned} P^x(T_y \leqslant t) &= 2P^x(X(t) > y) = 2\int_{y-x}^{\infty} (2\pi t)^{1/2} e^{-z^2/2t} dz \\ &= \int_0^t (y-x) \frac{e^{-(y-x)^2/2s}}{(2\pi s^3)^{1/2}} ds. \end{aligned}$$

This result is the first step in obtaining the hitting distribution of a hyperplane in $\mathbb{R}^d$. It should be clear that due to invariance under rotations, reflections on a plane and translations, we can assume that the initial point x in $\mathbb{R}^d$ has coordinates $x = (0, 0, \ldots, x_d)$

and that we are interested in $P^x(X(T) \in A)$ where $T = \inf\{t > 0 : X(t) \in H_0\}$, and $H_0 = \{y \in \mathbb{R}^d : y_d = 0\}$, and also $A \in H_0$. To begin with note that $T = \inf\{t > 0 : X_d(t) = 0\}$, and

$$X(T) = (X_1(T), \ldots, X_{d-1}(T), 0),$$

and since the components of the d-dimensional Brownian motion are independent, for any bounded function defined on H_0, we have

$$\begin{aligned} E^x[f(X(T))] &= E^x\big[E^x[f(X(T))|T]\big] \\ &= \int_0^\infty E^x[f(X(s))]\,\mathrm{d}P^x(T \leqslant s) \\ &= \int_0^\infty \mathrm{d}s \int_{\mathbb{R}^{n-1}} \mathrm{d}y f(y) \frac{\mathrm{e}^{-\|y-x\|_{d-1}^2/2s}}{(2\pi s)^{\frac{h-1}{2}}} x_d \frac{\mathrm{e}^{-x_d^2/2s}}{(2\pi s^3)^{1/2}} \mathrm{d}s \\ &= \int_{\mathbb{R}^{n-1}} f(y) \frac{\Gamma(d/2) x_d\,\mathrm{d}y}{\pi^{d/2}\|y-x\|_d^d}. \end{aligned}$$

Here $\|z\|_d = (\sum_{i=1}^d z_i^2)^{1/2} = (\|z\|_{d-1}^2 + z_d^2)^{1/2}$ and everything was written in such a way as to allow to be arbitrary with $x_n > 0$. From the above we have

$$P^x(X(T) \in \mathrm{d}y) = \frac{\Gamma(2) x_d\,\mathrm{d}y}{\pi^{d/2}\|x-y\|_d^{d/2}} \tag{2.14}$$

and we shall drop the subscript d from $\|\cdot\|_d$ when there is no confusion. The next example is important when we get to verify that the probabilistic representation for the exterior Dirichlet problem, that we provide below, satisfies the Sommerfeld radiation condition. Here we follow [5] but the same result had appeared earlier, hidden among other important results in [6]. The problem we want to solve is to compute the joint distribution of the hitting time and place of a sphere of radius R starting from a point x with $\|x\| > R$. We want to compute, for bounded f defined on $S = \{y : \|y\| = R\}$ and $\alpha > 0$

$$\begin{aligned} &E^x\big[f(X(T))\mathrm{e}^{-\alpha T}\big] \\ &\quad = \int \mathrm{e}^{\alpha s} f(\xi) P^x(X(T) \in \mathrm{d}\xi, T \in \mathrm{d}s), \end{aligned}$$

where $\mathrm{d}\xi$ denotes the "area" element on S. Again, due to rotational symmetry, this time around the axis passing through the origin and the starting point x, it suffices to consider

$$E^x\big[\mathrm{e}^{-\alpha T}\cos n\theta(T)\big],$$

where $\cos\theta(T) = (\hat{x}, X(T))/R$. Now, since $\cos n\theta$ can be written in terms of the Gegenbauer polynomials $C_n^h(\cos\theta)$, it suffices to compute

$$E^x\big[\mathrm{e}^{-\alpha T} C_n^h(\cos\theta)\big].$$

It is at this point that the identity $(\alpha - G)U^{\alpha} = I$ comes in. The operator U^{α} has the kernel $u_{\alpha}(|x-y|)$ given by (1.10) and it is now a simple exercise to verify that

$$F(t,\lambda) = \mathrm{e}^{-t\alpha^2/2}\frac{K_n(\alpha|x-y|)}{(\alpha|x-y|)^h}$$

satisfies

$$\frac{\partial F}{\partial t} + \frac{1}{2}\Delta F = 0$$

at $(t,x) \in (0,\infty) \times (\mathbb{R}^d - \{y\})$ and $h = (d-2)/2$. Bringing in the summation formula ([10])

$$\frac{K_h(Z)}{m^h} = 2^n\Gamma(h)\sum_{m=0}^{\infty}(h+m)\frac{K_{h+m}(Z_1)}{Z_1^h}\frac{I_{h+m}(Z_2)}{Z_2}C_m^h(\cos\theta),$$

where $Z^2 = Z_1^2 + Z_2^2 2Z_1Z_2\cos\theta$. Again, recur to (1.22) to verify that $F(t, X_t)$ is a martingale on $[0,T]$, where $T = \inf\{t > 0 : X(t) \in B(0,R)\}$, with respect to P^x, $|x| > R$. From the martingale property, and since the exponential becomes 0 when $T = \infty$, the optional sampling theorem yields that, for $|y| \leqslant B$ and $|x| > R$,

$$E^x\left\{\mathrm{e}^{-\frac{\alpha^2}{2}T}\frac{K_h(\alpha|B(T)-y|)}{(\alpha|B(T)-y|)}\right\} = \frac{K_h(\alpha|x-y|)}{(\alpha|x-y|)^h}.$$

Making use of the summation formula, and identifying coefficients of the same $I_{h+n}(\cdot)$, we obtain

$$E^x\left[\mathrm{e}^{-\frac{\alpha^2 T}{2}}\frac{K_{h+m}(\alpha|x(T))}{\alpha|x(T)|)^h}C_m^h(\cos\theta(T))\right] = \frac{K_{h+m}(\alpha|x|)}{(\alpha|x|)^h}C_m^h(\cos\phi);$$

here $\cos\phi = (x,y)/|x||y|$ and $|x(T)| = R$ for $T < \infty$, thus, replacing $\alpha^2/2$ by α, we have

$$E^x\left[\mathrm{e}^{-\alpha T}C_m^h(\cos\theta(T))\right] = \left(\frac{R}{|x|}\right)^h\frac{K_{h+m}(\sqrt{2}\alpha|x|)}{K_{h+m}(\sqrt{2}\alpha R)}$$

after we choose y parallel to x so that $\cos\phi = 1$ and normalize $C_m^h(1) = 1$.

2.1.4. Dirichlet problem for the Laplacian

This section is devoted to establishing the probabilistic representation of the solution to

$$(q(x) - \Delta/2)U(x) = f(x), \quad x \in D, \tag{P1-a}$$

$$u(x) = \varphi(x), \quad x \in \partial D, \tag{P1-b}$$

where (P1-b) is to mean $u(y) \to \varphi(x)$ as $y \in \partial D \to x \in D$.

Comment: We keep $1/2$ dividing the Laplacian for it facilitates the arithmetics when connecting with probability. It can be removed by redefining $\varphi(x)$ and $f(x)$. For (P1-b) to hold it is necessary that the points of ∂D satisfy

DEFINITION 2.1. A point $x \in \partial D$ is said to be *regular* (for D^c) whenever $P^x(T=0)=1$, where $T = \inf\{t > 0 : X_t \in \partial D^c\}$.

Comment: Note that as $x \in D$, $T = \inf\{t > 0 : X_t \in \partial D\}$ because of the continuity of the trajectories of X. We are ready to state

THEOREM 2.1. *Let D be a domain in $\mathbb{R}^n$ such that every point of ∂D is regular for D^c. Assume that $f : D \to \mathbb{R}$ is bounded and $\varphi : \partial D \to \mathbb{R}$ is continuous (in the induced topology of ∂D). Let $q : D \to \mathbb{R}$ be positive and continuous and let $u(x)$ be a solution to problem* (P1). *Then*

$$u(x) = E^x\left\{\int_0^T M_t f(X(t))\,\mathrm{d}t\right\} + E^x\big[M(T)\varphi(X(T)); T < \infty\big], \tag{2.15}$$

where $M(t) = \exp\{-\int_0^t q(X_s)\,\mathrm{d}s\}$.

Comment 1: When there is a Green function $G(x,y)$ satisfying $(q(x) - \Delta/2)G(x,y) = \delta(x-y)$ for $x \neq y$, it is clear that the first term on the right-hand side stands for

$$\int_D G(x,y)f(y)\,\mathrm{d}y,$$

whereas the second term stands for

$$-\int \frac{\partial G}{\partial n}(x,y)\varphi(y)\,\mathrm{d}\sigma(y)$$

with $\sigma(\mathrm{d}y)$ denoting the induced surface measure on ∂D.

Comment 2: When $q(x) \geqslant q_0 > 0$, the region $\{T < \infty\}$ can be removed from (2.15).

SKETCH OF PROOF. The basic idea for the proof of (2.15) is to consider the process $\widetilde{X}(t)$ obtained from $X(t)$ by killing at the rate $-\mathrm{d}M(t)/M(t)$, which has a generator $\widetilde{G} = (q - \Delta/2)$, as we saw in Subsection 7.3 of Chapter 1. Consider now the martingale

$$u\big(\widetilde{X}(t)\big) - U\big(\widetilde{X}(0)\big) + \int_0^t \widetilde{G}u(\widetilde{X}(s))\,\mathrm{d}s = Z(t)$$

with respect to $\widetilde{P}^x$ on Ω, and apply the optional sampling theorem at times $T \wedge n$, and then let $n \to \infty$, verifying that everything stays finite. Since $Z(0) = 0$ and $\widetilde{P}^x(\widetilde{X}(0) = x) = 1$, we thus obtain

$$u(x) = \widetilde{E}^x\big[U(\widetilde{X}(T)); T < \infty\big] + \widetilde{E}^x\left\{\int_0^T \widetilde{G}u(x(s))\,\mathrm{d}s; T < \infty\right\}$$

and note that, since $\widetilde{X}(T) = \lim \widetilde{X}(T \wedge n) = \lim X(T \wedge n) = X(T)$ and due to the definition of $\widetilde{P}^x$ and since $\widetilde{G}u = f$, the identity found becomes

$$u(x) = E^x\big[M(T)\varphi(X(T)); T < \infty\big] + E^x\left\{\int_0^T M(s)f(X(s))\,\mathrm{d}s; T < \infty\right\}$$

which is (2.15). □

2.2. Diffusions

We know how to define measures $P^\mu(\cdot)$ on $(\Omega, \mathcal{F})$, the sample space of continuous paths on $\mathbb{R}^n$ to answer questions about the statistical nature of microscopic motions behind

$$\frac{\partial u}{\partial t} = \frac{\sigma}{2}\Delta u.$$

Now, we would like to do the same, and be able to describe statistically the microscopic motions behind (or beneath)

$$\frac{\partial u}{\partial t} = Gu = \sum_y \frac{1}{2}a_{ij}(x)\frac{\partial^2}{\partial x_i \partial x_j}u + \sum b_j(x)\frac{\partial}{\partial x_j}u.$$

There are two complementary ways to achieve this. The first consists in taking some preassigned measure $m(\mathrm{d}y)$ on $\mathbb{R}^n$ and proving that

$$\frac{\partial \rho_t}{\partial t}(x, y) = G\rho_t(x, y), \quad t > 0, \tag{2.16}$$

with G acting on the x-variables, has a positive solution such that

$$\int \rho_t(x, y)m(\mathrm{d}y) = 1$$

for all x, with sufficient, smoothness properties. Usually $m(\mathrm{d}y) = \rho_e(y)\,\mathrm{d}y$ with $G^*\rho_e = \frac{1}{2}\sum \partial^2_{ij}a_{ij}(x)\rho_e(x) - \sum \partial_j(b_j\rho_e) = 0$, i.e., $\rho_e(y)$ is some equilibrium density, or perhaps a density $\rho(x)$ such that $\int f(x)Gg(x)\rho(x)\,\mathrm{d}x = \int (Gf)(x)g(x)\rho(x)\,\mathrm{d}x$, i.e., such that G is symmetric with respect to the scalar product $\int f(x)g(x)\rho(x)\,\mathrm{d}x$. With this preassigned ρ, there is an initial condition to be satisfied by $\rho_t(x, y)$, expressed as

$$\rho_{0+}(x, y) = \delta(x - y)/\rho(x). \tag{2.17}$$

With the aid of $\rho_t(x, y)$ one defines

$$P_t f(x) = \int \rho_t(x, y)f(y)m(\mathrm{d}y) \tag{2.18}$$

and then goes through the process of constructing the measures $P^\mu(\cdot)$ on $(\Omega, \mathcal{F}^0, \mathcal{F}_t)$, as specified in Chapter 1. In this fashion a strong Markov process, having continuous trajectories in $\mathbb{R}^n$, is constructed which has the transition semigroup P_t given by (2.2). Another approach, which could be thought of as the stochastic integration-martingale connection consists of two key ingredients.

(i) To prove that there exist measures $\{P^x : x \in \mathbb{R}^n\}$ on the space of all continuous on $\mathbb{R}^n$, such that for every f in $bC^2(\mathbb{R}^n)$

$$Z(t) = f\big(X(t)\big) - f\big(X(0)\big) - \int_0^t (Gf)X_s)\,\mathrm{d}s$$

is a $(\Omega, \mathcal{F}^0, \mathcal{F}_t^0, P^x)$-martingale, where $X(t) : \Omega \to \mathbb{R}^n$ is $X_t(w) = w(t)$ the coordinate map. Even though this theoretical existence result provides us with everything we need, we may want:

(ii) To explicitly construct the measures P^x that do the *job*. Above we showed such a construction that depended on being able to solve (2.1) and define (2.2). The second approach consists in transporting the Wiener measure P_0^x, defined in Section 1, by means of solution of the stochastic differential equation

$$\mathrm{d}X(t) = \sigma(X(t))\,\mathrm{d}B(t) + b(X(t))\,\mathrm{d}t \tag{2.19}$$

which we explain below.

2.2.1. Stochastic integration with respect to Brownian motion

This section is intended to provide a meaning to the first term on the right-hand side of (2.19), and to the resulting associated (stochastic) calculus. The obvious first step is to define $\int_0^t F(t)\,\mathrm{d}B_t$ for functions $F(t, w)$ that are constant $\int_0^t F(s)\,\mathrm{d}B_s$ on intervals $[t_i, t_{i+1})$. In order to do that, we assume our integrand $F(t, w)$ is of the type $F(t) = \sum_{i=0}^{\infty} F_i I_{[t_i, t_{i+1})}(t)$, where $0 = t_0 < t_1 < \cdots$ is an increasing sequence, and we choose $F_i \in \mathcal{F}_{t_i}$ for each i. Actually, when integrating with respect to a discontinuous integrator, we should consider $F_i \in \mathcal{F}_{t_i} = \lim_{s \uparrow t_i} \mathcal{F}_s = \bigcup_{s<t_i} \mathcal{F}_s$. But let us not get fancy for the time being. For such an $F(t)$ it is clear that

$$F \cdot B(t) := \int_0^t F(s)\,\mathrm{d}B(s) = \sum_i F_i(B_{t_{i+1}\wedge t} - B_{t_i \wedge t})$$

and it is clear that, if $t_n \leqslant t < t_{n+1}$, then the last term in the sum is $F_n(B_t - B_{t_n})$. Two facts about $F \cdot B$ are trivial. Assume that for each n, $\sum_{i=1}^n E^x(F_i^2) < \infty$ for any P^x, then

(i) $E^x(F \cdot B(t)) = 0$.

(ii) $E^x(F \cdot B(t))^2 = \sum_i E^x[F_i^2](t_{i+1} \wedge t - t_i \wedge t)$ and again, the last sum is finite. Moreover,

(iii) $E^x|F \cdot B(t)| \leqslant E^x[(F \cdot B(t))^2]$.

(iv) $E^x[F \cdot B(t)|\mathcal{F}_s] = F \cdot B(s)$, for $t \geqslant s$ which follows from the independence of the increments of $B(t)$. Furthermore:

(v) $F \cdot B(t)^2 - \sum (F_i)^2 (t_{t+1} \wedge t - t_i \wedge t)$ is a martingale.

The proof is again a simple computation and if relies on the independence of the increments of the Brownian motion $B(t)$ as well as our demanding that $F_i \in \mathcal{F}_{t_i}$. These results provide us with a proof of

LEMMA 2.3. *Assume that $\sum_{i=1}^n E^x[F_i^2] < \infty$ for each n and that $F_i \in \mathcal{F}_{z_i}$. Put $Z(t) = F \cdot B(t)$, then $Z(t)$ is a P^x-martingale for any P^x, and $Z^2(t) - \int_0^t (F(t)^2)\,\mathrm{d}t$ is again a P^x-martingale.*

The issue now is whether there exists an extension of the procedure outlined above to obtain $\int_0^t F(s)\,\mathrm{d}s$ for any $F(s, w)$, continuous in s and such that $F(s, w) \in \mathcal{F}_s$ and

$$E^x \int_0^t F(s)^2\,\mathrm{d}s = \int_0^t E^x (F(s)^2)\,\mathrm{d}s < \infty \tag{2.20}$$

for each x and each t. The obvious thing to do is to take a sequence of partitions of $[0, \infty)$; we take $t(n, i) = i/2^n$, $i = 0, 1, \ldots$ define the approximant $F_i^{(n)} = F(\frac{i}{2^n})$ to F. Note that the continuity of F and condition (2.20) imply that

$$E^x \int_0^t (F^n(s) - F(s))^2\,\mathrm{d}s = \int_0^t E(F^n(s) - F(s))^2\,\mathrm{d}s \to 0$$

as $n \to \infty$. Again, $F^{(n)}(s) := \sum F_i^n I_{[t(n,i),t(n,i+1))}(t)$. The *stochastic integral*

$$Z^{(n)}(t) = F^{(n)} \cdot B(t) = \int_0^t F^{(n)}(s)\,\mathrm{d}B(s) \tag{2.21}$$

was defined above and some of its basic properties collected in Lemma 2.6. In order to deal with the passage to the limit in (2.21), we have to take care of three aspects:

(a) The ever-present but never explicitly mentioned random element ω in $Z^{(n)}(t)$: The limit $Z(t)$ will be expected to be in $\mathcal{F}_t$.

(b) How the sequence converges, and

(c) To make sure that the convergence is true for the chosen sequence of partitions $\{[t(n, i), t(n, i+1);\ i = 0, 1, \ldots\}$ of $[0, \infty)$.

Certainly, getting (a) right will depend on (b). To see what the problem in (c) is, consider two consecutive values of n and assume for the moment that $F^n(s)$ is bounded. Consider now $|Z^n(t) - Z^{(n+1)}(t)|$. An elementary argument leads to

$$
\begin{aligned}
&E^x\left|Z^{(n)}(t) - Z^{(n+1)}(t)\right| \\
&\quad \leqslant 2\|f\|_\infty \sum E^x|B(t(n,i+1)\wedge t - B(t(n,i)\wedge t| \\
&\qquad \times \sqrt{\frac{2}{\pi}} 2\|f\|_\infty \sum |t(n,n+1)\wedge t - t(n,i)\wedge t|
\end{aligned}
$$

and it is easy to cook up a partition for which the sum diverges. What is at stake here is the fact that the trajectories of the standard Brownian notion are (almost surely) non-differentiable. But things go smoothly when we take convergence in the squared norm. For now

$$E^x\left\|Z^{(n)}(t) - Z^{(n+1)}(t)\right\|^2 \to 0$$

as $n \to \infty$, which provides us with a limit random variable $Z(t) \in \mathcal{F}_t$, such that $E^x|Z(t) - Z^{(n)}(t)|^2 \to 0$.

DEFINITION 2.2. We call $Z(t)$ the *stochastic integral* of $F(t)$ with respect of $\mathrm{d}B$ and write

$$Z(t) = \int_0^t F(s)\,\mathrm{d}B(s).$$

Again, the following properties of $Z(t)$ are expected and intuitive, given all the above, but their proof is lengthy.

PROPOSITION 2.1. *Let $F : [0,\infty) \times \Omega \to \mathbb{R}$ be such that $F(t) \in \mathcal{F}_t$, $t \to F(t,w)$ is continuous and $E^x \int_0^t F(s)^2\,\mathrm{d}s < \infty$. Define the stochastic integral $Z(t) = F \cdot B(t) = \int_0^t F(s)\,\mathrm{d}B(s)$ as above. Then*

(i) *$Z(t)$ is a martingale.*
(ii) *$Z^2(t) - \int_0^t F(s)^2\,\mathrm{d}s$ is a martingale.*
(iii) *If $G(t,w)$ is continuous in t, $\mathcal{F}_t$-measurable and for every t, $\sup_{w\in\Omega,\, s\leqslant t}|G(s,w)| < M(t)$, then*

$$Y(t) = \int_0^t G(s)\,\mathrm{d}Z(s) = \int_0^t G(s)F(s)\,\mathrm{d}B(s)$$

and

(iv) *$Y(t)$ is a martingale such that*

$$Y^2(t) - \int_0^t G^2(s)F(s)^2\,\mathrm{d}s$$

is a martingale.

(v) *If* $F_1(s)$ *and* $F_2(s)$ *are as* $F(t)$ *above, and* $Z_1(t)$ *and* $Z_2(t)$ *are correspondingly defined, then*

$$Z_1(t)Z_2(t) - \int_0^t F_1(s)F_2(s)\,\mathrm{d}s$$

is again a martingale.

The main result in this section, which contains the needed proofs to justify some computations in (1.5), is contained in

THEOREM 2.2 (Ito's formula). *Let* $f : \mathbb{R} \times \mathbb{R} \to \mathbb{R}$ *be a function that is once continuously differentiable in* t *and twice continuous differentiable in* $\mathbb{R}$. *Let* F *be as in the statement of Proposition* 2.6 *and* $g : [0, \infty) \times \Omega \to \mathbb{R}$ *be continuous in* t, $g(t) \in \mathcal{F}_t$, $\forall t \geqslant 0$ *and put* $A(t) = \int_0^t g(s)\,\mathrm{d}s$. *Consider* $f(A(t), Z(t))$. *Then*

$$\begin{aligned} f(A(t), Z(t)) &= f(0,0) + \int_0^t \frac{\partial f}{\partial t} g(s)\,\mathrm{d}s + \int_0^t \frac{\partial f}{\partial x} F(s)\,\mathrm{d}B(s) \\ &\quad + \frac{1}{2}\int_0^t \frac{\partial^2 f}{\partial x^2} F^2(s)\,\mathrm{d}s. \end{aligned} \tag{2.22}$$

SKETCH OF PROOF. Take a sequence $\{[t(m,i), t(n,i+1)),\ i \geqslant 0\}$ of partitions, with $\sup_i \Delta t(n,i) \to 0$ as $n \to \infty$; expand $f(t,x)$ up to the first order in t and second order in x to obtain

$$\begin{aligned} f(A(t), Z(t)) - f(0,0) &= \sum \frac{\partial f}{\partial t}(\xi_i(t))\big(A(t \wedge t(n,i+1)) - A(t \wedge t(n,i0))\big) \\ &\quad + \sum \frac{\partial f}{\partial x}(\xi_i(t))\big\{Z(t \wedge t(n,i+1)) - Z(t \wedge t(n,i))\big\} \\ &\quad + \frac{1}{2}\sum \frac{\partial^2 f}{\partial x^2}(\xi_i(t))\big\{Z(t \wedge t(n,i+1)) - Z(t \wedge t(n,i))\big\}^2 \\ &\quad + \text{remainder}. \end{aligned}$$

The rest (which is the hard part) of the proof consists of proving that each sum converges to the corresponding integral and the remainder tends to zero as the partition size tends to zero. □

Comment: Usually (2.22) is written in "differential form" as

$$\mathrm{d}f = \frac{\partial f}{\partial t} g(s)\,\mathrm{d}s + \frac{\partial f}{\partial x} F(s)\,\mathrm{d}B(s) + \frac{1}{2}\frac{\partial^2 f}{\partial x^2} F^2(s)\,\mathrm{d}s. \tag{2.23}$$

Let us go back to Chapter 1 and consider examples (1.21) and Lemma 1.1. Let $f(x)$ have two continuous derivatives. According to (2.23)

$$f\big(B(t)\big) - f\big(B(0)\big) - \int_0^t \frac{1}{2}\frac{\partial^2 f}{\partial x^2}(B(s))\,\mathrm{d}s = \int_0^t \frac{\partial f}{\partial x}(B(s))\,\mathrm{d}B_s.$$

Due to the continuity of the derivative $\partial f/\partial x$ we see that the right-hand side is a martingale, according to Proposition 2.6. Let us consider f to have two continuous derivatives and to be strictly positive. Then $\log f(x)$ is well defined and

$$\begin{aligned}\log f(B(t)) &= \log f(B(0)) + \int_0^t \left(\frac{f'}{f}\right)(B(s))\,\mathrm{d}B(s)\\ &\quad + \frac{1}{2}\int_0^t \left\{\frac{f''}{f}(B(s)) - \left(\frac{f'}{f}\right)^2(B(s))\right\}\mathrm{d}s,\end{aligned}$$

after exponentiating and rearranging, we obtain

$$\frac{f(B(t))}{f(B(0))}\mathrm{e}^{-\int_0^t(\frac{Gf}{f})(B(s))\,\mathrm{d}s} = \exp\left\{\int_0^t\left(\frac{f'}{f}\right)(B(s))\,\mathrm{d}B(s) - \frac{1}{2}\int_0^t\left(\frac{f'}{f}\right)^2(B(s))\,\mathrm{d}s\right\}$$

and we leave for the reader to apply (2.22) to the obvious thing to verify that the right-hand side satisfies

$$\mathrm{d}R(t) = R(t)\left(\frac{f'}{f}\right)(B(s))\,\mathrm{d}B(s)$$

if we denote it by $R(t)$. That is, the left-hand side is a martingale. Let us round up this section by stating the multidimensional extensions of Proposition 2.6 and Theorem 2.2.

PROPOSITION 2.2. *Let $B_1, \ldots, B_n$ be independent Brownian motions or, $(B_1, \ldots, B_n)$ be an n-dimensional Brownian motion. Let $\sigma(t, w)$ be a continuous (in t) $(k \times n)$-matrix-valued function, such that each entry $\sigma_{ij}(t) \in \mathcal{F}_t$. Then the $\mathbb{R}^k$-valued stochastic integral*

$$Z(t) = Z(0) + \int_0^t \sigma(s)\,\mathrm{d}B(s) \tag{2.24}$$

is perfectly well defined (when $Z(0) \in \mathcal{F}_0$), it is a martingale such that

$$(Z(t) - Z(0))(Z(t) - Z(0))^* - \int_0^t \sigma(s)\sigma(s)^*\,\mathrm{d}s \tag{2.25}$$

is a matrix valued martingale with the components

$$(Z_i(t) - Z_i(0))(Z_j(t) - Z_j(0)) - \int_0^t \sum_k \sigma_{ik}(s)\sigma_{jk}(s)\,\mathrm{d}s.$$

THEOREM 2.3 (Multidimensional Ito's formula). *Let the process $Z(t)$ be as in Proposition 13, and $g : [0,\infty) \times \Omega \to \mathbb{R}^m$ be continuous in t, and let $g_i(t) \in \mathcal{F}_t$ for $1 \leqslant i \leqslant k$. Put $A(t) = \int_0^t g(s)\,\mathrm{d}s$, and let $f : \mathbb{R}^m \times \mathbb{R}^k \to \mathbb{R}$ be once continuously differentiable in the first m variables and twice continuously differentiable in the last k variables. Let $A(0)$ and $Z(0)$ be $\mathbb{R}^k$-valued, $\mathcal{F}_0$-measurable random variables. Then*

$$\begin{aligned} &f\big(A(t) + A(0), Z(t) + Z(0)\big) \\ &\quad = f(A(0), Z(0)) + \sum_1^m \int_0^t \frac{\partial f}{\partial t_\ell}(s) g_i(s)\,\mathrm{d}s + \sum_{s=1}^k \sum_{k=1}^k \int_0^t \frac{\partial f}{\partial x_j} \sigma_{jk}(s)\,\mathrm{d}B_k \qquad (2.26) \\ &\qquad + \frac{1}{2} \sum_{\ell=1}^k \sum_{i,j,j=1}^k \int_0^t \frac{\partial^2 f}{\partial x_i \partial x_j} \sigma_{i\ell}\sigma_{j\ell}\,\mathrm{d}s. \end{aligned}$$

Homework: Deduce (2.15) from (2.17).

2.2.2. The Cameron–Martin–Girsanov–Maruyama transformation

As the usual Lebesque measure on $\mathbb{R}^n$ is invariant under translations and rotations, or the induced Lebesque measure on the n-dimensional sphere is invariant under rotations, the Wiener measures P_0^x on the space $(\Omega, \mathcal{F})$ are invariant under translations and rotations in the following sense. Let $S : \mathbb{R}^n \to \mathbb{R}^n$ be either a translation or a rotation, and again denote by the same symbol the induced map $S : \Omega \to \Omega$, $B_t(Sw) = S(w(t))$. It is easy to see, starting from cylindrical functions, that for any $H \in b\mathcal{F}$

$$E^x[H] = E^{S(x)}[H \circ S].$$

The question addressed in various forms by the authors cited in the title of the section is the following: Let $b : [0,\infty) \times \Omega \to \mathbb{R}$ be some continuous function such that $b(t) \in \mathcal{F}_t$, and let $a(t) = \int_0^t b(s)\,\mathrm{d}s$. Define $T_a : \Omega \to \Omega$ by $T_a(w) = w + a$ or $B_t(T_a w) = w(t) + a(t)$ and the question is how P^x is transported by T_a, or in simpler terms: keep P^x and a fixed and vary H in $E^x[H \circ T_a]$. Certainly one obtains P_a^x such that $E^x[H \circ T_a] = E_a^x[H] = \int H\,\mathrm{d}P_a^x$, and the question is: Does there exist $\mathrm{d}P_a^x/\mathrm{d}P^x$? What are its properties? The answer is contained in

THEOREM 2.4. *With the same notation and assumptions as above, if H is in $\mathcal{F}_t$*

$$E^x[H \circ T_a] = E^x[H M_t], \qquad (2.27)$$

where $M_t = \exp\{\int_0^t b(s)\,\mathrm{d}B(s) - \frac{1}{2}\int_0^t b^2(s)\,\mathrm{d}s\}$.

To see "why" this must be true and that there is nothing else involved than just completing squares, consider $H(w) = \exp(ikB_t(w))$, and assume that b is just a constant, so that $B_t(T_a(w)) = w(t) + bt$. Thus

$$E^x\left[e^{ikB_t}\circ T_a\right] = E^x\left[e^{ik(B_t+bt)}\right]$$
$$= e^{ikbt}E^x\left[e^{ik(B(t)-B(0))}e^{ikB(0)}\right] = e^{ikx}e^{ikbt}e^{-\frac{k^2t}{2}}.$$

Consider now

$$E^x\left[e^{ikB(t)}e^{\int_0^t b\,\mathrm{d}B(s)-\frac{1}{2}\int_0^t b^2\,\mathrm{d}s}\right] = E^x\left[e^{ikB(0)}e^{ik(1-ibk)(B(t)-B(0))}\right]e^{-\frac{1}{2}b^2t}$$
$$= e^{ikx}e^{-k^2(1-ib/k)^2t/2}e^{-b^2t/2} = e^{ikx}e^{ikbt}e^{-k^2t/2}.$$

Actually, using the continuity of $b(t)$ and the independence of the increments of B, the computation of (2.27) for H of the type $f(x_{t_1}, x_{t_2}, \ldots, x_{t_n})$ with $0 \leqslant t_1 < \cdots < t_n \leqslant t$ can be reduced to the above computation. Recall that we interpret the coordinate maps $X_t : \Omega \to \mathbb{R}^n$ as the positions of a randomly moving particle, and it is the probability laws on $(\Omega, \mathcal{F})$ that determine the statistical nature of randomness. Thus, a natural question is what the transition semigroup and infinitesimal generator are when P^x gets changed into P_a^x. Now $b(s, w)$ will be of the type $b(X_s)$ for some continuous $b(x)$. Note that the new transition semigroup P_t^a is given by

$$P_t^a f(x) = E_a^x[f(X_t)] = E^x[f(X_t)M_t]$$

with M_t as in (2.2). From Ito's formula we can rewrite the integrand above as follows:

$$f(X_t)M_t = f(X_0)M(0) + \int_0^t b(X_s)M(s)f(X_s)\,\mathrm{d}B(s) + \int_0^t M(s)f'(X_s)\,\mathrm{d}B(s)$$
$$+ \int_0^t b(X_s)M(s)f'(X_s)\,\mathrm{d}s + \frac{1}{2}\int_0^t M(s)f''(X_s)\,\mathrm{d}s,$$

and, since the two middle terms are martingales which vanish at $t = 0$, we have

$$P_t^a f(x) = E^x[f(X_t)M_t]$$
$$= f(x) + \int_0^t E^x[b(X_s)M_s f'(X_s)]\,\mathrm{d}s + \frac{1}{2}\int_0^t E^x[M(s)f''(X_s)]\,\mathrm{d}s$$

or, equivalently,

$$\frac{\partial}{\partial t}P_t^a f(x)\bigg|_{t=0} = \frac{1}{2}\frac{\partial^2}{\partial x^2}f(x) + b(x)\frac{\partial f}{\partial x}(x). \tag{2.28}$$

The moral thus being, that to add drift to a generator, it suffices to change measures in the path space.

2.3. Diffusions as solutions of stochastic equations

The aim of this section is to present diffusions as solutions to stochastic differential equations. As we have said above, we start with the standard Brownian motion $B(t)$ realized as coordinate maps in the sample space $(\Omega, \mathcal{F})$ of continuous maps in $\mathbb{R}^n$, on which we have put the Wiener measure P_0^x. Our goal will be to transport the measure P_0^x by means of the solution of

$$dX(t) = \sigma(X(t))\, dB(t) + b(X(t))\, dt$$

and to think of $X(t)$ as coordinate maps on $(\Omega, \mathcal{F})$. The coordinate maps will provide us with a diffusion with respect to the transported measure P^x.

2.3.1. *The basic stochastic differential equation*

The existence and uniqueness of solutions to

$$dX(t) = \sigma(X(t))\, dB(t) + b(X(t))\, dt, \qquad X(0) = x, \quad t > s$$

is well analyzed. As usual, the above equation is a shorthand for

$$X(t) = X(s) + \int_s^t \sigma(X(s))\, dB(s) + \int_s^t b(X_s)\, ds, \tag{2.29}$$

where $x \to \sigma(x)$ is an $(n \times n)$-matrix-valued function on $\mathbb{R}^n$, $x \to b(x)$ is an $\mathbb{R}^n$-valued function on $\mathbb{R}^n$ such that

$$|\sigma(x) - \sigma(y)| + |b(x) - b(y)| \leqslant k|x - y|$$

for some positive constant k. Furthermore we assume a restricted growth condition of $\sigma(x)$ and $b(x)$ of the type

$$|\sigma(x)|^2 + |b(x)|^2 \leqslant K^2(1 + |x|^2),$$

i.e., the growth in $|b(x)| = \sup |b_i(x)|$ and $|\sigma(x)| = \sup_y |\sigma_y(x)|$ is not faster than linear for large $|x|$. On $(\Omega, \mathcal{L}, P^0)$ we take $B(t)$ to denote the standard Brownian motion in $\mathbb{R}^n$ started at 0. Under the assumptions on $\sigma(x)$ and $b(x)$ stated above, for $s \in [0, T]$ the following iterative scheme

$$\begin{aligned} X_x^{(0)}(t, s) &= x \quad \text{for } t > s, \\ X_x^{(n+1)}(t, s) &= x + \int_s^t f(x^n(s))\, ds + \int_s^t \sigma(X^{n-1})\, dB \end{aligned}$$

provides us with a Cauchy sequence on $[s, T]$, i.e., such that

$$\sup_{s \leqslant t \leqslant T} \left| X_x^n(t) - X_x^{n+1}(t) \right| \to 0.$$

A systematic procedure allows us to verify that

$$X_x(t, s) = \lim X_x^{(n)}(t, s) \tag{2.30}$$

satisfies equation (1). To see why we began with P_0^0 as a measure on $(\Omega, \mathcal{F})$, note that when $\sigma = I$, the $n \times n$ identity matrix and $f \equiv 0$ and $s = 0$

$$X_x(t) = x + (B(t) - B(0))$$

and it is easy to see that for any $0 < t_1 < \cdots < t_n$, and any bounded $f(x_1, \ldots, x_n)$

$$E_0^0[f(X_1(t_1), \ldots, X_2(t_n)] = E_0^x[F(B(t_1), \ldots, B(t_n)],$$

i.e., we only shifted the origin.

2.3.2. The Markov nature of the solutions

We describe here how the independence of the increments of the Brownian motion B determine the Markov property of the solutions to (2.29). What we want to prove is the following: if we denote $\widehat{\mathcal{F}}_t = \sigma(X_x(u) : u \leqslant t)$, and we let $t > s \geqslant 0$, then for bounded $f(x)$,

$$E\big[f(X_x(t))|\widehat{\mathcal{F}}_s\big] = E^0\big[f(X_x(t))|X_x(s)\big] \tag{2.31}$$

where for $t > 0$, $X_x(t)$ satisfies

$$X_x(t) = x + \int_0^t \sigma(X_x(u))\,\mathrm{d}B(u) + \int_0^t (X_x(u))\,\mathrm{d}u \tag{2.32}$$

and it is clear that for $s < t$

$$X_x(t) = X_x(s) + \int_s^t \sigma(X_x(u))\,\mathrm{d}B(u) + \int_s^t b(X_x(u))\,\mathrm{d}u, \tag{2.33}$$

and that the two integrals on the right-hand side are independent of $X_x(s)$. To see it, just approximate by sums in such a way that s always correspond to a partition point. This type of argument shows several things:

(a) $X_x(t) \in \mathcal{F}_t$ for all t.
(b) For any t, $\sigma(X_x(u) : u \leqslant t) := \widehat{\mathcal{F}}_t \subset \mathcal{F}_t$.
(c) $X_x(s) \in \mathcal{F}_s$ is independent of $\int_s^t \sigma(X_x(u))\,\mathrm{d}B_u + \int_s^t b(X_x(u))\,\mathrm{d}u$.

Taking all this into account and making use of (1.2.2), we can easily convince ourselves that for $t > s$ and bounded f

$$\begin{aligned} E^0\big[f(X_x(t))|\widehat{\mathcal{F}}_s\big] &= E^0\big\{E^0\{f(X_x(s) + \Delta)|\mathcal{F}_s\}|\widehat{\mathcal{F}}_s\big\} \\ &= E^0\big\{E^0\{f(\xi + \Delta)\}|_{\xi = X_x(s)}|\widehat{\mathcal{F}}_s\big\} \\ &= E^0\{f(\xi + \Delta)\}|_{\xi = X_x(s)}, \end{aligned}$$

where, of course, Δ stands for the last two integrals in (2.33), and since $X(s) \in \widehat{\mathcal{F}}_s$, the last step clearly implies (2.31). If we define the family of mappings

$$K_x : \Omega \to \Omega, \qquad w \to \gamma_x, \tag{2.34}$$

where γ_x satisfies (2.33) for $s = 0$, we can define on $(\Omega, \mathcal{F}^0)$ the family of measures

$$E^x[H] = E_0^0[H \circ K_x], \tag{2.35}$$

where P_0^0 is the Wiener measure on $(\Omega, \mathcal{F}^0)$ started at 0, and H is a bounded, $\mathcal{F}^0$-measurable function. In terms of these P^x, $x \in \mathbb{R}^n$ one can define the family of operators

$$P_{t+s} f(x) := E^x f(X_t) = E_0^0 f(X_x(t)), \tag{2.36}$$

and, if one verifies the intuitively obvious identity $X_x(t+s) = X_{X_x(t)}(s)$, then, making use of the Markov property described above, one obtains the Chapman–Kolmogorov identity for $P_t f(x)$

$$P_{t+s} f(x) = P_t P_s f(x).$$

To finish this list of properties, the Ito formula in (1.17) yields

$$\begin{aligned} \frac{\partial}{\partial t} P_t f(x)\bigg|_{t=0} &= \frac{\partial}{\partial t} E_0^0\{f(X_x(t))\}|_{t=0} \\ &= \sum_i b_i(x) D_i f(x) + \frac{1}{2} \sum_{ij} a_{ij}(x) \partial_{ij}^2 f(x) \end{aligned}$$

for any f with two continuous bounded derivatives. The matrix $a(x)$ is related to $\sigma(x)$ by $a(x) = \sigma(x)\sigma(x)^+$, $\sigma(x)^+$ denoting the transpose of $\sigma(x)$.

2.4. Reflected diffusions

This section is designated to recalling some results about the solutions to

$$\begin{aligned} &\frac{\partial u}{\partial t}, (t, x) = (\mathcal{L} - q)u(t, x), \quad (t, x) \in (0, \infty) \times D, \\ &u(0+, x) = f(x), \quad x \in D, \\ &(\operatorname{grad} u(t, x), \gamma(x)) = 0, \quad t > 0, x \in \partial D, \end{aligned} \tag{2.37}$$

where D is a compact domain in $\mathbb{R}^3$, with a smooth boundary ∂D (meaning that the tangential derivatives of the inner normal $\gamma(x)$ to ∂D exist at least up to third order), and $f(x)$ is assumed bounded in D. The operator $\mathcal{L}$ being given by

$$\mathcal{L} = \frac{1}{2} \sum_{ij} a_{ij}(x) \partial_{ij} + \sum b_i(x) \partial_i \tag{2.38}$$

with $x \to a_{ij}(x)$ being at least twice continuously differentiable, with Lipschitz continuous derivatives of order k, $0 < k \leqslant 1$, and $x \to b_i(x)$ being at least continuously differentiable, with Lipschitz continuous derivatives of order k, $0 < k \leqslant 1$. Also, $x \to q(x)$ is assumed to be Lipschitz continuous. Recall that this means that there exists positive A such that $|q(x) - q(y)| \leqslant A|x-y|^k$. The matrix a is assumed to be strictly positive definite at each $x \in \overline{D}$ and $g_{ij} = (a^{-1})_{ij}$ defines a geodesic distance $d(x, y)$. We denote $\det g_{ij}(x)$ by $g(x)$. We assume the existence of a strictly positive solution to $\frac{1}{2}\sum_j \partial_{ij} a_{ij}(x)\rho = b_i(x)\rho$ with the aid of which $\mathcal{L}$ can be rewritten either as

$$\mathcal{L} = \frac{1}{2}\frac{1}{\rho(x)}\sum \partial_i(\rho(x)a_{ij}(x)\partial_j) \tag{2.39}$$

or as

$$\mathcal{L} = \frac{1}{2}\frac{1}{\sqrt{g}}\sum_j \partial_j(a_{ij}\sqrt{g}\partial_i) + \sum c_i\partial_i = \mathcal{L}_0 + \sum c_i\partial_i, \tag{2.40}$$

where

$$c_i(x) = \frac{1}{2}\sum_j a_{ij}(x)\partial_j \log\left(\frac{\rho(x)}{\sqrt{g(x)}}\right).$$

In the former case, $\mathcal{L}$ is symmetric with respect to the scalar product $(f, g)_\rho = \int f(x)g(x)\rho(x)\,\mathrm{d}x$, while in the latter, $\mathcal{L}_0$ is symmetric with respect to the scalar product $(f, g)_{\sqrt{g}} = \int f(x)g(x)\sqrt{g(x)}\,\mathrm{d}x$. Defining $m(A) = \int_A \sqrt{g(x)}\,\mathrm{d}x$, we can induce a measure m on ∂D as follows: let $D(\varepsilon) = \{x \in D : d(x, \partial D) < \varepsilon\}$ and $F(x) : \overline{D(\varepsilon)} \to \mathbb{R}$ be any continuous function. Then

$$\int_{\partial D} F(\xi)\widetilde{m}(\mathrm{d}\xi) = \lim_{\varepsilon \downarrow 0}\frac{1}{\varepsilon}\int_{D(\varepsilon)} F(x)m(\mathrm{d}x).$$

It is known that there exist positive, continuous, transition densities

$$\rho_0(t, x, y), k_\beta(t, x, y),$$

with respect to $\rho(y)\,\mathrm{d}y$ and $m(\mathrm{d}y)$, respectively, such that the solution to each of the following problems

$$\begin{aligned}
&\text{(a)}\quad \frac{\partial u}{\partial t} = \mathcal{L}u, \quad (t, x) \in (0, \infty) \times D,\\
&\text{(b)}\quad u(0+, x) = f(x), \quad x \in D,\\
&\text{(c)}\quad (\operatorname{grad} u(t, x), \gamma(x)) = 0, \quad t > 0, x \in \partial D
\end{aligned} \tag{2.41}$$

and for $\beta > 0$

$$\begin{aligned}
&\text{(a)}\quad \frac{\partial v}{\partial t} = (\mathcal{L} - \beta q)v, \quad (t, x) \in (0, \infty) \times D,\\
&\text{(b)}\quad v(0+, x) = f(x), \quad x \in D,\\
&\text{(c)}\quad (\operatorname{grad} v(t, x), \gamma(x)) = 0, \quad t > 0, x \in \partial D
\end{aligned} \tag{2.42}$$

can, respectively, be represented by

$$u(t,x) = \int \rho_0((t,x,y) f(y) \rho(y)\, \mathrm{d}y \tag{2.43}$$

and also,

$$v(t,x) = \int k_\beta(t,x,y) \sqrt{a(y)} f(y)\, \mathrm{d}y. \tag{2.44}$$

Note that as $\beta = 0$, (2.41) and (2.42) are coincidental, so the uniqueness would require that

$$k_0(t,x,y) \frac{\sqrt{a(y)}}{\rho(y)} = \rho_0(t,x,y)$$

and be symmetric in x and y since $\mathcal{L}$ is symmetric with respect to $(f,g)_\rho$. The two choices of reference measure are motivated by the fact that the drift term in (2.40) may be vanishing, or compensated by an appropriate change of measure, leaving us with $\mathcal{L}_0$ which is symmetric with respect to $(f,g)_a$. It follows from the Feynmann–Kac formula that

$$\begin{aligned} u(t,x) &= E^x\left[f(X(t)) \mathrm{e}^{-\beta \int_0^t q(X_s)\, \mathrm{d}s}\right] = Q_t^\beta f(x) \\ &= \int k_\beta(t,x,y) f(y) \sqrt{a(y)}\, \mathrm{d}y, \end{aligned} \tag{2.45}$$

where $X(t)$ is the process having the transition probability $k_0(t,x,y)$. Writing

$$\mathrm{e}^{-\beta \int_0^t q(X_s)\, \mathrm{d}s} = 1 - \beta \int_0^t q(X(s)) \mathrm{e}^{-\beta \int_0^s q(X(u))\, \mathrm{d}u}\, \mathrm{d}s,$$

substituting it in (2.45) and using the Markov property, one obtains

$$Q_t^\beta f(x) = Q_t^0 f(x) - \beta \int_0^t \mathrm{d}s\, Q_t^\beta q Q_s^0 f$$

or in terms of the densities

$$k_\beta(t,x,y) = k_0(t,x,y) - \beta \int_0^t \mathrm{d}s \int_D k_\beta(t-s,x,z) q(z) k_0(s,z,y) \sqrt{a(z)}\, \mathrm{d}z.$$

Denoting $k_\beta(t,x,y)\sqrt{a(y)}/\rho(y)$ by $\bar{\rho}_\beta(t,x,y)$, we obtain from the above

$$\bar{\rho}_\beta(t,x,y) = \rho_0(t,x,y) - \beta \int_0^t \mathrm{d}s \int_D \bar{\rho}_\beta(t-s,x,z) q(z) \rho_0(t,z,y) \rho(z)\, \mathrm{d}z$$

from which it follows by iteration that, if $\rho_0(t,x,y) = \rho_0(t,y,x)$, then $\rho_\beta(t,x,y) = \bar{\rho}_\beta(t,y,x)$. A different, slicker, and more comprehensive proof, is as follows: Let (f,g) be a scalar product such that $(f, \mathcal{L}, g) = (\mathcal{L}f, g)$ for f and g in the domain of $\mathcal{L}$. We prove now that $(U^\alpha f, g) = (f, U^\alpha g)$ from which

$$\int_0^\infty \mathrm{e}^{-\alpha t}\big(Q_t^\beta f, g\big)\,\mathrm{d}t = \int_0^\infty \big(f, Q_t^\beta g\big)\mathrm{e}^{-\alpha t}\,\mathrm{d}t$$

follows. Since (see Chapter 1, Section 7.3) for a dense class of f's and g's, $U^\alpha g$ and $U^\alpha f$ satisfy $(\alpha - (\mathcal{L} - \beta q))U^\alpha f = f$ and $\{\alpha - (\mathcal{L} - \beta q)U^\alpha g = g$, we have

$$\begin{aligned}(U^\alpha f, g) &= \big(U^\alpha f, \{\alpha - (\mathcal{L} - \beta a)\}U^\alpha g\big) = \big(\{\alpha - (\mathcal{L} - \beta q)\}U^\alpha f, U^\alpha g\big)\\ &= (f, U^\alpha g),\end{aligned}$$

where the second step follows from the symmetry of the operator $\mathcal{L}$. Now Fubini's theorem and the uniqueness of the Laplace transform operation imply that

$$\big(Q_t^\beta f, g\big) = \big(f, Q_t^\beta g\big)$$

from which the desired symmetry of the transition density drops out. Actually, instead of (2.42), we know that the density $k_1(t, x, y)$ satisfies

$$\begin{aligned}&\frac{\partial k}{\partial t}(t, x, y) = (\mathcal{L} - q)k(t, x, y), \quad (t, x) \in (0, \infty) \times D \times \overline{D},\\ &k(0+, x, y) = \delta(x - y)/\sqrt{a(x)},\\ &(\operatorname{grad} k(t, x, y), \gamma(x)) = 0.\end{aligned} \tag{2.46}$$

With this knowledge, it is easy to verify that the solution to the nonhomogeneous Neumann problem

$$\begin{aligned}&\text{(a)}\quad \frac{\partial v}{\partial t} = (\mathcal{L} - q)v, \quad (t, x) \in (0, \infty) \times D,\\ &\text{(b)}\quad v(0+, x) = f(x),\\ &\text{(c)}\quad (\operatorname{grad} v(t, x), \gamma(x)) + \varphi(t, x) = 0,\end{aligned} \tag{2.47}$$

where $\varphi(t, x)$ is continuous in $(0, \infty) \times \partial D$, is given by

$$v(t, x) = \int_D k_1(t, xy) f(y) m(\mathrm{d}y) + \int_{\partial D} k_1(t, x, y)\varphi(t, y)\widetilde{m}(\mathrm{d}y). \tag{2.48}$$

If the consistency of the initial conditions at ∂D and the boundary conditions were required, we would need to have

$$\lim_{x \in D \to y \in \partial D} f(x) = \lim_{t \downarrow 0} \varphi(t, y) \quad \text{at } y \in \partial D.$$

If, instead of (2.37), had we wanted to solve

$$\begin{aligned} &(\alpha - (\mathcal{L} - q))v(\alpha, x) = -f(\alpha, x), \quad \alpha > 0, x \in D, \\ &(\operatorname{grad} v(\alpha, x), \gamma(x)) + \tilde{\varphi}(\alpha, x) = 0, \quad \alpha > 0, x \in \partial D \end{aligned} \tag{2.49}$$

using (2.48), the solution could be represented as

$$v(\alpha, x) = \int_D G_\alpha(x, y) f(x, y) m(\mathrm{d}y) + \int_{\partial D} G_\alpha(x, y) \tilde{\varphi}(\alpha, y) \widetilde{m}(\mathrm{d}y) \tag{2.50}$$

where, of course,

$$G_\alpha(x, y) = \int_0^\infty k_\alpha(t, x, y) \mathrm{e}^{-\alpha t} \, \mathrm{d}t.$$

To close this section, we are going to write down the stochastic differential equation that is to be satisfied by a diffusion in $\mathbb{R}^n$ which is reflected at the boundary ∂D of D as soon as it gets there. For that we need to find (X, L) in (the Skorohod equation)

$$\begin{aligned} &\mathrm{d}X(t) = \sigma(X(t)) \, \mathrm{d}B(t) + b(X(t)) \, \mathrm{d}t + I_{\partial D}(X) \gamma(X(t)) \, \mathrm{d}L_t, \\ &X(0) = x, \end{aligned}$$

where, as above, $\sigma\sigma^+ = a$, and L_t is an increasing additive functional such that

$$L_t = \int_0^t I_{\partial D}(X_s) \, \mathrm{d}L_s,$$

i.e., L_t grows only when the diffusion X visits ∂D. Note that the last term describes the "random kick" that the particle diffusing, according to the first two terms, gets back into D as soon as it reaches the boundary. Using (X, L), the solution $v(\alpha, x)$ to (2.49) can be represented by means of

$$\begin{aligned} v(\alpha, x) = &\int_0^\infty \mathrm{e}^{-\alpha t} E^x \left\{ \mathrm{e}^{-\int_0^t q(X(u)) \, \mathrm{d}u} f(\alpha, X(t)) \right\} \mathrm{d}t \\ &+ E^x \left\{ \int_0^\infty \mathrm{e}^{-\alpha t} \mathrm{e}^{-\int_0^t q(X(u)) \, \mathrm{d}u} \tilde{\varphi}(\alpha, X(t)) \, \mathrm{d}L_t \right\}. \end{aligned}$$

2.5. Killed diffusions and some fundamental identities

We are interested in representing probabilistically the solution to

$$\begin{aligned} &\frac{\partial u}{\partial t}(t, x) = \mathcal{L}u(t, x) + s(t, x), \quad (t, x) \in (0, \infty|x|), \\ &u(0, x) = f(x), \quad x \in D, \\ &u(t, x) = h(t, x), \quad (t, x) \in (0, \infty)_x \partial D. \end{aligned} \tag{2.51}$$

Again, D is a domain in $\mathbb{R}^d$, bounded or unbounded, such that every point $x \in \partial D$ is regular for D^c. Besides, both $f(x)$ and $h(t,x)$ are bounded continuous in their domains. If the consistency is wanted at $t=0$, we must assume $f(x)$ extensible to $\overline{D}$ and $h(0+,x)=f(x)$ for $x \in \partial D$. If we take the Laplace transform of problem (2.51), we obtain

$$\begin{aligned} &(\alpha - \mathcal{L} \sim u(\alpha,x) = \sim s_1(\alpha,x) = \sim s(\alpha,x) + \alpha f(x), \quad x \in D, \alpha > 0, \\ &\sim u(\alpha,x) = \sim h(\alpha,x), \quad x \in \partial D_0. \end{aligned} \tag{2.52}$$

In both (2.51) and (2.52), the boundary conditions are to hold in the limit. For example, for $x \in \partial D$, $\sim u(\alpha,x) = \lim_{y \in D \to x \in \partial D} \sim u(\alpha,y) = \sim h(\alpha,x)$. If $X(t)$ denotes the diffusion on $\mathbb{R}^d$ with an infinitesimal generator $\mathcal{L} = \sum a_i \partial_{ij} + \sum_i \partial_i$, we assume that it has a symmetric transition density with respect to a measure $m(\mathrm{d}x) = m(x)\,\mathrm{d}x$ on $\mathbb{R}^n$, where $m(x) > 0$. That is,

$$P_t F(x) = \int \rho(t,x,y)\big[(y)n(y)\,\mathrm{d}y = E^x Lf(x(H))\big] \tag{2.53}$$

for every bounded F. If we consider problem (2.51) with $D = \mathbb{R}^d$, we have the following representation theorem.

THEOREM 2.5. *Let $u(t,x)$ be a bounded solution to* (2.51), *which is continuously differentiable, once with respect to t and twice with respect to $x \in \mathbb{R}^d$. Then*

$$u(t,x) = E^x[f(x(t))] + \int_0^t E\big[st', X((t-t'))\big]\,\mathrm{d}t'. \tag{2.54}$$

For details refer to Vol. 2 of Dynkin's [2-4] or to Freidlin's [5-4]. To obtain similar representations, for (2.51) as stated, we have to consider the process killed upon leaving D which was introduced in Section 1.73. Define $T = \inf\{t > 0 \mid X(t) \notin D\} = \inf\{t > 0 \mid x(t) \in \partial D\}$, and for $F(x)$ bounded on D, put

$$Q_t F(x) = E^x\{F(X_t); t > t\}. \tag{2.55}$$

Again, Q_t is a positive semigroup on $L_\infty(D)$, and if F is continuous $Q_t F(x) \to F(x)$ as $t \to 0$ for each $x \in D$ and if F has a compact support contained in D, and at least two continuous derivatives, then

$$\frac{\partial Q_t}{\partial t} F(x) = \mathcal{L} Q_t F(x), \quad t > 0, x \in D. \tag{2.56}$$

The analogue of 2.5 is now

THEOREM 2.6. *Let $u(t,x)$ be a bounded solution to* (2.51) *which is once continuously differentiable with respect to t and twice with respect to $x \in D$, then*

$$u(t,x) = E^x\big[f(X(t)); t < T\big] + E^x\big[H(t-T, X(T), T \geqslant t\big] + E^x\left\{\int_0^{t\wedge T} s(t-t', X(t'))\,dt'\right\}. \tag{2.57}$$

SKETCH OF PROOF. The idea is to use the established routin. Consider $u(t-t', X(t'))$ on $(0,t)$ and use Ito's formula:

$$u(t-t', X(t')) = u(t, X(0)) + \int_0^{t'} \left(\frac{\partial u}{\partial t} + \mathcal{L}u\right)(t-t'', X(t''))\,dt'' + M(t'),$$

where $M(t')$ is a martingale that vanishes at $t'=0$. From (2.51) we have

$$u(t, x(0)) = u(t-t', X(t')) + \int_0^{t'} s(t-t'', X(t''))\,dt'' - M(t').$$

Now compute at $t' = t \wedge T$, and invoke the optional sampling theorem (Theorem 1.1) to obtain, after computing expected valves with respect to P^x, that

$$u(t,x) = E^x\big[u(t-t\wedge T, X(t\wedge T))\big] + E^x \int_0^{t\wedge T} s(t-t', X(t'))\,dt'.$$

And now split the first term into two, according to whether $t < T$ or $T \geqslant t$ as

$$E^x\big[u(t-r\wedge T, X(t\wedge T))\big] = E^x\big[f(x_t); t < T\big] + E^x\big[h(t-T, X(T)); T \geqslant t\big],$$

which completes the proof. □

Another way of obtaining (2.56) would be to invert Laplace transforms in the probabilistic representation of the solution to (2.52) which goes as follows.

THEOREM 2.7. *For any $\alpha > 0$, let $\tilde{u}(\alpha, x)$ be a solution to* (2.52) *which has two continuous derivatives with respect to $x \in D$. Then*

$$\tilde{u}(\alpha, x) = E^x \int_0^T e^{-\alpha t}\tilde{s}_1(\alpha, X(t))\,dt + E^x\big[e^{-\alpha T}\tilde{h}(\alpha, X(T))\big]. \tag{2.58}$$

PROOF. The idea is pretty much the same as the one sketched above. Start from $\tilde{u}(\alpha, X(t))$, invoke Ito's formula and (2.52), compute at T and take an expected valve with respect to P^x. For further details consult in [5-4], for example.

That is all well known. Let us now obtain some identities from which the reciprocity identities for wave propagation drop out.

Starting from the assumption that the density $\rho(t,x,y)$ introduced in (2.5.3) is symmetric in (x,y), one can prove that the semigroup $\{Q_t : t \geqslant 0\}$ introduced in (2.55) has a symmetric density $P_D(t,x,y)$. With these one can define the symmetric α-potentials

$$g(\alpha,x,y) = \int_0^\alpha \rho(t,x,y)\,\mathrm{d}t, \qquad g_0(\alpha,x,y) = \int_0^\infty \rho_0(t,x,y)\,\mathrm{d}t$$

which are the integral kernels of the operators

$$U^\alpha f(x) = E\left\{\int_0^T \mathrm{e}^{-\alpha t} f(X(t))\,\mathrm{d}t\right\}; \qquad V^\alpha f(x) = E^x\left\{\int_0^T \mathrm{e}^{-\alpha t} f(X(t))\,\mathrm{d}t\right\}.$$

Consider now the following sequence of identities

$$\begin{aligned} U^\alpha f(x) &= E^x\left\{\int_0^T \mathrm{e}^{-\alpha t} f(X(t))\,\mathrm{d}t\right\} + E^x\left\{\int_0^T \mathrm{e}^{-\alpha t} f(X(t))\,\mathrm{d}t\right\} \\ &= V^\alpha f(x) + E^x\left\{\mathrm{e}^{-\alpha T}\left(\int_0^\infty \mathrm{e}^{-\alpha t} f(X(t)) \circ \theta_T\,\mathrm{d}t\right)\right\} \\ &= V^\alpha f(x) + E^x\left[\mathrm{e}^{-\alpha T} U^\alpha f(X(T))\right]. \end{aligned}$$

Again, and for emphasis, as $\alpha > 0$ we can forget about the possibility that $\rho^x\{T = \infty\} > 0$.

Introducing the kernels $h^\alpha(x,\mathrm{d}z) = E^x[\mathrm{e}^{-\alpha T}; X(T) \in \mathrm{d}z]$ which, for each $x \in \mathbb{R}^d$, are measures on $B(\partial D)$ (because $X(T)$ is contained in ∂D), we can rewrite the previous chain as

$$\begin{aligned} &\int g(\alpha,x,y) f(y) m(\mathrm{d}y) \\ &\quad = \int_D g_D(\alpha,x,y) f(y) m(\mathrm{d}y) + \int\left\{\int_{\partial D} h^\alpha(x,\mathrm{d}z) g(\alpha,z,y)\right\} f(y) m(\mathrm{d}y) \end{aligned}$$

which, being valid for f positive measurable f supported in D, implies that

$$g(\alpha,x,y) - g_D(\alpha,x,y) = \int_{\partial D} h^\alpha(x,\mathrm{d}z) g(\alpha,z,y) \tag{2.59}$$

which is called "the fundamental identity" for α-potentials in Chapter 2 of [1-6]. From the symmetry of $g(\alpha,x,y)$ and $g_D(\alpha,x,y)$ in x, y it follows that

$$E^x\left[\mathrm{e}^{-\alpha T} g(\alpha, X(T), y)\right] = E^y\left[\mathrm{e}^{-\alpha T} g(\alpha, X(T), x)\right]$$

or, if you prefer,

$$\int_{\partial D} h^\alpha(x,\mathrm{d}z) g(\alpha,z,y) = \int h^\alpha(y,\mathrm{d}z) g(\alpha,z). \tag{2.60}$$

We shall make use of this in Chapters 3 and 9. □

References

Brownian motion, diffusions and stochastic calculus are household items in many quarters. Beside references in Chapter 1, the following contain interesting and useful material.

1. Karlin, S. and Taylor, H. M.: *A Course in Stochastic Processes*, Academic Press, New York (Vol. I, 1975; Vol. II, 1981).
2. Arnold, L.: *Stochastic Differential Equations. Theory and Applications*, Wiley-Interscience, New York, 1974.
3. Rogers, L. C. G. and Williams, D.: *Diffusions, Markov Processes and Martingales*, Wiley, Chichester, 1987.
4. Varadham, S. R. S.: *Diffusion Problems and Partial Differential Equations*, Springer-Verlag, Berlin, 1980.

A computation of the joint distribution of hitting time and place of a sphere, by a Brownian motion starting from its exterior and its application to Sommerfeld's radiation condition appeared in

5. Betz, C. and Gzyl, H.: Hitting spheres with Brownian motion and Sommerfeld's radiation condition, *J. Math. Anal. Appl.* **182** (1994), 301–308.

Similar argument to compute the joint distribution had appeared earlier in the interesting

6. Horowitz, J.: Measure-valued random processes, *Z. Wahrsch Verv. Geb.* **70** (1985), 213–216.

And an interesting recent reference containing the basics about diffusions and third boundary value problems is

7. Papanicolau, V. G.: The probabilistic solution of the third boundary problem for second order elliptic equations, *Probab. Theory Related Fields* **87** (1990), 27–77.

But the reader would fare well checking some of the references there, in particular, the classic by

8. Sato, K. and Ueno, T.: Multidimensional diffusions and Markov processes on the boundary, *J. Math. Kyoto Univ.* **3**(4) (1965), 529–605.

as well as the recent

9. Hsu, P.: Probabilistic approach to the Neumann problem, *Comm. Pure Appl. Math.* **38** (1989), 445–472.

CHAPTER

THREE

WAVES

Introduction

The general problem we shall be addressing throughout consists of solving the wave equation

$$\partial_t^2 u(t,x) = Lu(t,x) + s(t,x), \quad (t,x) \in (0,\infty) \times D. \tag{3.1}$$

Here D is an open domain in $\mathbb{R}^d$ for some $d \geqslant 1$ and $s(t,x)$ describes the sources producing the waves. As above

$$L = \sum a_{ij}(x)\partial_{ij}^2 + \sum b_j(x)\partial_j - q(x). \tag{3.2}$$

In order to connect well with the theory developed in Chapter 2, we place here the same assumptions about the coefficients $a_{ij}(x), b_j(x)$ and $q(x)$ as we did in Chapter 2. Problem (3.1) must be supplemented with initial conditions

$$u(0+,x) = f(x), \qquad u_t(0+,x) = g(x), \quad x \in D \tag{3.3}$$

as well as with boundary conditions, which we take to be either of the Dirichlet type

$$u(t,x) = \lim_{y\in D\to x\in\partial D} = u(t,y) = h(t,x), \quad t>0,\ x\in\partial D \tag{3.4}$$

or of the Neumann type

$$\left\langle \gamma(k), \lim_{y\in D\to x\in\partial D} \operatorname{grad} u(t,x)\right\rangle + h(t,x) = 0, \quad t>0,\ x\in\partial D,$$

where $\gamma(x)$ denotes the inner normal at $x \in \partial D$ which is assumed to be at least continuously differentiable. When consistency between the initial data and the boundary data is required, the necessary constraints should be established. In (3.3) the data functions are assumed to be bounded, or at least positive, and both in (3.4) or (3.5) the function $h(t,x)$ is assumed to be continuous and bounded.

This chapter is devoted to recalling some classical results on the solution of the wave equation. Since we compiled some probability theory for applied wave theorists, this chapters is to reciprocate and provide applied probabilists with the very basics about waves. We shall consider waves of constant velocity in $\mathbb{R}^d$, $d \geqslant 1$, after which we consider waves in heterogeneous media. To close this section, let us record some relationships between the solutions to (3.1) subject to different inhomogeneous initial conditions.

Let $s(t,x)$, be given and let $s_1(t,x) = \int_0^t s(t',x)\,\mathrm{d}t'$ and $u(t,x)$ be the unique solution to

$$u_{tt} = Lu + s_1, \qquad u(0+,x) = 0, \qquad u_t(0+,x) = g(x).$$

Then $v(t,x) = u_t(t,x)$ satisfies

$$v_{tt} = \mathcal{L}v + s, \qquad u(0+,x) = g(x), \qquad v_t(0+,x) = 0$$

and a simple integration provides us with a converse. With the aid of the Laplace transform, it is easy to see that the solution to the problem

$$\frac{\partial^2 v}{\partial t^2} = \mathcal{L}v + s(t,x) + \delta(t)g(x) + \delta'(t)f(x),$$
$$v(0+,x) = 0, \qquad v_t(0+,x) = 0$$

is the same as that of (3.1). This is due to the fact that after a Laplace transformation both solutions satisfy the same elliptic problem.

$$(\alpha^2 - L)\widetilde{w}(\alpha,x) = \tilde{s}(\alpha,x) + g(x) + \alpha f(x)$$

plus the Laplace transformed boundary conditions, where, of course,

$$\widetilde{w}(\alpha,x) = \int_0^\infty w(t,x)\mathrm{e}^{-\alpha t}\,\mathrm{d}t.$$

3.1. Waves of constant speed in $\mathbb{R}^d$

This section is meant for solving the equation describing the propagation of acoustic (scalar) waves in a full homogeneous isotropic space; i.e., waves in $\mathbb{R}^d$ which propagate with a constant speed v, that is, we are interested in (3.1) with $a_{ij} = v^2\delta_{ij}, b_i = 0$ and $q = 0$.

3.1.1. *Waves in one-dimensional space*

Let us consider, to begin with, the equation

$$\partial_t^2 u = v^2 \partial_x^2 u, \quad t > 0,\ x \in \mathbb{R} \tag{3.5}$$

subject to

$$u(0+, x) = f(x), \qquad u_t(0+, x) = g(x), \quad x \in \mathbb{R}. \tag{3.6}$$

The standard trick is to introduce auxiliary variables $\xi = x + vt$, $\eta = x - vt$, in terms of which $u^\sharp(\xi, \eta) = u(\frac{\xi - y}{2v}, \frac{\xi + y}{2})$ satisfies

$$u^\sharp_{\xi\eta} = 0$$

having the general solution

$$u^\sharp(\xi, \eta) = F(\xi) + G(\eta),$$

where F and G are true arbitrary continuously differentiable functions. Reverting to the original variables, we have

$$u(t, x) = F(x + vt) + G(x - vt)$$

and, after arranging for F and G to satisfy the initial conditions, we obtain

$$u(t, x) = \frac{1}{2}\{f(x + tv) + f(x - tv)\} + \frac{1}{2}\int_{x - vt}^{x + vt} g(\xi)\, \mathrm{d}\xi. \tag{3.7}$$

PROPOSITION 3.1. *The solution to* (3.5), *subject to the initial conditions* (3.6), *with* $f \in C^2(\mathbb{R})$ *and* $g \in C^1(\mathbb{R})$ *is given by* (3.7).

We did the one-dimensional case using the (covariant) factorization technique to anticipate, and for the sake of comparison with a (noncovariant) factorization of the wave equation to be carried out below. To anticipate the techniques to be used below, consider the equation

$$\frac{\partial^2 \hat{u}}{\mathrm{d}t^2}(t, k) + \omega^2 \hat{u}(t, k) = \hat{s}(t, k), \quad k \in \mathbb{R},$$

for the spatial-Fourier transform $\hat{u}(t, k) = \int \mathrm{e}^{ikx} \hat{u}(t, x)\, \mathrm{d}x$ of $u(t, x)$. Here $\omega^2 = (kv)^2$. Certainly, $\hat{u}(t, k) = \hat{u}_h(t, k) + \hat{u}_p(t, x)$, where, naturally, $\hat{u}_h(t, x)$ is the Fourier transform of the solution to the homogeneous equation, as given by (3.7). It is an elementary exercise to verify that

$$\begin{aligned}\hat{u}_p(t,k) &= \int_0^t \frac{\sin\omega(t-t')}{\omega}\hat{s}(t',k)\,\mathrm{d}t' \\ &= \int_0^t \mathrm{d}t' \int_{-(t-t')}^{(t-t')} \mathrm{d}u\, \mathrm{e}^{iwu}\hat{s}(t',k) \\ &= \int_0^t \mathrm{d}t' \int_{-vt'}^{vt'} \frac{\mathrm{d}y}{v}\mathrm{e}^{iyk}\hat{s}(t-t',k).\end{aligned}$$

Taking antitransforms we obtain

$$\begin{aligned}u_p(t,x) &= \int_0^t \mathrm{d}t' \int_{-vt'}^{vt'} \mathrm{d}u\, s(t-t',x-u) \\ &= \int_0^t \mathrm{d}t' \int_{x-vt'}^{x+vt'} s(t-t',\xi)\,\mathrm{d}\xi.\end{aligned} \tag{3.8}$$

PROPOSITION 3.2. *Let $s(t,x)$ be $C^1(\mathbb{R}^2)$. Then* (3.8) *provides a solution to the nonhomogeneous wave equation $\partial_t^2 u(t,x) = v^2\partial_x^2 u(t,x) + s(t,x)$ with the homogeneous initial condition.*

Exercise: Recast the initial conditions as sources and apply (3.8) to obtain $u_h(t,x)$.

Comment: The differentiability requirements on the data and source terms could be dispensed with using "weak solutions". We shall just manipulate these solutions in the distribution sense as if they were strong solutions.

3.1.2. Waves in spaces of dimension $d > 1$: Averaging methods

This will be the first of several approaches to the problem of obtaining integral representations of the solutions to the equation

$$\partial_t^2 u(t,x) = v^2\Delta u(t,x) + s(t,x), \quad t > 0,\ x \in \mathbb{R}^d$$

with initial data

$$u(0+,x) = f(x), \qquad u_t(0+,x) = 0.$$

As above, we first assume that there are no sources and begin with an approach consisting of transforming problem (3.1) into a problem in 1 spatial dimension:

The method of spherical means: Let $v(x)$ be any continuous function defined on $\mathbb{R}^d$. Its average over a sphere of radius r centered at x is defined by

$$I(v,x,r) = \frac{1}{\omega(d)}\int v(x+r\xi)\,\mathrm{d}\sigma(\xi),$$

where $\omega(d)$ is the area of the $(d-1)$-dimensional unit sphere

$$S_{d-1} = \{y : \|y - x\| = 1\}$$

and $d\sigma(\xi)$ denotes the Lebesque measure on S_{d-1}. Simple computations plus hypotheses of continuity on $v(x)$ and its derivatives yield

$$\left(\frac{\partial^2}{\partial r^2} + \frac{d-1}{r}\frac{\partial}{\partial r}\right) I(v,x,r) = \Delta I(v,x,r), \tag{3.9}$$

$$I(v,x,0+) = v(x), \qquad \left(\frac{\partial}{\partial r} I_v\right)(x,0+) = 0. \tag{3.10}$$

These imply that, if $u(t,x)$ satisfies the wave equation, then $I(u,t,x,r)$, the spherical average of $u(t,\cdot)$, satisfies

$$\partial_t^2 I(u,t,x,r) = v^2\left(\partial_r^2 + \frac{d-1}{r}\partial_r\right) I(u,t,x,r) \tag{3.11}$$

plus the initial conditions

$$I(u,0+,x,r) = I(t,x,r), \qquad \partial_t I(u,0+,x,r) = I(g,x,r).$$

Assume, before going on, that we have rescaled the time variable to $vt = \tau$, but we denote it by t again, so that in (3.9) we are setting $v = 1$. To begin with, note that any twice continuously differentiable function $F(t)$ defined on $\mathbb{R}$ can be thought of as a function on $\mathbb{R}^d$ depending only on the first coordinate. In this case,

$$(AF)(t,r) := I(f,(t,0,\ldots,0),r) = \frac{\omega(d-1)}{\omega(d)} \int_{-1}^{1} F(t+sr)(1-s^2)^{(d-3)/2}\,ds$$

and another simple computation, like that leading to (3.9), leads to

$$\partial_r^2(AF)(t,r) + \frac{d-1}{r}\partial_r(AF)(t,r) = (AF)''(t,r),$$

where $(AF)''$ denotes second derivatives of (AF) with respect to the time variable. Denoting $(AF)(0,r)$ by $(AF)(r)$, the previous identities become

$$(AF)(r) = \frac{\omega(d-1)}{\omega(d)} \int_{-1}^{1} F(rs)(1-s^2)^{(d-3)/2}\,ds \tag{3.12}$$

and

$$\partial_r^2(AF)(r) + \frac{d-1}{r}\partial_r(AF)r = (AF)''(r). \tag{3.13}$$

Note that A acts on the first variable in $u(t,x)$. Since $u(t,x)$ satisfies the wave equation, we obtain from (3.13) that

$$\partial_r^2(Au)(r,x)+\frac{d-1}{r}\partial_r(Au)(r,x)=\Delta(Au)(r,x)$$

subject to $(Au)(0+,x)=u(0+,x)=f(x)$, $(A\partial_r)(0+,x)=0$; that is $(Au)(r,x)$ satisfies *Darboux's* equation (3.9), therefore

$$(Au)(r,x)=\frac{1}{\omega(d)}\int f(x+r\xi)\,\mathrm{d}\sigma(\xi)=I(f,x,r).$$

To solve it, set $r=\sqrt{\tau}, rs=\sqrt{\sigma}$ in (3.12) applied to $u(t,x)$ this time, and obtain after rescaling

$$\tau^{(d-2)/2}I(f,x,\sqrt{\tau})=\frac{\omega(d-1)}{\omega(d)}\int_0^t\frac{u(\sqrt{\sigma},x)}{\sqrt{\sigma}}(\tau-\sigma)^{(d-3)/2}\,\mathrm{d}\sigma. \tag{3.14}$$

Note that for all this it is convenient to make use of the form of the initial data and the invariance of the wave equation under time reversal to assume that $u(t,x)$ is even in t. The final steps depend on whether d is odd or even. When it is odd, we can differentiate (3.14) $(d-1)/2$ times with respect to τ to obtain

$$\frac{\omega(d-1)}{\omega(d)}\left(\frac{d-3}{2}\right)!\frac{u(\sqrt{\tau},x)}{\sqrt{\tau}}=\left(\frac{d}{\mathrm{d}\tau}\right)^{(d-1)/2}\left\{I(f,x,\sqrt{\tau})\tau^{(d-2)/2}\right\}$$

and now, undo the change of variables setting $\sqrt{\tau}=t$, to obtain

$$u(t,x)=\frac{\omega(d)}{\omega(d-1)}\left(\frac{1}{2}\right)^{(d-1)/2}\left(\frac{1}{t}\frac{\mathrm{d}}{\mathrm{d}t}\right)^{(d-1)/2}\left\{I(f,t)t^{d-2}\right\}. \tag{3.15}$$

When n is even, the inversion is performed by fractional differentiation of order $(d-1)/2$ to obtain

$$\frac{\omega(d-1)}{\omega(d)}\frac{u(\sqrt{\tau},x)}{\sqrt{\tau}}=\frac{1}{\Gamma\pi\Gamma(\frac{d-1}{2})}\left(\frac{\mathrm{d}}{\mathrm{d}t}\right)^{d/2}\int_0^2\frac{I(f,x,\sqrt{\sigma})}{\sqrt{\tau-\sigma}}\sigma^{(d-2)/2}\,\mathrm{d}\sigma$$

or, equivalently, setting $t=\sqrt{\tau}$

$$u(t,x)=\frac{2t}{\sqrt{\pi}\Gamma(\frac{d-1}{2})}\frac{\omega(d)}{\omega(d-1)}\left(\frac{1}{2}\right)^{d/2}\left(\frac{1}{t}\frac{\mathrm{d}}{\mathrm{d}t}\right)^{d/2}\int_0^t\frac{(I(f,x,s))s^{d-1}}{\sqrt{t^2-s^2}}\,\mathrm{d}s. \tag{3.16}$$

To unclutter (3.15) and (3.16) a bit, use $\omega(d)/\omega(d-1)=\sqrt{\pi}\Gamma((d-1)/2)/\Gamma(d/2)$. We can sum up as

PROPOSITION 3.3. *Depending on whether d is odd or even, the solution to the wave equation with f in $C_0^2(\mathbb{R}^d)$ is given by (3.15) or (3.16) respectively.*

Comment: When $u_t(0,x)$ is not identically 0, we must add a term to (3.15) or (3.16) according to the comments in Section 1. Let us consider now the following nonhomogeneous equation

$$\partial_t^2 u(t,x) = \Delta u + s(t,x), \quad t > 0,\ x \in \mathbb{R}^d$$

with homogeneous boundary conditions

$$u(0+,x) = 0, \qquad u_t(0+,x) = 0$$

and let us explain *Duhamel's method* to take care of the source term. Denote by $U(g,t,x)$ the solution to

$$\partial_t^2 u = \Delta u, \quad t > 0,\ x \in \mathbb{R}^d, \tag{3.17}$$

$$u(0+,x) = 0, \qquad u_t(0+,x) = g. \tag{3.18}$$

Then, for each fixed $t' \geqslant 0$, $U(s_{t'}, t-t', x)$ denotes the solution to (3.17) for $t > t'$, such that at $t = t'$ it satisfies

$$u(t'+,0) = 0, \qquad u_t(t',u) = s_{t'}(x) = s(t',x).$$

With this notation define

$$u(t,x) = \int_0^t U(s_{t'}, t-t', x)\,\mathrm{d}t' \tag{3.19}$$

and it is a simple exercise to verify that (3.19) satisfies the nonhomogeneous wave equation (3.17) with homogeneous initial condition (3.18).

3.1.3. *Waves in spaces of dimension $d > 1$. The Fourier transform approach*

When dealing with the propagation of waves in isotropic and homogeneous spaces, one can take Fourier transforms in the time variable, and spatial variables or in both; corresponding, respectively, to the considered waves of definite time frequency, a definite spatial wave vector or plane waves. Actually, since there are differential operators on the right- and left-hand sides of the wave equation, from the abstract point of view all this corresponds to expansion in terms of eigenfunctions of the corresponding differential operators.

3.1.4. *Fourier transform in time*

If we consider equation (3.1) with the initial conditions included in the source term, and therefore with homogeneous initial conditions, and set

$$\hat{u}(\omega,x) = \int \mathrm{e}^{i\omega t} u(t,x)\,\mathrm{d}t \quad \text{and} \quad u(t,x) = \int \mathrm{e}^{-i\omega t}\hat{u}(\omega,x)\frac{\mathrm{d}\omega}{2\pi},$$

and then transform both members of (3.1), we will obtain

$$(\Delta+\omega^2)\hat{u}(\omega) = -(\hat{s}(\omega,x)+i\omega f(x)+g(x)), \tag{3.20}$$

where we assume that both data and sources vanish outside some compact set, and we leave for the reader to change things a bit to cover the case $v \neq 1$. The Green function for $\Delta+\omega^2$ was computed in Section 1 of Chapter 2 for $\omega=-i\alpha$. In what follows we shall deal only with the case $d=3$ for the sake of concreteness and ease of computation. It is a standard exercise to verify that

$$g_\pm(|x-y|) = \frac{\mathrm{e}^{\pm i\omega|x-y|}}{4\pi\,|x-y|}$$

both satisfy

$$(\Delta+\omega^2)g_\pm = -\delta(x-y) \tag{3.21}$$

and the signs being such that

$$u_\pm(t,x) = \int \mathrm{e}^{-i\omega t}\,\frac{\mathrm{e}^{\pm i\omega|x-y|}}{4\pi\,|x-y|}\frac{\mathrm{d}\omega}{2\pi} = \frac{\delta(t\pm|x-y|)}{4\pi\,|x-y|} \tag{3.22}$$

represent "outgoing" and "incoming" waves produced at y by a "shot" at $t=0$. Physical demands constrain us to consider outgoing waves and we only keep $g_+(|x-y|)$. Assume now that the solution $\hat{u}(\omega,x)$ to (3.20) and its derivatives decay rapidly enough at infinity, i.e., faster than $1/|x|$, and consider for each x

$$\int_{B(R)} \left\{\hat{u}(\omega,y)\Delta g_+(|x-y|) - \Delta\hat{u}(\omega,y)g_+(|x-y|)\right\}\mathrm{d}y \tag{3.23}$$

$$= \int_{S(R)} \left\{\hat{u}(\omega,y)\frac{\partial}{\partial n}g_+(|x-y|) - g_+(|x-y|)\frac{\partial}{\partial n}\hat{u}(\omega,y)\right\}\mathrm{d}y, \tag{3.24}$$

where $B(R)$ denotes a ball of radius R centered at x and $s(R)$ its surface, etc. From our assumptions on the support of the data and about the decay of $\hat{u}(\omega,x)$ and $\partial\hat{u}/\partial\eta$; we see that, after letting $R\to\infty$ in (3.23), we obtain

$$\hat{u}(\omega,x) = \frac{1}{4\pi}\int \frac{\mathrm{e}^{i\omega|x-y|}}{|x-y|}\left\{\hat{s}(\omega,y) - i\omega f(y) + g(y)\right\}\mathrm{d}y. \tag{3.25}$$

To obtain $u(t,x)$ from $\hat{u}(\omega,x)$ given by (3.24) we must identify the Fourier transforms involved. This is an easy task recurring to (3.21). Begin with the last term on the right-hand side of (3.25).

$$\frac{1}{2\pi}\int e^{i\omega t}\int \frac{e^{i\omega|x-y|}}{4\pi(x-y)}g(y)\,dy = \int \frac{\delta(t-|y|)}{4\pi|y|}g(x+y)\,dy$$
$$= \int_0^\infty \frac{r\delta(t-r)}{4\pi}\left(\int g(x+\xi r)\,d\sigma(\xi)\right)dr$$
$$= tI(g,x,t).$$

To invert the second term is equally easy, just note that

$$\int i\omega\frac{e^{-i\omega t}}{2\pi}\left(\frac{e^{i\omega|x-y|}}{4\pi|x-y|}f(y)\,dy\right)dw = \frac{\partial}{\partial t}(tI(t,x,t)).$$

To take care of the first term, it suffices to note that it is Fourier transform of the convolution of $\delta(t-|x-y|)/4\pi|x-y|$ with $s(t,y)$ integrated with respect to y over $\mathbb{R}^3$, that is, of

$$\int_{-\infty}^{-\infty} dt' \int \frac{\delta(t-|x-y|)}{4\pi|x-y|}s(t',y)\,dy.$$

If we add to the condition (natural after all) that $s(t,x)=0$ for $t<0$, the above integration reduces to

$$\int_0^t (t-t')I\big(s(t'),x,t\big)\,dt'$$

as it should.

3.1.5. *Fourier transform in space*

This time we consider the wave equation (1.1) but do not add the initial conditions. We rather drag them along. Set

$$u^\sharp(t,k) = \int e^{-i(k,x)}u(t,k)\,dx, \qquad u(t,x) = \frac{1}{(2\pi)^3}\int e^{i(t,x)}u^\sharp(t,k)\,dk$$

and take Fourier transform of the wave equation and the initial conditions to obtain

$$\frac{d^2u^\sharp}{dt^2}(t,k) = -k^2u^\sharp(t,k)+s^\sharp(t,k), \qquad u^\sharp(0+,k)=f^\sharp(k), \qquad u^\sharp_t(0+,k)=g^\sharp(k)$$

which can be integrated to yield

$$u^\sharp(t,k) = f^\sharp(k)\cos kt + \frac{\sin kt}{k}g^\sharp(k) + \int_0^t \frac{\sin k(t-t')}{k}s(t',k)\,dt'.$$

We invert the transform corresponding to the first term just to emphasize an issue: that of the incoming and outgoing waves. Note that

$$\int f^{\sharp}(k)\cos kt e^{i(k,x)}\frac{dk}{(2\pi)^3}=\int f(y)\left\{\int \cos k e^{ik\cdot(x-y)}\frac{dk}{(2\pi)^3}\right\}dy.$$

Choose in $k=(k_1,k_2,k_3)$ space a system of coordinates with the k_3-axis along $(x-y)$, pass to spherical coordinates, integrate with respect to the angle, and symmetrize the domain of integration with respect to k to obtain

$$\frac{1}{8\pi^2}\frac{\partial}{\partial t}\int_{-\infty}^{\infty}\sin kt\left\{\int\left(\frac{e^{ik|x-y|}-e^{-ik|x-y|}}{i|x-y|}\right)dy\right\}dk.$$

Now write $\sin kt=(e^{ikt}-e^{-ikt})/2i$, carry out the products, exchange integrals (or invert the resulting transforms) to obtain

$$\frac{\partial}{\partial t}\int\frac{\delta(t-|x-y|)-\delta(t+|x-y|)}{4\pi|x-y|}f(y)\,dy$$

from which we see that, while we consider evolution into the future (after $t=0$, or $t>0$, only), the second term does not contribute to the wave front. So, neglecting waves incoming from infinity was sensible.

3.1.6. *Expansion of the solution in plane waves*

Just a few words to explain the different choice of signs for the temporal and spatial Fourier transforms. The expansion of the solution $u(t,x)$ to the wave equation in plane waves is defined by

$$u(t,x)=\int e^{-i\{\omega t-(k,x)\}}\tilde{u}(\omega,k)\frac{d\omega\,dk}{(2\pi)^4} \tag{3.26}$$

which can be thought of as a "superposition" of the methods developed in (3.24) and (3.25). The important thing to notice there is that the phase $\omega t-(k,x)$ is invariant under the same group of symmetries that leaves the equation $\partial_t^2 w-v^2\Delta w=0$ invariant whenever $w^2=v^2k^2$. If we take a combined Fourier transform of the wave equation with the initial conditions added as sources, we obtain

$$(\omega^2-v^2k^2)\tilde{u}(\omega,k)=-\frac{(\tilde{s}(\omega,k)+i\omega f^{\sharp}(k)+g^{\sharp}(k))}{\omega^2-v^2k^2}$$

which, substituted into (6), reproduces the results we have obtained above.

3.1.7. *The (de)complexification approach*

There are two ways of turning problem (1.1) into an elliptic problem in $d+1$ variables. One of them will be described below in Chapter 7 devoted to transmutation operators, the other one, briefly described here, is taken from Garabedian's book and consists in letting t be a complex variable, which is then considered to vary along the imaginary axis. Thus, setting $x_0 = -it$, instead of (3.1), we obtain

$$\left(\frac{\partial^2}{\partial x_0^2} + \Delta\right)\omega(x_0, x) = 0.$$

This equation, when considered in the whole space with the initial conditions correspondingly rotated, can be integrated, because there is an explicit representation for the corresponding Green function. By means of a clever choice of contours in the complex x_0-plane, the following representations are obtained in Garabedian's book: When n is even

$$\begin{aligned} u(t,x) = \frac{1}{2.4\ldots(n-2)}\Bigg\{ & \frac{\partial}{\partial t}\left(\frac{\partial}{t\partial t}\right)^{(n-2)/2}\int_0^t I(f,x,\rho)\frac{\rho^{n-1}d\rho}{\sqrt{t^2-\rho^2}} \\ & + \left(\frac{\partial}{\partial t}\right)^{(n-2)/2}\int I(g,x,\rho)\frac{\rho^{n-1}d\rho}{\sqrt{t^2-\rho^2}}\Bigg\}. \end{aligned}$$

And when n is odd, the result is

$$\begin{aligned} u(t,x) = \frac{1}{1.3\ldots(n-2)}\Bigg\{ & \frac{\partial}{\partial t}\left(\frac{1}{t}\frac{\partial}{\partial t}\right)^{(n-3)/2}\left(t^{n-2}I(t,x)\right) \\ & + \left(\frac{1}{t}\frac{\partial}{\partial t}\right)^{(n-3)/2}\left(t^{n-2}I(g,t,x)\right)\Bigg\}. \end{aligned}$$

We leave for the reader to verity that they are coincidental with the representation obtained above by means of the averaging method. These will be re-obtained below by a different method.

3.1.8. Factorization

This section is designated to a variation on the theme developed to solve the wave equation in a space of dimension one. There we wrote $(\partial_t^2 - c^2\partial_x^2)u = s$ as $(\partial_t - c\partial_x)(\partial_t + c\partial_x)u = s$ and integrated in terms of auxiliary variables. This factorization is interesting, for it preserves the symmetries of the wave equation, and can be extended to higher dimensions leading to all sorts of "relativistic wave equations". But there is another procedure which involves the extraction of square roots, by means of which (3.1) can be factorized. To describe it, we have to recall a few facts. Recall, first of all, that we want to solve

$$\begin{aligned} &\partial_t^2 u = Lu + s, \quad t > 0,\ x \in \mathbb{R}^d, \\ &u(0+,x) = f(x), \qquad u_t(0+,x) = g(x), \quad x \in \mathbb{R}^d, \end{aligned}$$

where $L = \sum a_{ij}(x)\partial_i\partial_h + \sum b_i(x)\partial_i q(x)$, the coefficients of which are usually bounded and reasonably smooth. We need to assume the existence of a positive solution to

$\frac{1}{2}\sum \partial_j a_{ij}(x)\rho(x) = b_i(x)\rho(x)$ on $\mathbb{R}^n$, determining a measure $\rho(x)\,\mathrm{d}x = m(\mathrm{d}x)$ with respect to which the semigroup $\{Q_s; s \geqslant 0\}$, having L as an infinitesimal generator, has a symmetric transition density $q(s, x, y)$; i.e.,

$$\int (Q_s f(x))h_2(x)\rho(x) = \int h_1(x)(Q_s h_2(x))\rho(x)\,\mathrm{d}x$$

for all positive $h_1(x), h_2(x)$. We consider the semigroup to be acting on $\mathbf{B} = C_0(\mathbb{R}^d)$ and denote by $\mathbf{D}(\mathcal{L})$ the domain of its infinitesimal generator. We saw in Chapter 3 that, for any initial datum $w_0(x)$ in $\mathbf{B}$, $Q_s w_0(x) = w(s, x)$ solves

$$\frac{\partial w}{\partial s}(s, x) = \mathcal{L}w(s, x), \qquad w(0+, x) = w_0(x).$$

Also, if we change time Q_s according to

$$v(\tau, x) = K_\tau w_0(x) = \int_0^\infty \frac{\tau e^{-\tau \frac{2}{4s}}}{(4\pi s^3)^{1/2}} Q_s w_0(x)\,\mathrm{d}s, \tag{3.27}$$

we obtain a new semigroup acting on $\mathbf{B}$, with an infinitesimal generator $-\sqrt{-\mathcal{L}} = G$ defined on $\mathbf{D}(G) \supset \mathbf{D}(\mathcal{L})$, that solves

$$\frac{\partial v}{\partial \tau}(\tau, x) = Gv(\tau, x), \qquad v(0+, x) = w_0(x). \tag{3.28}$$

This is all standard. To understand the definition of G, one only has to recall that for $h_1(x), h_2(x)$ having at least two continuous derivatives and compact support

$$\int h_1(x)\mathcal{L}h_2(x)\rho(x)\,\mathrm{d}x \leqslant 0,$$
$$\int h_1(x)\mathcal{L}h_2(x)\rho(x)\,\mathrm{d}x = \int (\mathcal{L}h_1(x))h_2(x)\rho(x)\,\mathrm{d}x.$$

Thus, loosely speaking, $\sqrt{-\mathcal{L}}$ is well defined and has a positive spectrum, and $G = -\sqrt{-\mathcal{L}}$ has a negative spectrum and generates a contraction semigroup K_t. To factor the wave equation, note that since we assume static media (time-independent coefficients) we can write the wave equation as

$$(\partial_t - iG)(\partial_t + iG)u(t, x) = s(t, x) \tag{3.29}$$

which we can rewrite as a system

$$\begin{aligned}(\partial_t - iG)u(t, x) &= v(t, x),\\ (\partial_t + iG)v(t, x) &= s(t, x)\end{aligned} \tag{3.30}$$

each of which can be simply integrated. The only formal difficulty is to pass from $K_\tau = \exp(-\tau G)$ to $U_t = \exp iG$, which is covered by

PROPOSITION 3.4. *Let G be a self-adjoint operator in a complex Hilbert space and G be bounded below. Then the semigroup $\{K_\tau : \tau \geqslant 0\}$ generated by G extends to a holomorphic semigroup $\{P(z) : \Re z > 0\}$ such that the unitary group $\{U(t) : r \in \mathbb{R}\}$ generated by iG satisfies*

$$U(t)f = \left\{\lim_{z\to\infty} P(z)f\ \forall f \in X\right\}.$$

For us $\mathcal{H}$ is the Hilbert space obtained from the scalar product $(f,g)_\rho = \int \bar{f}(x)g(x)\rho(x)\,\mathrm{d}x$. Before going on, *assume* that we have incorporated the initial conditions for the wave equation problem into the source term, and that we want to solve (3.30) with homogeneous initial conditions.

A simple integrating factors technique applied twice yields

$$\begin{aligned} u(t,x) &= U(s-t)\left(\int_0^s U(s-s')f(s',x)\,\mathrm{d}s'\right)\mathrm{d}s \\ &= \int_0^t \mathrm{d}s' \int_{s'}^t U(2s-(t+s'))f(s',x)\,\mathrm{d}s. \end{aligned}$$

Setting $\sigma = 2s-(t+s')$, we obtain

$$\begin{aligned} u(t,x) &= \frac{1}{2}\int_0^t \mathrm{d}s' \int_{-(t-s')}^{(t-s')} U(\sigma)f(s',x)\,\mathrm{d}\sigma \\ &= \frac{1}{2}\int_0^t \mathrm{d}s' \left\{\int_0^{t-s} U(\sigma)f(s',x)\,\mathrm{d}\sigma - \int_{-t+s'}^0 U(\sigma)f(s',x)\,\mathrm{d}\sigma\right\}. \end{aligned}$$

Making $\sigma' = t-\sigma$ in the first integral and $\sigma' = \sigma + t$ in the second one, after renaming variables we obtain

$$u(t,x) = \frac{1}{2}\int_0^t \mathrm{d}s \left\{\int_s^t U(t-\sigma)f(s,x)\,\mathrm{d}\sigma + \int_s^t U(-(t-\sigma))f(s,x)\,\mathrm{d}\sigma\right\} \tag{3.31}$$

and, exchanging order of integration once more, we have

$$u(t,x) = \frac{1}{2}\int_0^t \mathrm{d}\sigma \int_0^u \left\{U(t-\sigma)f(s,x) + U(-(t-\sigma))f(s,x)\right\}\mathrm{d}s. \tag{3.32}$$

Comment: First of all, note that u is real when f is real. Also, differentiating twice with respect to t, we obtain

$$\partial_t^2 u = \mathcal{L}u + f.$$

Let us particularize $f(t,x)$ a bit. With the notation introduced above, when $f(t,x) = \Phi(t)g(x)$

$$u(t,x) = \frac{1}{2}\int_0^t \mathrm{d}\sigma \left\{U(t-\sigma)g(x) + U(-(t-\sigma))g(x)\right\} \int_0^\sigma \Phi(s)\,\mathrm{d}s \tag{3.33}$$

which for $\Phi(t) = \delta(t)$ yields

$$u(t,x) = \frac{1}{2}\int_0^t \mathrm{d}\sigma \left\{U(t-\sigma)g(x) + U(-(t-\sigma))g(x)\right\}. \tag{3.34}$$

As $\Phi(t) = \delta'(t)$, the integration by parts, which is easier when carried out starting from (3.29), yields

$$u(t,x) = \frac{1}{2}(U(t)g(x) + U(-t)g(x)). \tag{3.35}$$

If G^{-1} exists, writing $U(t-\sigma) = -\frac{-G^{-1}}{i}\frac{\mathrm{d}}{\mathrm{d}\sigma}U(t-\sigma)$ and substituting in (3.30) we obtain instead

$$u(t,x) = \frac{1}{2i}G^{-1}\int_0^t \left\{U(t-s)f(s,x) - U(-(t-s))f(s,x)\right\}\mathrm{d}s, \tag{3.36}$$

which could be formally written as

$$u(t,x) = G^{-1}\int_0^t \sin G(t-s)f(s,x)\,\mathrm{d}s,$$

as expected.

3.2. Propagators and Green functions

The propagator, that is the operator mapping the data at time t' to the data at time t, is defined to be the solution to

$$\partial_t^2 K(x,t;x',t') = \mathcal{L}K(x,t;x',t') \tag{3.37}$$

subject to the initial conditions

$$\begin{aligned} K(x,t'+;x',t') &= 0, \\ K_t(x,t'+;x',t') &= A\delta(x-x'), \end{aligned} \tag{3.38}$$

where the constant A is related to the density $\rho(x')$ which makes $\mathcal{L}$ a symmetric operator. With the aid of $K(x,t;x',t')$ it is easy to see that

$$u(t,x) = \int K(x,t,;x',t')g(x')\rho(x')\,\mathrm{d}x'$$

satisfies

$$\partial_t^2 u = \mathcal{L}u, \qquad u(0+, x) = 0, \qquad u_t(0+, x) = g(x)$$

if $A = \rho(x')^{-1}$ (or $A = \rho(x)^{-1}$; It does not matter). The Green function for the wave equation is defined to be the solution of

$$\partial_t^2 G(x, t; x', t') = \mathcal{L}G(x, t; x', t') + A\delta(t - t')\delta(x - x') \tag{3.39}$$

satisfying homogeneous initial conditions at $t = t'$ as well as

$$G(x, t; x', t') = 0 \quad \text{for } t < t' \tag{3.40}$$

which is a causality condition.

We should add that, if we were considering (3.35) or (3.37) in a domain D, we would have had to supplement (3.36) and (3.38), respectively, with appropriate boundary conditions, either of the Dirichlet or Neumann types. For the time being, let us verify the

LEMMA 3.1. *With the notation introduced above*

$$G(x, t; x', t') = \theta(t - t')K(x, t; x', t').$$

PROOF. Recall that $\partial_t \theta(t - t') = \delta(t - t')$ and differentiate $\theta(t - t')K$ twice with respect to t to obtain

$$\partial_t^2 G = \delta'(t - t')K + 2\delta(t - t')K_t + \theta\mathcal{L}K.$$

Now $\theta\mathcal{L}K = \mathcal{L}G$ since we are considering static media only and by making use of test functions φ

$$\int \varphi\delta'(t - t')K = -\int \delta(t - t')(k\varphi)' = -\int \delta(t - t')(K_t\varphi + K\varphi')$$

since the last term vanishes on account of (3.37), which also implies that

$$-\delta(t - t')K_t = A\delta(t - t')\delta(x - x').$$

Let us now recall some simple consequences of the invariance of wave equation (3.1) under time translations and time reflections. Since the solution to (3.35)–(3.36) is unique, and for any s, $K(x, t + s; x', t' + s)$ satisfies both the equation and the initial data, we have

$$K(x, t + s; x', t' + s) = K(x, t; x', t' + s).$$

Take now $s = -t'$ and define

$$K(x, x'|t - t') = K(x, t - t'; x', 0).$$

It is a trivial exercise to verify that

$$K(x,-t';x',-t) = K(x,t;x',t). \tag{3.41}$$

Since the wave equation involves a second derivative with respect to time, $K(x,x'|-\tau)$ also satisfies

$$\partial_\tau^2 K = \mathcal{L}K, \qquad K(x,x'|0+) = 0, \qquad K_\tau(x,x'|0+) = -A\delta(x-x')$$

from which we conclude that

$$K(x,x'|-\tau) = -K(x,x',\tau). \tag{3.42}$$

□

PROPOSITION 3.5. *The reciprocity identity. Under the assumptions and notation introduced above, we have the symmetry relationship*

$$K(x,t;x',t') = K(x',t';x,t) = K(x',-t';x,-t) \tag{3.43}$$

plus the reciprocity identify

$$G(x,t;x',t') = G(x',t;x,t'). \tag{3.44}$$

PROOF. Clearly, it suffices to verify (3.41). For that consider both $K(x,t;x',t')$ and $K(x,t;x'',t'')$, which satisfy

$$(\partial_t^2 - \mathcal{L})K(x,t;x',t') = 0,$$
$$(\partial_t^2 - \mathcal{L})K(x,t;x'',t'') = 0$$

plus the respective initial condition at times t'' and t', respectively. Multiply the first of these identities by $K(x,t;x'',t'')$, the second by $K(x,t;x',t')$, subtract and integrate over $\mathbb{R}^d$ with respect to $\rho(x)\,\mathrm{d}x$ and with respect to t over the interval (t_1,t_2), to obtain

$$\int_{t_1}^{t_2} \mathrm{d}t \int \big\{K(x,t;x'',t'')\partial_t^2 K(x,t,x',t') - K(x,t;x't'\partial_t^2 K(x,t;x'',t'')\big\}\rho(x)\,\mathrm{d}x$$
$$= \int_{t_1}^{t_2} \mathrm{d}t \int \big\{K(x,t;x'',t'')\mathcal{L}K(x,t;x',t') - K(x,t;x',t')\mathcal{L}K(x,t;x't'')\big\}\rho(x)\,\mathrm{d}x.$$

Note now, that, due to the symmetry of $\mathcal{L}$, the right-hand side vanishes identically, and, integrating by parts the first one, we obtain

$$0 = \int \big\{K(x,t;x'',t'')K_t(x,t;x',t') - K(x,t;x',t')K_t(x,t;x'',t'')\big\}\Big|_{t=t_1}^{t=t_2}\rho(x)\,\mathrm{d}x.$$

Now, let $t_2 \downarrow t'$ and $t_1 \downarrow t''$, and use the initial condition

$$\int \{K(x,t';x'',t'')\delta(x-x') - K(x,t;x't')\delta(x-x'')\}\,\mathrm{d}x$$

which yields

$$K(x',t';x'',t'') = K(x'',t'';x',t') = 0,$$

or from (3.39) and (3.40) we obtain (3.41), after setting $x'' = x$ and $t'' = t$. □

Comment: It is (3.42) that realizes the physically intuitive notion of switching "source" with "observer", and hence the name of the identity. If we insisted in using the scalar product with respect to the usual Lebesque measure, i.e., $(f,g) = \int f(x)g(x)\,\mathrm{d}x$; then $\mathcal{L}$ would have the adjoint

$$\mathcal{L}^* = \sum a_{ij}\partial_{ij} + \sum b_i^*\partial_i - q^*,$$

where $b_1^* = b_i(x) - \sum_j \partial_j a_{ij}(x)$, $q^*(x) = q(x) + \sum \partial_i b_i(x) + \frac{1}{2}\sum_{ij}\partial_{ij}a_{ij}(x)$. If we let $K^*(x,t;x',t')$ and $G^*(x,t;x',t')$ be, respectively, the propagator and Green's function, corresponding to $\partial_t^2 - \mathcal{L}^*$, satisfying the same initial (and boundary) conditions as $K(x,t;x',t')$ and $G(x,t;x',t')$, respectively, then we would have

PROPOSITION 3.6. *With the notation introduced above, we have*

$$K^*(x,t;,x',t') = K(x',t';x,t) \tag{3.45}$$

as well as

$$G^*(x,t;x',t') = G(x',t;x,t'). \tag{3.46}$$

Let us close this section with the derivation of a representation formula involving the Green function. Let $G(x,t;x',t')$ be the solution to

$$(\partial_t^2 - \mathcal{L})G = A\delta(t-t')\delta(x-x'), \quad x,x' \in D,\ t,t' \geqslant 0 \tag{3.47}$$

such that $G(x,t;x',t') = 0$ for $t' > t$ and it satisfies the homogeneous boundary conditions of the Dirichlet type

$$G(x,t;x',t') = 0 \quad \text{for } x \in \partial D,\ x' \in D \tag{3.48}$$

or of the Neumann type

$$\frac{\partial}{\partial n}G(x,t;x',t') = 0 \quad \text{for } x \in \partial D,\ x \in D. \tag{3.49}$$

Even if we did not prove it, the reciprocity theorem is valid for this $G(x,t;x',t')$ as well, the proof of this assertion is as above. Let now $u(t,x)$ be the solution to the problem

$$(\partial_t^2 - \mathcal{L})u(t,x) = s(t,x), \quad x \in D,\ t \geqslant 0 \tag{3.50}$$

in which initial conditions are incorporated into the source, and $u(t,x)$ satisfies either the Dirichlet conditions

$$u(t,x) = h(t,x), \quad x \in \partial D,\ t > 0 \tag{3.51}$$

or the Neumann conditions

$$\frac{\partial u}{\partial n}(t,x) + h(t,x) = 0, \quad x \in \partial D,\ t > 0, \tag{3.52}$$

where, as above, $\frac{\partial}{\partial n}u = \sum n_i(x)a_{ij}(x)\partial_j u(t,x)$. If we insist on consistency at $t=0$ between the boundary and initial data, appropriate restrictions should be included. Consider now the domain $\overline{D \times (0,\infty)}$ with a boundary $D \times \{0\} \cup \partial D_x[0,\infty)$ and the integral

$$I = \int_0^\infty \int_{\bar{D}} \mathrm{d}t' \rho(x')\,\mathrm{d}x' \big\{u(t',x')(\partial_t^2 - \mathcal{L})G(x,t';x',t') \\ - G(x,t;x',t')(\partial_t^2 - \mathcal{L})u(t',x')\big\}.$$

Using the equations satisfied by each of the pieces in it, we obtain

$$I = u(t,x) - \int_0^t \mathrm{d}t' \int_{\bar{D}} G(x,t,x',t')s(x',t')\rho(x')\,\mathrm{d}x'.$$

Now, Green's identity, applied to the two differential operators, yields

$$I = \int \rho(x')\big\{u(t',x)\partial_t G(x,t;x',t') - \partial_t u(t',x')G(x,t;x',t')\big\}\big|_0^\infty \mathrm{d}x' \\ + \int_0^t \mathrm{d}t' \int_{\partial D} \left\{u(t'x)\frac{\partial}{\partial n}G(x,t;x',t') - G(x,t;x',t')\frac{\partial}{\partial h}u(t,x')\right\}\widetilde{m}(\mathrm{d}x'),$$

where $\widetilde{m}(\mathrm{d}x')$ is the surface measure on ∂D as explained in [2–4]. Here and there we have made use of the fact that $G(x,t;x',t') = 0$ whenever $t' > t$. Note as well that the first integral is zero since we are dealing with homogeneous initial conditions and because of the previous remarks. To write down the dependence on the initial data explicitly, consider

$s(t,x) = s_0(t,x) + \delta(t)g(x) + \delta'(t)f(x)$, and substitute in the second term of the first identity for I, it is easy to see that

$$\begin{aligned}
&\int_0^t \mathrm{d}t' \int_{\bar{D}} G(x,t;x',t')s(t',x')\rho(x')\,\mathrm{d}x' \\
&\quad = \int_0^t \mathrm{d}t' \int_{\bar{D}} G(x,t;x',t')s_0(t',x')\rho(x')\,\mathrm{d}x' \\
&\qquad + \int_{\bar{D}} G(x,t;t',x')g(x')\rho(x')\,\mathrm{d}x' + \frac{\partial}{\partial t}\int_{\bar{D}} G(x,t;x',t')f(x')\rho(x')\,\mathrm{d}x'.
\end{aligned}$$

Putting it all together we obtain

$$\begin{aligned}
u(t,x) = &\int_0^t \mathrm{d}t' \int_D G(x,t;x',t')s(t',x')\rho(x')\,\mathrm{d}x' \\
&+ \int_0^t \mathrm{d}t' \int_{\partial D}\Big\{u(t',x')\frac{\partial}{\partial n}G(x,t;x',t') \\
&\qquad - G(x,t;x',t')\frac{\partial}{\partial n}u(t',x')\Big\}\widetilde{m}(\mathrm{d}x').
\end{aligned} \tag{3.53}$$

3.3. Geometrical optics

3.3.1. Basics

The results below serve several purposes: First they provide means to describe the propagation of "wave fronts" or discontinuities in the initial data. These computations will be useful when describing the propagation in heterogeneous media. We want to construct a function $S(t,x)$ such that the wave field $u(t,x)$ vanishes in the region where $S(t,x) < 0$, and is different from zero in the region, where $S(t,x) > 0$. By analogy with a power series expansion, the following expansion

$$u(t,x) = \sum_{n\geqslant 0} \Phi_n(t,x) f_n(S) \tag{3.54}$$

is proposed for the solution of (1.1). Again, by analogy with the simple powers $f_n(u) = U^n/n!$, we impose the restriction

$$f_n'(u) = f_{n-1}(u) \tag{3.55}$$

and, since the wave equation involves second derivatives, we will be forced to assign $f_{-2}(u)$ a priori. Or we could assign say, $f_0(u)$ and then set $f_{-1} = f_0'$ and $f_{-2} = f_0''$ the substitution of (3.54) into (1.1) leads to the following identity

$$
\begin{aligned}
& f_{-2}\Big\{(\partial_t S)^2 - \sum a_{ij}\partial_t S\partial_j S\Big\}\Phi_0 + \Phi_1 f_{-1}\Big\{(\partial_t S)^2 - \sum a_{ij}\partial_i S\partial_j S\Big\} \\
& \quad + f_{-1}\Big\{\Phi_0\partial_t^2 S + \partial_t S\partial_t\Phi_0 + \sum a_{ij}\partial_{ij}^2 S - \sum b_i\partial_i S\Phi_0\Big\} \\
& \quad + \sum_{n\geqslant 2} f_{n-2}\Big\{(\partial_t S)^2 - \sum a_{ij}\partial_i S\partial_j S\Big\}\Phi_n \\
& \quad + \sum_{n\geqslant 1} f_{n-1}\Big\{\Phi_n\partial_t^2 S - 2\partial_t S\partial_t\Phi_n - \Phi_n\sum a_{ij}\partial_{ij} S \\
& \qquad\qquad - \sum a_{ij}\partial_i\Phi_n\partial_j S - \sum b_i\Phi_n\partial_i S\Big\} \\
& \quad + \sum_{n\geqslant 0} f_n\Big\{\partial_t^2\Phi_n - \sum a_{ij}\partial_{ij}^2\Phi_n - \sum b_i\partial_i\Phi_n\Big\} = 0.
\end{aligned}
$$

Invoking the independence of $f_n(u)$, or arguing that the equation is satisfied whenever the coefficients of $f_n(s)$ are chosen to vanish identically, we obtain the following family of recursive equations. The first item is

$$
(\partial_t S)^2 - \sum a_{ij}\partial_i S\partial_j S = 0 \tag{3.56}
$$

for the function $S(t,x)$, and for Φ's we obtain

$$
\begin{aligned}
& 2\Big(\partial_t\Phi_0\partial S_t - \sum a_{ij}\partial_i S\partial_j\Phi_0\Big) + \Phi_0\Big(\partial_t^2 S - \sum a_{ij}\partial_{ij}^2 S - \sum b_i\partial_i S\Big) = 0, \\
& 2\partial_t\Phi_{n+1}\partial_t S_t - \sum a_{ij}\partial_i S\partial_j\Phi_{n+1} + \Phi_{n+1}\Big(\partial_t^2 S - \sum a_{ij}\partial_{ij}^2 S - \sum b_i\partial_i S\Big) \\
& \quad + \partial_t^2\Phi_n - \sum a_{ij}\partial_{ij}^2\Phi_n - \sum b_i\partial_j\Phi_n = 0
\end{aligned} \tag{3.57}
$$

for all $n \geqslant 0$. If we introduce the D'Alambert operator $\Box = \partial_t^2 - \sum a_{ij}\partial_{ij}^2 - \sum b_i\partial_j$, we can rewrite the last two equations as

$$
2\nabla\Phi_0\cdot\nabla S + \Phi_0\Box S = 0, \qquad 2\nabla\Phi_{n+1}\cdot\nabla S + \Phi_{n+1}\Box S + \Box\Phi_n = 0 \tag{3.58}
$$

which clearly are a recursive set for the coefficient functions $\Phi_n(t,x)$. Above we are using the obvious notation $\nabla\Phi\cdot\nabla S = \partial_t S\partial_t\Phi - \sum a_{ij}\partial_i S\partial_j\Phi$.

3.3.2. *Solving the eikonal equation*

The eikonal equation (3.55) is the mathematical counterpart of the intuitive or phenomenological notion of propagation of disturbances along rays. Beside the dependence on the shape of the initial data, the solution depends on the class of functions within which we search for it. Instead of solving (3.55) using the method of characteristics in the space time variables (t,x), we are going to look for solutions of the type

$$
S(t,x) = t - \tau(x) \tag{3.59}
$$

in which the dependence of $\tau(x)$ on other variables is to be specified later. Note that (3.58) forces both S and τ to have dimensions of time. Substituting (3.58) into (3.55), we obtain for τ the equation

$$\sum a_{ij}(x)\partial_i\tau\partial_j\tau = 1 \tag{3.60}$$

and, if we define independent variables p_i, which along the trajectories of the characteristics are to be $p_i = \partial_i\tau$, the equations of the bi-characteristics, for $i = 1, \ldots, n$ are:

$$\frac{\mathrm{d}x_i}{\mathrm{d}t} = \sum a_{ij}(x)p_j = \frac{\partial}{\partial p_i}\frac{1}{2}\Big(\sum a_{ij}p_jp_k\Big), \tag{3.61}$$

$$\frac{\mathrm{d}p_i}{\mathrm{d}t} = -\frac{1}{2}\sum_{jk} p_jp_k\partial_j a_{jk}(x) = -\frac{\partial}{\partial x_i}\left(\frac{1}{2}\sum a_{jk}p_jp_k\right), \tag{3.62}$$

$$\frac{\mathrm{d}\tau}{\mathrm{d}t} = \sum a_{ij}(x)p_ip_j = \sum p_i\frac{\partial}{\partial p_i}\left(\sum \frac{1}{2}a_{jk}p_jp_k\right) = 1. \tag{3.63}$$

We wrote (3.60) and (3.61) in Hamiltonian form, where the Hamiltonian function is $H(x, p) = \frac{1}{2}\sum a_{ij}(x)p_ip_j$. If we differentiate (3.60) and make use of (3.61), we obtain

$$\frac{\mathrm{d}^2x_i}{\mathrm{d}t^2} + \sum \Gamma^i_{jk}\frac{\mathrm{d}x_j}{\mathrm{d}t}; \qquad \frac{\mathrm{d}x_k}{\mathrm{d}t} = 0, \quad 1, \ldots, n, \tag{3.64}$$

where $\Gamma^i_{jk}(x) = \frac{1}{2}\sum a_{ie}(x)\{\partial_j g_{ek}(x) + \partial_k g_{ej}(x) - \partial_e g_{jk}(x)\}$, where, as above, $g_{ij}(x)$ denotes the inverse to the matrix $a_{ij}(x)$, and Γ^i_{jk} denotes the Christoffel symbol associated to g_{ij}.

Equation (3.62) asserts that the spatial components $x(t)$, or the bi-characteristic $(x(t), p(t))$ are geodesics in the metric $g_{ij}(x)$. These equations also appear as the Euler–Lagrange equations describing the trajectories minimizing the action

$$A(\gamma) = \int_\gamma L(\gamma(s), \gamma(s))\,\mathrm{d}s, \qquad L(x', x) = \frac{1}{2}\sum g_{ij}(x)x_ix_j, \tag{3.65}$$

where γ is any continuously differentiable trajectory with fixed points x and x^0. The Euler–Lagrange equations can be rewritten as

$$\frac{\mathrm{d}}{\mathrm{d}t}\left(\frac{\partial L}{\partial \dot{x}}\right) - \frac{\partial L}{\partial x_i} = 0, \quad i = 1, \ldots, n. \tag{3.66}$$

It is actually easy to see that (3.60) implies $H = \frac{1}{2}\sum a_{ij}(x)p_ip_j = \frac{1}{2}\sum g_{ij}(x)\dot{x}_j\dot{x}_j$ and that either the set (3.60)–(3.61) or (3.65) imply $H = L$ is a constant, which, according to (3.59), must have the value $1/2$. This restricts the choices of the initial velocities $\dot{x}(0)$ or moments $p(0)$ to an initial cone. Note that $L = \frac{1}{2}\sum g_{ij}(x)\dot{x}_i\dot{x}_j$ is dimensionless, but

$\sqrt{g_{ij}}(x)\,\mathrm{d}x_i\,\mathrm{d}x_j$ has dimensions of time. To tie things up a bit, we begin with the following observations. From (3.65) it follows readily that

$$\frac{\mathrm{d}L}{\mathrm{d}t} = 0. \tag{3.67}$$

Thus, if $f(\lambda)$ is a continuously differentiable function, multiplying (3.65) by $f'(L)$ we see that $x(t)$ solves

$$\frac{\mathrm{d}}{\mathrm{d}t}\left(\frac{\partial}{\partial \dot{x}_i} f(L)\right) - \frac{\partial f(L)}{\partial x_i} = 0, \quad i = 1, 2, \ldots, n \tag{3.68}$$

which is the Euler–Lagrange set corresponding to the action

$$A_f(\gamma) = \int_\gamma f(L(s))\,\mathrm{d}s. \tag{3.69}$$

If we wanted the trajectories that minimize (3.68) and satisfy (3.66), to minimize (3.64) and satisfy (3.65) as well, we should check the analogue of (3.65), namely,

$$\frac{\mathrm{d}}{\mathrm{d}t} f(L) = \sum x'(L)\frac{\partial L}{\partial \dot{x}_i}\ddot{x}_i + f'(L)\frac{\partial L}{\partial x_i}\dot{x}_i = \frac{\mathrm{d}}{\mathrm{d}t}\left(\sum \frac{\partial f(L)}{\partial \dot{x}_i}\dot{x}_i\right)$$

or, equivalently,

$$\frac{\mathrm{d}}{\mathrm{d}t}\left(\sum \frac{\partial f(L)}{\partial \dot{x}_i}\dot{x}_i - f(L)\right) = 0. \tag{3.70}$$

Note (curious coincidence or is there something behind it?) that if $f(L) = \sqrt{L}$, the quantity within the parentheses above is just $\sqrt{L}$!. Thus for $f(L) = \sqrt{L}$, the minimum of

$$T(\gamma) = \int_\gamma \sqrt{L(\dot{\gamma}(s), \gamma(s))}\,\mathrm{d}s \tag{3.71}$$

occurs along curves $\xi(s)$ satisfying (3.67), and because of (3.69), we have

$$\frac{\mathrm{d}\sqrt{L}}{\mathrm{d}t} = 0 \tag{3.72}$$

which implies that the curves satisfy both (3.67) and (3.65). Note that the integral in (3.68) is independent of parametrization and can be written as

$$T(\gamma) = \int_\gamma \sqrt{A_{ij}(x)\,\mathrm{d}x_i\,\mathrm{d}x_j} \tag{3.73}$$

and, since a_{ij} has dimensions of (velocity)^2, $\sqrt{g_{ij}\,\mathrm{d}x_i\,\mathrm{d}x_j}$ comes out in time units. Thus (3.70) or (3.72) is the time along curve γ, and what the above analysis asserts is that geodesics minimizes the travel time. To particularize to the case, where $a_{ij}(x) = c^2(x)\delta_{ij}$ and $g_{ij}(x) = \delta_{ij}c^2(x)$, note that (3.59) and (3.72) become

$$c^2(x)(\nabla\tau)^2 = 1 \quad \text{and} \quad T(\gamma) = \int_\gamma |\mathrm{d}x|/c(x),$$

where certainly $|\mathrm{d}x| = (\sum(\mathrm{d}x_i)^2)^{1/2}$. If we set $\hat{n}(x)$ to be the unit normal to the surfaces of $\tau =$ constant, to find how fast the wave front $\{(t,x) : s(t,x) = 0\}$ is moving, the classical trick is to look at

$$S(t_0, x_0) = 0 \quad \text{and} \quad S(t_0 + \delta t, x_0 + \hat{n}\delta\ell) = 0$$

from which one obtains

$$\frac{\delta\ell}{\delta t} = \frac{1}{\hat{n}\cdot\nabla\tau_{(x_0)}} = \frac{1}{|\nabla\tau(x_0)|} = c(x_0).$$

To obtain an interesting relationship between the travel time along geodesics and the solutions to (3.63), note that, if we denote by

$$x(t) = F(t, x(0), \dot{x}(0)) \tag{3.74}$$

the solution to (3.63) with initial data $(X(0), X(0))$, since (3.63) is invariant under the transformation $t \to \bar{t} = t\alpha$ and $x(t) \to \bar{x}(\bar{t}) = x(\bar{t}/\alpha)$, we can write

$$x = \varphi(x(0), T(x,x)\dot{x}(0))$$

when $x = x(T(x, x_0))$. If we assume that $\varphi(x_0, \xi) = x$ or, equivalently, $F(t, x_{(0)}, \xi)) = x$ can be solved for ξ to obtain

$$\xi = \Psi(x, x(0)), \tag{3.75}$$

then, from definition (3.70) (or (3.72) and the constancy of $L(x, x)$, we obtain

$$T(x, x_0)^2 = \sum g_{ij}(x_0)\Psi_i(x, x_0)\Psi_j(x, x_0).$$

3.3.3. Solving the transport equations

Equations (3.55) and (3.56) are called transport equations, for they describe the evolutions of whatever is measured by the "amplitudes" Φ_n along the times of the field tangent to ∇S. Let us begin with (3.54) in its general form

$$\frac{\partial S}{\partial t}\frac{\partial \Phi}{\partial t} - \sum_{ij} a_{ij}(x)\partial_j\Phi\partial_j S = -\frac{\Phi}{2}\Box S \tag{3.76}$$

and note that, as we set $S(t,x)=t-\tau(x)$, the equation above becomes

$$\frac{\partial \Phi}{\partial t}+\sum a_{ij}(x)\partial_i\Phi\partial_j\tau=-\Phi\Delta\tau/2.$$

Consider the system

$$\frac{\mathrm{d}x_i}{\mathrm{d}s}(x)=-\sum a_{ij}(x(s))\partial_j\tau(x(s)),\qquad x_i(0)=x_i,\quad i=1,\ldots,n, \tag{3.77}$$

and the evaluation $F(s)=\ell_n\phi(s,x(t-s)), 0\leqslant s\leqslant t$ along the trajectories of (3.75). Note that

$$\frac{\mathrm{d}F}{\mathrm{d}s}=-\frac{1}{2}\Delta\tau(x(t-s))$$

which, integrated with respect to s on $(0,t)$, yields

$$\Phi(t,x)=\Phi(0,x(t))\mathrm{e}^{-\int_0^t\frac{1}{2}\Delta\tau(x(s))\,\mathrm{d}s}. \tag{3.78}$$

Comment: Extending the metric $a_{ij}(x)$ to a space time metric $a_0=1$, $\bar{a}_{0j}=0$, $\bar{a}_{ij}=a_{ij}$, $i,j\geqslant 1$, we would have obtained

$$\Phi(t,x)=\Phi(0,x(t))\mathrm{e}^{\frac{1}{2}\int_0^t\Box S(x(s))\,\mathrm{d}s}, \tag{3.79}$$

where $x_i(s)$, $i=0,1,\ldots,n$ solves the analogue of (3.75) with $\bar{a}_{ij}$ replacing a_{ij}. To solve the rest of (3.56), we follow the same routine. Rewrite it as

$$\begin{aligned}&\frac{\mathrm{d}}{\mathrm{d}s}\left(\Phi_{n+1}(s,x(t-s))\mathrm{e}^{\frac{1}{2}\int_0^s\Delta\tau(x(\sigma))\,\mathrm{d}\sigma}\right)\\&\quad=\mathrm{e}^{-\frac{1}{2}\int_0^s\Delta\tau(x_{(\sigma)})\,\mathrm{d}s}\Box\Phi_n(x_{(s)})\end{aligned}$$

and now integrate both sides over $(0,t)$ to obtain

$$\begin{aligned}\Phi_{n+1}(t,x)&=\Phi_n(0,x(t))\mathrm{e}^{-\frac{1}{2}\int_0^t\Delta\tau(\sigma)\,\mathrm{d}\sigma}\\&\quad+\int_0^t\mathrm{e}^{-\frac{1}{2}\int_s^t\Delta\tau(x(\sigma))\,\mathrm{d}\sigma}\Box\Phi_n(x(s))\,\mathrm{d}s.\end{aligned} \tag{3.80}$$

Note that this family of solutions inserted in (3.53) would yield solutions to the wave equation in as much as $f_n(s)$ can be chosen appropriately.

3.4. General representation of solutions

This section will deal with outlining a general representation of the solutions of the wave equation

$$(\partial_t^2 - \mathcal{L})u(t,x) = s(t,x), \quad t > 0,\ x \in D \tag{3.81}$$

skipping the details concerning the regularity of the solutions. We follow Romanov's presentation addressing the reader there for details and further references. As above,

$$\mathcal{L}u = \sum a_{ij}(x)\partial_i\partial_j u + \sum b_i(x)\partial_i u + c(x)u$$

and we assume that our source term is of the form

$$s(t,x) = f(x)h(t,x). \tag{3.82}$$

We assume that $a_{ij}(x) \in \mathcal{C}^{\ell+4}(D), b_i(x) \in \mathcal{C}^{\ell+2}$ and $c(x) \in \mathcal{C}^e(D)$ for ℓ large enough. If we denote, as above, by $g_{ij}(x)$ the matrix inverse of $a_{ij}(x)$, and consider the geodesics defined by it on D, we assume that, for each pair x_1, x_2, there exists a unique geodesic totally contained in D joining x_1 and x_2. Denote it by $\Gamma(x_1, x_2)$. Instead of solving (3.79) for any (3.80), we will look for solutions when

$$s(t,x) = \delta(t - t_0)\delta(x - x_0),$$

which vanish for $t < t_0$. Since the coefficients in (3.79) are time independent, we can obtain fundamental solutions of the type $H(t - t_0, x, x_0)$. So, fix $x_0 \in D$, and we saw in Subsection 6.2 that at x_0 we can introduce special (Riemann) coordinates $\zeta = (\zeta_1, \zeta_2, \zeta_3)$ so that $x \in D$ is described by $x = \varphi(\zeta, x_0)$. The interpretation of ζ becomes clear when written as $\zeta = \tau(x, x_0)(\alpha_1, \alpha_2, \dots, \alpha_n)$, where $(\alpha_1, \alpha_2, \dots, \alpha_n)$ is tangent to the geodesic $\Gamma(x_0, x)$ at x_0, and has a unit length. Recall that $x = \varphi(\zeta, x_0)$ can be solved for ζ to obtain $\zeta = \Psi(x, x_0)$ and

$$\tau^2(x_0, x) = \sum g_{ij}(x^0)\psi_i(x, x_0)\psi_j(x, x_0).$$

It can be proved that $\psi_i(x, x_0) \in \mathcal{C}^{\ell+3}(D \times D)$, therefore $\tau^2(x_0, x) \in \mathcal{C}^{\ell+3}(D \times D)$. The following result embodies a generalization of (3.51); we state it for $d = 3$.

THEOREM 3.1. *If $\ell \geqslant 2N + 7$, $N \geqslant -1$, the fundamental solution to* (3.78) *is of the form*

$$H(t, x, x_0) = \frac{1}{2}\theta(t)\sum_{k=1}^{N}\sigma_k(x, x^0)\theta_k(t^2 - \tau^2(x, x_0)) + v_N(t, x, x^0). \tag{3.83}$$

Here $\theta(t)$ is the Heaviside function, $\theta_{-1}(t)=\delta(t)$ and for $k\geqslant 0$, $\theta_k(t)=\theta(t)t^k/k!$. The functions $\sigma_k(x,x_0)$ are calculated according to the recursion relations

$$\begin{aligned}\sigma_{-1}(x,x_0)=&\left(\frac{1}{|A(x_0)|}\left|\frac{\partial}{\partial x}\right||(x,x_0)|\right)^{1/2}\\&\times\exp\left\{-\frac{1}{2}\int_0^1\sum_k\left[b_i(\xi)-\sum_j\frac{\partial}{\partial\xi_j}a_{ij}(\xi)\right]g_{kj}^{ke}(x_0)\psi_k(x,x_0)\right.\\&\left.\times\frac{\partial}{\partial\xi_j}\psi_\ell(\xi,x^0)\bigg|_{\xi=\varphi(t\psi(x,x_0),x_0)}\mathrm{d}t\right\},\\\sigma_k(x,x_0)=&\frac{1}{4}\sigma^{-1}(x,x_0)\int_0^1[\sigma_{-1}(\xi,x_0)]^{-1}t^k\mathcal{L}_\xi\sigma^{k-1}(\xi,x_0)\bigg|_{\xi=\varphi(t\psi(x,x_0),x_0)}\mathrm{d}t\end{aligned}$$

for $k\geqslant 0$. Above $\mathcal{L}_\xi$ denotes the operator $\mathcal{L}$ acting on the ξ variable of the indicated function. The coefficients $\sigma_k(x,x_0)\in C^{\ell-2k}(D\times D)$. For $N=-1$, $v_N(t,x,x_0)$ is continuous in $G=\{(t,x,x^0):x\in D,\ x_0\in D,\ t>\tau(x,x_0)\}$ and piecewise continuous in $G_0=G\cup G'$, with

$$G'=\{(t,x,x_0):x,\ x_0\in D,\ 0\leqslant t\leqslant\tau(x,x_0)\}.$$

For $N=0$, $v_N(t,x,x_0)$ is continuous in G_0, and for $N\geqslant 1$, $v_n(t,x,x_0)$ is continuous in G_0, together with its derivatives $\partial_x^\alpha\partial_t^\beta\partial_{x_0}^\gamma v_N$, $\alpha=(\alpha_1,\alpha_2,\alpha_3)$ etc., $|\alpha|=\alpha_1+\alpha_2+\alpha_3$ etc., and $|\alpha|+|\gamma|\leqslant\ell-2s-3$, $|\alpha|+\beta+|\gamma|\leqslant s-1$. Moreover, for each $s\geqslant -1$,

$$v_N(t,x,x_0)=\theta_{N+1}\big(t^2-\tau^2(x,x^0)\big)v_N^0(t,x,x_0),\tag{3.84}$$

where $v_N^0(t,x,x_0)$ is bounded in bounded domains contained in G_0.

PROOF. Pick up your N, and set

$$u_N:=\frac{1}{2a}\theta(t)\sum_{k=1}^N\gamma(x,x^0)\theta_k\big(t^2-\tau^2(x,x^0)\big).\tag{3.85}$$

Once we have verified that

$$\big(\partial_t^2-\mathcal{L}\big)u_N=-\frac{1}{2\pi}\theta(t)\theta_N\big(t^2-\gamma^2(x,x^0)\big)\mathcal{L}\sigma_N+\delta(x-x^0)\delta(t),\tag{3.86}$$

the problem reduces to verifying that the equation

$$\big(\partial_t^2-\mathcal{L}\big)v_N=\frac{1}{2\pi}\theta(t)\theta_N\big(t^2-\tau^2(x,x_0)\big)\mathcal{L}\sigma_N\tag{3.87}$$

subject to

$$v_N(t,x,x_0)=0 \quad \text{for } t<0 \tag{3.88}$$

has a solution with the desired properties. When applying $\partial_t^2-\mathcal{L}$ to u_N given by (3.82), in order to verify (3.83), we obtain equations to determine σ_k's. Introducing the notation $\Gamma(t,x,x_0)=t^2-\tau^2(x,x_0)$, we obtain

$$\begin{aligned}
&(\partial_2^2-\mathcal{L})u_N\\
&=\frac{1}{2\pi}\sum_{k=-1}^{N}\{\delta'(t)\theta_{k+1}(\Gamma)+2\delta(t)\partial_t\theta_k(\pi)\}\sigma_k(x,x_0)\\
&\quad+\frac{1}{2\pi}\theta(t)\sum_{k=1}^{N}\Big\{\sigma_k\Big[\theta_k''(\Gamma)\Big(\Gamma_t^2-\sum_{ij}a_{ij}(x)\partial_i\Gamma\partial_j\Gamma\Big)+\theta_k^1(\Gamma)\partial_t^2\Gamma\Big]\\
&\quad-\theta_k(\Gamma)\mathcal{L}\sigma_k-\theta_k'(\Gamma)\Big[2\sum a_{ij}\partial_j\Gamma\partial_j\sigma_k+\sigma_k\sum b_i\partial_i\Gamma\\
&\qquad+\sigma_k\sum_{ij}a_{ij}(x)\partial_i\Gamma\partial_j\Gamma\Big]\Big\}.
\end{aligned}\tag{3.89}$$

To verify this computation, just use the following simple verifiable identities

$$\begin{aligned}
&(\partial_t\Gamma_t)^2-\sum a_{ij}(x)\partial_i\Gamma\partial_j\Gamma=4\Gamma, \qquad \partial_t^2\Gamma=2,\\
&\Gamma\theta_t''(\Gamma)=(k-1)\theta_{k-1}(\Gamma), \qquad \theta_k^1(\Gamma)=\theta_{k-1}(\Gamma),\\
&\delta'(t)\theta_k(\Gamma)+2\delta(t)\partial_t\theta_k(\Gamma)=2\partial_t[\delta(t)\theta_k(\Gamma)],\\
&-\delta'(t)\theta_k(\Gamma)=\delta(t)\theta(\Gamma)/t.
\end{aligned}$$

To compute the left-hand side of the last one, note that for any infinitely differentiable function $\psi(t,x)$

$$\int\frac{1}{t}\delta(t)(\Gamma)\psi(t,x)\,\mathrm{d}t\,\mathrm{d}x=\lim_{t\to0}\frac{1}{t}\int_{\mathbb{R}^3}\theta_k(\Gamma)\psi(t,x)\,\mathrm{d}x,$$

which when introducing the Riemann coordinates τ,θ,ϕ for x and using

$$x=\phi(\zeta,x^0), \quad \zeta=\zeta'B^{-1/2}(x^0),\ \zeta'=\tau(\sin\theta\cos\phi,\sin\theta\sin\phi,\cos\phi),$$

the sequence continues as

$$\begin{aligned}
&\lim_{t\to0}\frac{1}{t}\int_0^{2\pi}\int_0^{\pi}\int_0^{\infty}(t+\tau)^k\theta_k(t-\tau)\psi(t,x')\left|\frac{\partial\phi}{\partial\zeta}\right|\left|B(x^0)\right|^{-1/2}\tau^2\sin\theta\,\mathrm{d}\tau\,\mathrm{d}\phi\,\mathrm{d}\theta\\
&\quad=2\pi\psi(0,x^0)\left|A(x^0)\right|^{1/2}\begin{cases}1, & k=-1,\\ 0, & k\leqslant 0.\end{cases}
\end{aligned}$$

□

Comment: It was here where the three spatial dimensions simplify the involved computations. Substituting in the first term on the right-hand side of (3.89), we obtain

$$\begin{aligned}(\partial_t^2 - \mathcal{L})u_N &= |A(x_0)|^{1/2}\sigma_{-1}(x_0^0, x_0)\delta(x - x_0)\delta(t) \\ &\quad + \frac{1}{2\pi}\theta(t)\sum_{k=-1}^{N}\Big\{\theta_{k-1}(\Gamma) - \Big[2\sum_{ij} a_{ij}\partial_i\Gamma\partial_j\sigma_k \\ &\quad - \sigma_k\sum_{ij} a_{ij}\partial_i\partial_j\Gamma + \sum_i b_i\partial_i\Gamma + (4k-2)\sigma_K\Big] - \theta_k(\Gamma)\mathcal{L}\sigma_k\Big\}.\end{aligned}$$

Above, $A(x_0)$ denotes the matrix $a_{ij}(x_0)$ and $B = A^{-1}$, and $|A(x_0)| = \det A_1$. Now choose $\sigma_{-1}(x_0, x_0) = |A(x_0)|^{-1/2}$ and the coefficients $\sigma_k(x, x_0)$ should be determined so that (3.83) obtains, i.e.,

$$2\sum_{ij} a_{ij}\partial_i\Gamma\partial_i\sigma_k + \sigma_k(\mathcal{L}'\Gamma - 4k + 2) = \begin{cases} 0, & k = -1, \\ -\mathcal{L}\sigma_{k-1}, & k \geqslant 0, \end{cases} \tag{3.90}$$

where $\mathcal{L}'\Gamma = \sum a_{ij}\partial_{ij}^2\Gamma + \sum b_i\partial_i\Gamma$. To see that these equations yield the required σ_k, note that the equations of the characteristics $\mathrm{d}x_i/\mathrm{d}\tau = \sum_j a_{ij}\partial_j\tau$ and $\partial\Gamma/\partial\tau = -2\tau$ imply.

$$-2\tau\frac{\partial\sigma}{\partial\tau}k = \sum_{ij} a_{ji}(x)\partial_j\Gamma\frac{\partial\sigma_k}{\partial x},$$

where clearly we are using the length of the geodesic to parametrize it. Using these identities, it can be proved that

$$\sum_{ij}\partial_i(a_{ij}\partial_j\Gamma) = -6 + 2\tau\frac{\partial}{\partial\tau}\ell n\left|\frac{\partial\psi}{\partial x}(x, x^0)\right|. \tag{3.91}$$

With this result we can write $\mathcal{L}'\Gamma$ as

$$\sum\frac{\partial}{\partial x_i}\left(a_{ij}\frac{\partial}{\partial x_j}\Gamma\right) + \sum_i\left(b_i - \sum_{j=1}^{3}\frac{\partial}{\partial x_j}a_{ij}\right)\frac{\partial\Gamma}{\partial x_i}.$$

From this and (3.87) we rewrite (3.86) as

$$\begin{aligned}&\frac{\partial\sigma_k}{\partial\tau} + \sigma_k\left[\frac{k+1}{\tau} - \frac{1}{2}\frac{\partial}{\partial\tau}h_i\left|\frac{\partial\psi}{\partial\tau}(x, x^0)\right| - \frac{1}{4\tau}\sum_i\left(b_j - \sum_j\frac{\partial}{\partial x_j}a_{ij}\right)\frac{\partial\Gamma}{\partial x_i}\right] \\ &\quad = \frac{1}{4\tau}(\mathcal{L}\sigma_{k-1})\begin{cases} 0, & k = -1, \\ 1, & k > -1\end{cases}\end{aligned}$$

or, equivalently,

$$\frac{\partial}{\partial \tau}\left(\tau^{k+1} M \sigma_k\right) = \frac{1}{4\tau}\left(\tau^k M \mathcal{L}\sigma_{k-1}\right)\begin{cases} 0, & k=-1, \\ 1, & k>-1 \end{cases} \tag{3.92}$$

and, for $k > -1$, where

$$M(x, x_0) = \left|\frac{\partial \psi}{\partial x}(x, x^0)\right|^{-1/2} \exp\frac{1}{2}\left\{\int_{\Gamma(x,x^0)} \sum_i \left(b_i - \sum_j \frac{\partial \tau}{\partial x_j}\right)\frac{\partial \tau}{\partial x_i}\, \mathrm{d}\ell\right\},$$

ℓ varying between 0 and $\tau(x, x^0)$ along $\Gamma(x, x^0)$. A simple integration of both sides of (3.89) and the appropriate choice of the integration constant yields

$$\sigma_{-1}(x, x^0) = \left[|A(x_0)|^{1/2} M(x, x_0)\right]^{-1}, \quad k = -1,$$

$$\sigma_k(x, x_0) = \left[\frac{\tau(x, x_0)^{1(1+k)}}{4M(x, x_0)}\right]\int_{(x,x_0)} \ell^k M(\xi, x_0)\mathcal{L}_\xi \sigma_{k-1}(\xi, x^0)\, \mathrm{d}\ell, \quad k > -1.$$

To obtain an exact match with the formulae in the statement of the theorem, change ℓ to $\tau(x, x_0)$ as an integration variable. We direct the reader to the original sources for the study of the solution $v_N(t, x, x_0)$ of

$$\left(\partial_t^2 - \mathcal{L}\right)v_N = \frac{1}{2\pi}\theta(t)\theta(t^2 - \tau^2)\mathcal{L}\sigma_n.$$

Comment: When all the coefficients involved are C^∞, we can let $N \to \infty$ and obtain the representation for the solution of (3.81), as mentioned at the beginning of Section 6. Also, when $a_{ij}(x)$ are independent of $x, b_i(x) = e(x) = 0$, these results yield $\sigma_1(x, x_0) = \text{constant}$, $\sigma_0 = 0$ and

$$H(t, x, x^0) = \frac{1}{2\pi}\theta(t)\delta(t^2 - \tau^2(x, x_0))\sigma_{-1},$$

$$\sigma_{-1} = |Ar|^{-1/2}, \quad \tau^2(x) = \sum_{ij} g_{ij}(x - x_0)_i(x - x_0)_j.$$

References

Two complementary, basic textbooks are

1. John, F.: *Partial Differential Equations*, Springer-Verlag, Berlin, 1978.
2. Treves, F.: *Basic Linear Partial Differential Equations*, Academic Press, New York, 1975.

For the method of spherical means we followed

3. Courant, R.: *Methods of Mathematical Physics*, Wiley-Interscience, New York, 1962.

   But a nice version of the method appeared in

4. Folland, G. B.: *Introduction to Partial Differential Equations*, Princeton Univ. Press, Princeton, 1976.

The decomplexification approach is outlined in

5. Garabedian, P. R.: *Partial Differential Equations*, Wiley, New York, 1964.

Among the books devoted to applications of wave equations we have

6. Bleistein, N.: *Mathematical Methods for Wave Phenomena*, Academic Press, New York, 1989.
7. Aki, K. and Richards, P. G.: *Quantitative Seismology*, Vol. I, W. H. Freeman Co, New York, 1980.
8. Witham, G. B.: *Linear and Nonlinear Waves*, Wiley-Interscience, New York, 1974.

From this last one we got more than inspiration for the material in Section 4. For the material in Section 5 we submit the reader to

9. Romanov, V. G.: *Inverse Problems of Mathematical Physics*, VNU Science Press, Utrecht, 1987.

A whole lot about oscillatory phenomena can be seen in

10. Rabinovich, M. I. and Trubetskow, D. I.: *Oscillations and Waves in Linear and Nonlinear Systems*, Kluwer Acad. Publ., Dordrecht, 1989.

And a recent, modern presentation of the classical stoff is contained in

11. Shatah, J. and Strowe, M.: *Geometric Wave Equations*, Amer. Math. Soc. Lecture Notes, Providence, RI, 2000.

CHAPTER

FOUR

WAVES AND BROWNIAN MOTIONS

This chapter is designated to waves in isotropic homogeneous spaces in which the waves propagate at constant speed. The aim is to re-obtain classical results by representing the appropriate Green functions for wave propagation problems in terms of path integrals with respect to the Wiener measure associated with some Brownian motion.

4.1. Waves in full space

As usual, we begin with

$$\partial_t^2 u(t,x) = c^2 \Delta u(t,x) + s(t,x), \quad t > 0, x \in \mathbb{R}^d \tag{4.1}$$

subject to

$$u(0,x) = f(x), \qquad u_t(0,x) = g(x), \quad x \in \mathbb{R}^d, \tag{4.2}$$

where c is a positive constant, with no extra conditions on $f(x), g(x), s(t,x)$ until needed. Assuming that a solution exists and that it does not behave too bad in time, we put

$$\tilde{u}(\alpha, x) = \int_0^\infty e^{-\alpha t} u(t,x)\, dt \tag{4.3}$$

to transform problem (4.1) into

$$\left(\frac{1}{2}\left(\frac{\alpha}{c}\right)^2 - \frac{1}{2}\Delta\right)\tilde{u}(\alpha,x) = \widetilde{S}(\alpha,x), \tag{4.4}$$

where $\widetilde{S}(\alpha, x) = (\tilde{s}(\alpha, x) + \alpha f(x) + g(x))/2c^2$.

The reason we divide by 2 is to think of $\Delta/2$ as the infinitesimal generator of the transition semigroup for Brownian motion. We saw in Chapter 1, (1.13) and (1.14) that the solution to (4.4) is given by

$$\begin{aligned}\tilde{u}(\alpha, x) &= U^{\beta}\widetilde{S}(\alpha, x) = \int_0^{\infty} e^{-\beta t} P_t \widetilde{S}(\alpha, x) \\ &= E^x \int_0^{\infty} e^{-\beta t} \widetilde{S}(\alpha, B(t))\, dt\end{aligned} \tag{4.5}$$

with $\beta = \frac{1}{2}(\frac{\alpha}{c})^2$, and $B(t)$ being an $\mathbb{R}^d$-dimensional Brownian motion.

When $d = 3$, thinks are easy to compute explicitly, for in this case,

$$\int_0^{\infty} e^{-\beta t} \frac{e^{-(x-y)^2/2t}}{(2\pi t)^{3/2}}\, dt = \frac{e^{-\sqrt{2\beta}|x-y|}}{2\pi|x-y|}.$$

This being so, it is clear which assumptions to impose on $f(x), g(x), \tilde{s}(\alpha, x)$ for the exchange of integrals to explicitly work out (4.6). If $f, g, \tilde{s}(\alpha, \cdot)$ are bounded in $\mathbb{R}^d$, then

$$\tilde{u}(\alpha, x) = \int \frac{e^{-\frac{\alpha}{c}|x-y|}}{2\pi|x-y|} \widetilde{S}(\alpha, y)\, dy \tag{4.6}$$

is well defined as $\alpha > 0$.

To complete the exercise, we split $\widetilde{S}(\alpha, x)$ to three pieces and consider to begin with

$$\int \frac{e^{-\frac{\alpha}{c}|x-y|}}{|x-y|} \tilde{s}(\alpha, y)\, dy = \int_0^{\infty} e^{-\alpha t} \int \frac{\theta\left(t - \frac{|x-y|}{c}\right)}{2\pi|x-y|} s\left(t - \frac{|x-y|}{c}, y\right) dy.$$

Thus, the sources $s(t, x)$ contribute the following term to the full solution

$$\begin{aligned}u_s(t, x) &= \frac{1}{2c^2} \int \frac{\theta(t - |x-y|/c)}{2\pi|x-y|} s(t - |x-y|/c)\, dy \\ &= \frac{1}{4\pi c^2} \int \frac{\theta(t - |y|/c)}{|y|} s(t - |y|/c, y)\, dy \\ &= \frac{c^2}{4\pi c^2} \int_0^t r \int_{|\xi|=1} s(t - r, cr\xi)\, d\sigma(\xi)\, dr,\end{aligned}$$

where $d\sigma(\xi)$ denotes the (Lebesque) area measure on the unit sphere $\{\xi : |\xi| = 1\}$. Rewritten a bit this becomes

$$u_s(t, x) = \frac{1}{4\pi} \int_0^t \frac{d\sigma}{t - \sigma} \int_{|\xi|=c(t-\sigma)} S(\sigma, x + \xi)\, d\Sigma(\xi)$$

with $\mathrm{d}\Sigma(\xi) = \xi^2\,\mathrm{d}\sigma(\xi)$. This is the standard Duhamel' result. Consider now

$$\int_0^\infty \alpha \frac{\mathrm{e}^{-\frac{\alpha}{c}|x-y|}}{|x-y|} f(y)\,\mathrm{d}y = \int_0^\infty \mathrm{e}^{-\alpha t}\left(\frac{\partial}{\partial t}\int \sigma\left(t - \frac{|x-y|}{c}\right)\frac{f(y)}{|x-y|}\,\mathrm{d}y\right)\mathrm{d}t$$

which contributes to the solution with a term

$$\begin{aligned} u_f(t,x) &\frac{\partial}{\partial t}\int \frac{\delta(t-|y|/c)}{4\pi|y|c^2} f(x+y)\,\mathrm{d}y \\ &= \frac{1}{4\pi}\frac{\partial}{\partial t} t \int_{|\xi|=1} f(x+et\xi)\,\mathrm{d}\sigma(\xi) \end{aligned}$$

and we leave for the reader to verify that the remaining term is

$$u_g(t,x) = \frac{t}{4\pi}\int_{|\xi|=1} f(x+ct\xi)\,\mathrm{d}\sigma(\xi).$$

Collecting these, we obtain the classical solution

$$\begin{aligned} u(t,x) = {} & \frac{1}{4\pi c^2 t}\int_{|\xi|=ct} g(x+\xi)\,\mathrm{d}\Sigma(\xi) + \frac{1}{4\pi c^2}\frac{\partial}{\partial t}\frac{1}{t}\int_{|\xi|=ct} f(x+\xi)\,\mathrm{d}\Sigma(\xi) \\ & + \frac{1}{4\pi c^2}\int_0^t \frac{\mathrm{d}\sigma}{t-\sigma}\int_{|\xi|=c(t-\sigma)} s(\sigma, x+\xi)\,\mathrm{d}\Sigma\xi. \end{aligned} \tag{4.7}$$

Enough of this and back to (4). Using

$$\int \frac{\mathrm{e}^{-(x-y)^2/2t}}{(2\pi t)^{d/2}}\mathrm{e}^{-\beta t} = \frac{2}{(2\pi)^{d/2}}\left(\frac{\sqrt{2\beta}}{|x-y|}\right)^{\frac{d}{2}-1} K_{\frac{d}{2}-1}\left(\sqrt{2\beta}|x-y|\right) \tag{4.8}$$

we could obtain implicit analogues of (5) for $d \neq 3$.

Actually, the cases $d = 1$ and $d = 2$ can be worked out explicitly as well, but we leave them as an exercise for the reader.

The following are exercises for the reader as well:

(i) Use

$$\int \mathrm{e}^{-ik\cdot x}\frac{\mathrm{e}^{-|x-y|^2/2t}}{(2\pi t)^{d/2}}\,\mathrm{d}y = \mathrm{e}^{-\frac{k^2 t}{2}}\mathrm{e}^{ik\cdot x}$$

to obtain the usual Fourier representation of the solutions to (1).

(ii) Look up the integral

$$\int_{|\xi|=1} e^{ik\cdot\xi}\, d\sigma(\xi)$$

over the sphere $\{|\xi|=1\}$ in $\mathbb{R}^d$ to go from the Fourier representation to what she/he would get using (4.5) and (4.7).

4.2. Dirichlet problems

We consider here the problem

$$\partial_t^2 u(t,x) = c^2 \Delta u(t,x) + s(t,x), \quad t>0, x\in D \tag{4.9}$$

with initial conditions

$$u(0,x) = f(x), \qquad u_t(0+,x) = g(\alpha), \quad x\in D \tag{4.10}$$

as well as the boundary condition

$$u(t,x) = h(t,x), \quad t>0,\ x\in\partial D. \tag{4.11}$$

When the consistency of boundary conditions and initial conditions are required, we must assume that $f(x)$ and $g(x)$ can be extended to $\overline{D}$ and $\lim_{z\in D\to x\in\partial D} f(z) = \lim_{t\downarrow 0} h(0,x)$, and that $\frac{\partial h}{\partial t}(t,x)$ is defined on $(0,t)$ and continuous on $[0,\infty)$ for every x in ∂D, and, of course, $h_t(0+,x) = \lim_{z\in D\to x\in\partial D} g(z)$.

When we take the Laplace transform of problem (1), we obtain, with $\widetilde{S}(\alpha,x)$ as above,

$$\left\{\frac{1}{2}\left(\frac{\alpha}{c}\right)^2 - \frac{1}{2}\Delta\right\}\tilde{u}(\alpha) = \widetilde{S}(\alpha,x), \tag{4.12}$$

$$\lim_{z\in D\to x\in\partial D} \tilde{u}(\alpha,z) = \tilde{h}(\alpha,x), \quad \alpha>0, \tag{4.13}$$

at each $x\in\partial D$ for which $\tilde{h}(\alpha,\cdot)$ is continuous.

The domain D which we consider is any connected, open set $D\subset\mathbb{R}^n$, bounded or not, and such that every point $x\in\partial D$ is regular (for $\overline{D}^c$), i.e., such that $P^x(T_D=0)=1$, where $T_D = \inf\{t>0 : B(t)\in\overline{D}^c\}$.

Invoking the results of Section 3 of Chapter 2, we know that the solution to (4.12) can be represented by

$$\begin{aligned}\tilde{u}(\alpha,x) &= E^x\big[M(T_D)\tilde{h}(\alpha,B(T_D));\, T_D<\infty\big] \\ &\quad + E^x\left[\int_D^{T(D)} M(t)\widetilde{S}(\alpha,B(T))\,dt\right],\end{aligned} \tag{4.14}$$

where $M(t) = \exp -\frac{1}{2}(\frac{\alpha}{c})^2 t$. Note that for $\alpha > 0$, when $\tilde{h}(\alpha, x)$ is a bounded function of x, it does not matter whether we restrict the integral in (4.14) to the set $\{T_D < \infty\}$ or not.

We shall now present two examples that use the results obtained in Section 1.3 of Chapter 2: propagation of waves in a half-space and propagation of waves in a sphere.

4.2.1. *Waves in half space with Dirichlet boundary conditions*

Thus, consider $D = \{x \in \mathbb{R}^d : x_d > 0\}$ and $\partial D = \{x \in \mathbb{R}^d : x_d = 0\}$. Obviously,

$$T_D = \inf\{t > 0 : B^d(t) = 0\}$$

depends only on the last component of $B(t) = (B^1(t), \ldots, B^d(t))$, and for any $x \in D$, $P^x(T_D < \infty) = 1$. This time the first term of (4.14) is

$$E^x\big[M(T_D)\tilde{h}(\alpha, B(T_D))\big] = E^x\big[\mathrm{e}^{-\beta T_D}\tilde{h}(\alpha, B'(T_D), \ldots, B^{d-1}(T_D)\big],$$

where $\beta = \frac{1}{2}(\frac{\alpha}{c})^2$. Let us write $x = (x', \zeta)$, for $x \in D$, $\zeta \geqslant 0$, and $y = (y', 0)$ for $y' \in \partial D$. Then

$$\begin{aligned} &E^x\big[M(T_D)\tilde{h}(\alpha, B(T_D))\big] \\ &= \int_0^\infty \mathrm{e}^{-\beta t} \frac{\zeta \mathrm{e}^{-\zeta^2/2t}}{(2\pi t^3)^{1/2}} \int_{\partial D} \frac{\mathrm{e}^{-(x'-y')^2/2t}}{(2\pi t)^{(d-1)/2}} \tilde{h}(\alpha, y')\,\mathrm{d}y' \\ &= \frac{2\zeta}{(2\pi)^{d/2}} \int_{\partial D} \left(\frac{\alpha}{c|x-y|}\right)^d 2K_{\frac{d}{2}}\left(\frac{\alpha}{c}|x-y|\right)\tilde{h}(\alpha, y')\,\mathrm{d}y', \end{aligned} \tag{4.15}$$

where $|x-y|^2 = (|x'-y'| + \zeta^2)$, as expected.

To compute the second term in (4.14), it is easier to use the fact that the transition density of the Brownian motion, killed upon hitting ∂D, is

$$\begin{aligned} p_t((x', \zeta), (y', \eta)) &= p_t(x', y')(p_t(\xi, \eta) - p_t(\xi, -\eta)) \\ &= p_t(x, y) - p_t(x, \bar{y}), \end{aligned}$$

where the appropriate meaning of the symbol $p_t(\cdot, \cdot)$ is determined by the dimension of space of the labels, for example,

$$p_t(x', y') = (2\pi t)^{-(d-1)/2} \exp\left\{-\frac{1}{2\pi}(x'-y')^2\right\}.$$

Also $\bar{y} = (y', -\eta)$. This killed semigroup has a λ-resolvent with the density $u^\lambda(x, y)$ given by

$$\frac{2}{(2\pi)^{(d-1)/2}}\left\{\left(\frac{\sqrt{2\lambda}}{|x-y|}\right)^{\frac{d}{2}-1} K_{\frac{d}{2}-1}\left(\sqrt{2\lambda}|x-y|\right) - \left(\frac{\sqrt{2\lambda}}{|x-\bar{y}|}\right)^{\frac{d}{2}-1} K_{\frac{d}{2}-1}\left(\sqrt{2\lambda}|x-\bar{y}|\right)\right\}.$$

Taking this into account, and performing a simple change of variables, the second term in (4.14) is

$$E^x\left\{\int_0^{T(D)} M(t)\widetilde{S}(\alpha, B(t))\,\mathrm{d}t\right\}$$
$$= \frac{2}{(2\pi)^{(d-1)/2}}\int\left(\frac{\alpha}{c|x-y|}\right)^{\frac{d}{2}-1} K_{\frac{d}{2}-1}\left(\frac{\alpha}{c}|x-y|\right)\left(\tilde{s}(\alpha, y)-\tilde{s}(\alpha,\bar{y})\right)\mathrm{d}y. \tag{4.16}$$

Comment: We leave for the reader to apply the well known identities, relating the Neumann function to its derivative, to verify that the integral kernel in (4.15) is the normal derivative (at ∂D) of the kernel in (4.16), i.e., the usual representation

$$\tilde{u}(\alpha, x) = \int_D G(x,y)\widetilde{S}(\alpha, y)\,\mathrm{d}y + \int_{\partial D}\frac{\partial G}{\partial n}(x,y)\tilde{h}(\alpha, y')\,\mathrm{d}y'$$

is satisfied.

When $d = 3$, the integral in (4.16) can be broken down as in Section 1 to obtain the time dependence of the waves produced by the volume sources. For fun, let us invert the Laplace transform in (4.15). Denote the left-hand side there by $\tilde{u}_b(\alpha, x)$. Thus,

$$\begin{aligned}\tilde{u}_b(\alpha, x) &= \frac{2\zeta}{(2\pi)^{3/2}}\int_{\partial D}\left(\frac{\alpha}{c|x-y|}\right)^{3/2} K_{\frac{3}{2}}\left(\frac{\alpha}{c}|x-y|\right)\tilde{h}(\alpha, y')\,\mathrm{d}y' \\ &= -\frac{2}{(2\pi)^{3/2}}\frac{\partial}{\partial\zeta}\int_{\partial D}\left(\frac{c}{\alpha|x-y|}\right)^{1/2} K_{\frac{1}{2}}\left(\frac{\alpha}{c}|x-y|\right)\tilde{h}(\alpha, y')\,\mathrm{d}y'\end{aligned}$$

and since $K_{\frac{1}{2}}(\rho) = \sqrt{\frac{\pi}{2}\frac{1}{\rho}}\mathrm{e}^{-\rho}$, we have

$$\tilde{u}_b(\alpha, x) = \frac{-1}{2\pi}\frac{\partial}{\partial\zeta}\int_{\partial D}\frac{c}{\alpha|x-y|}\mathrm{e}^{-\frac{\alpha}{c}}\tilde{h}(\alpha, y')\,\mathrm{d}y'$$

or, equivalently,

$$u_b(t, x) = -\frac{c}{2\pi}\frac{\partial}{\partial\zeta}\int_D^t \mathrm{d}s\int_{\partial D} h\left(s - \frac{|x-y'|}{c}; y'\right)\frac{\theta\left(s-\frac{|x-y|}{c}\right)}{|x-y|}\,\mathrm{d}y',$$

and we should not forget that $|x-y|^2 = |x'-y'|^2 + \zeta^2$. Things can be rewritten a bit to obtain, say

$$u_b(t, x) = -\frac{c^2}{2\pi}\frac{\partial}{\partial\zeta}\theta(t-\zeta/c)\int_{\zeta/c}^t \mathrm{d}\sigma\int_\sigma^t \langle h(S-\sigma, x', c\sqrt{\sigma^2-\zeta)^2})\rangle\,\mathrm{d}s,$$

where $\langle h(\tau, x', \rho)\rangle = \int_0^{2\pi} h(\tau, x' + \rho(\cos\theta, \sin\theta)\,\mathrm{d}\theta$.

4.2.2. Waves in a sphere with Dirichlet boundary conditions

The results in this section are chosen to exemplify the use of martingale theory. The same results could be obtained by separation of variables and eigenfunction expansions.

Consider how to simplify the problem of solving

$$\partial_t^2 u = c^2 \Delta u, \quad t > 0, x \in B(0, R) \tag{4.17}$$

with vanishing initial conditions $u(0+, x) = 0$, $u_t(0+, x) = 0$ and the boundary condition

$$u(t, x) = h(t, x), \quad t > 0, x \in \partial D. \tag{4.18}$$

According to (4.14), the solution to the corresponding version of equation (4.12) is this time

$$\tilde{u}(\alpha, x) = E^x\left[e^{-\beta T} \tilde{h}(\alpha, B(T))\right]$$

with $T = \inf\{t > 0 : |B(T)| = R\}$. Again $\beta = \alpha^2/2c^2$. Expanding $\tilde{h}(\alpha, x)$ in terms of spherical harmonics

$$\tilde{h}(\alpha_1; x) = \sum_{n=0}^{\infty} \sum_{k=0}^{n} a_{n,k}(R) P_{n,k}(\cos\theta) e^{ik\phi},$$

where $x = R(\sin\theta\cos\varphi, \sin\theta\sin\varphi, \cos\theta)$, we see that it suffices to know what is

$$E^x\left[e^{-\beta T} e^{ik\phi(T)} P_{n,k}(\cos\theta(T))\right]. \tag{4.19}$$

To begin with, note that due to the symmetry of rotation around the $\overline{0x}$ axis, the hitting distribution of

$$B(T) = R(\sin\theta(T)\cos\phi(T), \sin\theta(T)\sin\phi(T), \cos\theta(T))$$

is invariant around that axis, implying that the non-vanishing terms of type (4.19) are those with $k = 0$, thus reducing our worries to compute

$$E^x\left[e^{-\beta T} P_n(\cos\theta(T))\right]$$

for $n \geqslant 0$. To accomplish that, we recur to our exponential martingale $e^{a \cdot B(t) - \frac{a^2 t}{2}}$ for any vector $a = (a_1, a_2, a_3)$. Using the Optional Sampling Theorem, since the hitting time of the sphere is finite and has a finite expectation, we have

$$E^x\left[e^{a \cdot B(T) - a^2 \frac{T}{2}}\right] = e^{a \cdot x}.$$

If we take a coordinate system with z-axis along a, and invoking

$$e^{v\cos\theta} = \sum_{r=0}^{\infty}(2n+1)\sqrt{\frac{\pi}{2v}}I_{n+1/2}(v)P_n(\cos\theta)$$

we obtain, after identifying the coefficients of $I_{n+1}(v)$, that

$$E^x\big[e^{-a^2T/2}P_n(\cos\theta(T))\big] = \left(\frac{|x|}{R}\right)^{1/2} P_n(\cos\gamma)\frac{I_{n+1/2}(a|x|)}{I_{n+1/2}(aR)},$$

where $\gamma =$ angle $(a0x)$, which can now be equated to zero.

Putting all of this together, we have

$$\tilde{u}(\alpha, x) = \sum_{n=0}^{\infty} a_{n,0}(R)\left(\frac{|x|}{R}\right)^{1/2}\frac{I_{n+1/2}\left(\frac{\alpha}{c}|x|\right)}{I_{n+1/2}\left(\frac{\alpha}{c}R\right)}.$$

How to invert this is described in the references.

4.3. Neumann type boundary conditions

This time we consider the problem of solving

$$\partial_t^2 u(t,x) = c^2\Delta u(t,x) + s(t,x), \quad t>0, x\in D, \tag{4.20}$$

where $D = \{(x',\zeta)\in\mathbb{R}^n : x'\in\mathbb{R}^{n-1}, \zeta>0\}$ under the initial conditions

$$u(0+,x) = f(x), \qquad u_t(0+,x) = g(x), \quad x\in D \tag{4.21}$$

as well as the boundary conditions

$$\langle\nabla u(t,x), \gamma(x)\rangle + h(t,x) = 0, \quad t>0, x\in\partial D, \tag{4.22}$$

where $\gamma(x)$ is a vector field forming an acute angle with the inner normal $(0,0,\ldots,1)$ to ∂D.

Again, assuming the solution is well behaved as t grows and taking the Laplace transform in time, we transform problem (4.20) into the following problem

$$\left(\beta - \frac{1}{2}\Delta\right)\tilde{u}(\alpha,x) = S(\alpha,x), \quad x\in D, \tag{4.23}$$

$$\langle\nabla(\alpha,x), \gamma(x)\rangle + \tilde{h}(\alpha,x) = 0, \quad x\in\partial D, \alpha>0 \tag{4.24}$$

and we saw in Section 4 of Chapter 2 how to write down the solution to this problem, whenever the transition semigroup of the reflected diffusion is known, or in terms of path integrals over the reflected diffusion.

In this example, to get a hold of the reflected Brownian motion is rather easy: If $B(t)$ denotes the d-dimensional Brownian motion, then $X(t) = (B'(t), |B^d(t)|)$ is the Brownian motion in D, where the last component gets reflected on ∂D. It is also very easy to see that the transition semigroup is

$$p_t((x', \zeta), (y', \eta)) = p_t(x', y')(p_t(\zeta, \eta) + p_t(\zeta, -\eta)) = p(x, y) + p(x, \bar{y}), \tag{4.25}$$

where, of course,

$$p_t(x', y') = (2\pi)^{\frac{d-1}{2}} \exp\{-(x' - y')^2/2t\},$$

and

$$p_t(\zeta, \eta) = (2\pi t)^{1/2}\{\exp -(\zeta - \eta)^2/2t\}.$$

Again, the dimension of the space in which the points lie determine the meaning of $p_t(\cdot, \cdot)$. Also we use the same notation as above $\bar{y} = (y', -\eta)$ if $y = (y', \eta) \in D$.

The λ-resolvent of the reflected semigroup being

$$\begin{aligned} v_\gamma^\lambda(x, y) &= \frac{1}{(2\pi)^{\frac{d-1}{2}}} \left\{ \left(\frac{\sqrt{2\lambda}}{|x - y|} \right)^{\frac{d}{2}-1} K_{\frac{d}{2}-1}(\sqrt{2\lambda}|x - y|) \right. \\ &\quad \left. + \left(\frac{\sqrt{2\lambda}}{|x - \bar{y}|} \right)^{\frac{d}{2}-1} K_{\frac{d}{2}-1}(\sqrt{2\lambda}|x - \bar{y}|) \right\} \end{aligned}$$

in terms of which (2.4. 13.a) becomes, with $\beta = \frac{1}{c}(\frac{\alpha}{c})^2$

$$\tilde{u}(\alpha, x) = \int_D v_\gamma^\beta(x, y)\widetilde{S}(\alpha, y)\, dy + \int_{\partial D} v_\gamma^\beta(x, y)\tilde{h}(\alpha, y')\widetilde{m}(dy'), \tag{4.26}$$

where, of course, $\widetilde{m}(dy') = dy_1, \ldots, dy_{d-1}$ is the surface measure on ∂D. These are variations on the theme of Section 2.1 above. Indeed, to obtain the solution to problem (4.20) from the solution in the whole space by symmetrization with respect to the boundary would have been just as easy. This is the standard way of solving that problem.

Using the probabilistic representations of the involved Green functions, namely (2.2.14), instead of (4.26) we would have

$$\widetilde{U}(\alpha, x) = E^x \int_0^\infty e^{-\beta\gamma} \widetilde{S}(\alpha, X(t))\, dt + E^x \int_0^\infty e^{-\beta t} \tilde{h}(\alpha, X(t))\, dL_t, \tag{4.27}$$

where $X(t) = (B'(t), \ldots, B^{d-1}(t), |B^d(t)|)$ and L_t is, in this simple case, the local time of the one-dimensional Brownian motion $B^d(t)$ at $x = 0$.

Computing these would yield (4.26) again.

4.4. Existence results

An interesting aspect of the probabilistic representations (4.5), (4.14) and (4.27) is that they can be used to prove the existence of solutions to the respective wave problems. For example, we could start from (4.5) and prove that the solution $\tilde{u}(\alpha, x)$ (1.3) is the Laplace transform of a function $u(t, x)$ and then to use (1.4) to verify that $u(t, x)$ satisfies (1.1) indeed.

This method will be particularly interesting when considering propagation problems in heterogeneous media, where the standard representation formulae obtained above, for constant c^2, do not apply.

To exemplify, consider (4.3) and (4.4) which can be explicitly computed as (4.5). Clearly, if for each y, $\tilde{s}(\alpha, y)$ is the Laplace transform of some measure on $\mathbb{R}$ and since multiplying by $\exp\{-(\alpha|x-y|/c)\}$ amounts to take Laplace transforms of the time shifted (by $|x-y|/c$) measure on $[0, \infty)$, then (4.6) is a Laplace transform of something.

Moreover, representation (1.4) provides us with a weak solution to (4.3), which becomes a classical solution when some assumptions on the smoothness in x are imposed on the terms of $\tilde{s}(\alpha, x)$.

So let $\tilde{s}(\alpha, x)$ be bounded in x independently of $\alpha > 0$, and for each $\alpha >$ let it be the Laplace transform of a measure on $[0, \infty)$, and $U(\mathrm{d}s, x)$ be family measures on $[0, \infty)$ such that

$$\tilde{u}(\alpha, x) = \int_0^\infty \mathrm{e}^{-\alpha t} U(\mathrm{d}s, x), \tag{4.28}$$

where $\tilde{u}(\alpha, x)$ solves (4.3). Rewrite (4.28) as

$$\alpha^2 \tilde{u}(\alpha, x) - \alpha f(x) - g(x) = \int_0^\infty \mathrm{e}^{-\alpha t} V(\mathrm{d}t, x), \tag{4.29}$$

where, of course, for each interval $I \subset (0, \infty)$, the measures V and U are related by

$$V(I, x) = \Delta U(I, x) + \int_I S(t, x)\,\mathrm{d}t. \tag{4.30}$$

If we introduce

$$U_1(t, x) = \int_0^t V(\mathrm{d}s, x) + g(x),$$

$$\overline{U}(t, x) = \int_0^t U_1(s, x)\,\mathrm{d}s + f(x),$$

it is clear that $\overline{U}(0,x) = f(x)$, $\frac{\partial \overline{U}}{\partial t}(0,x) = g(x)$ and an integration by parts leads to

$$\int_0^\infty \mathrm{e}^{-\alpha t} \mathrm{d}_t \frac{\partial \overline{U}}{\partial t} = \alpha^2 \widetilde{\overline{U}}(\alpha, x) - \alpha f(x) - g(x). \tag{4.31}$$

Comparing (4.29) and (4.31), we obtain that

$$\alpha^2 \widetilde{U}(\alpha, x) - \alpha f(x) - g(x) = \alpha^2 \widetilde{\overline{U}}(\alpha, x) - \alpha f(x) - g(x)$$

and therefore

$$\widetilde{U}(\alpha, x) = \widetilde{\overline{U}}(\alpha, x)$$

from which it follows that $U(\mathrm{d}t, x)$ has a density, i.e., $U(\mathrm{d}t, x) = u(t, x)\,\mathrm{d}t$. From the relationship between V and $\overline{U}$ given by (4.30) it follows that $V(\mathrm{d}t, x)$ has a density, denoted by $v(t, x)$, satisfying

$$v(t, x) = \Delta u(t, x) + s(t, x)$$

from which, by (4.31) again, $\mathrm{d}_t \frac{\partial \overline{U}}{\partial t}$ is the function $v(t, x)$, that is

$$\frac{\partial^2 u}{\partial t^2}(t, x) = v(t, x) = \Delta u(t, x) + s(t, x).$$

The initial conditions

$$u(0+, x) = f(x), \qquad u_t(0+, x) = g(x).$$

Certainly, these identifications happen to be a.e. with respect to $\mathrm{d}t$ on $(0, \infty)$, but if we check the definitions of $\overline{U}(t, x)$ and $U_1(t, x)$, we see that they are continuous, thus the identifications can be made for all values of t.

4.5. Problems of Dirichlet type in unbounded domains: from the Markov property to the Huygens condition and the Sommerfeld radiation condition

We consider $d = 3$ throughout this section.

4.5.1. The Dirichlet problem in unbounded domains

One of the usual presentations of this problem is as follows: it is required to find the solution to

$$\partial_t^2 u = c^2 \Delta u \quad \text{for } t > 0, x \in D \tag{4.32}$$

if it is known that $u(t,x)=v(t,x)+I(t,x)$, and that on ∂D the resultant wave field is to vanish, i.e., on ∂D

$$0=v(t,x)+I(t,x). \tag{4.33}$$

The field $I(t,x)$ is called an incoming field, and we put $\frac{\partial^2 I}{\partial t}-c^2\Delta I=-s(t,x)$. Therefore we have to find $v(t,x)$ in D, which for $t>0$ satisfies

$$\partial_t^2 v=c^2\Delta v+s(t,x) \tag{4.34}$$

plus the boundary condition (4.33). We assume that D^c is a bounded, simply connected domain with a smooth enough surface ∂D.

Again, by taking Laplace transforms we obtain, with $\beta=\frac{1}{2}(\frac{\alpha}{t})^2$ as above,

$$\left(\beta-\frac{1}{2}\Delta\right)\tilde{v}(\alpha,x)=\tilde{s}(\alpha,x)=\frac{1}{2c^2}\widetilde{S}(\alpha,x) \tag{4.35}$$

together with

$$v(\alpha,x)=\tilde{h}(\alpha,x)=-\tilde{I}(\alpha,x). \tag{4.36}$$

We are now back to our favourite game: the solutions to (3) are represented by

$$v(\alpha,x)=E^x\left[e^{-\beta T}\tilde{h}(\alpha,B(T));T<\infty\right]+E^x\left\{\int_0^T e^{-\beta I}\widetilde{S}(\alpha,B(T))\,dt\right\}. \tag{4.37}$$

Here the second term is the usual $\int_D G_D^\beta(x,y)\widetilde{S}(\alpha,y)\,dy$ and the first one is $-\int_{\partial D}\frac{\partial G_D^\beta}{\partial n}(x,y)\tilde{h}(\alpha,y)\,d\sigma(y)$, where $G^\beta(x,y)$ denotes the Green function for $(\beta-\frac{1}{2}\Delta)$ in D, when it can be found.

If we had no incident field, and the boundary ∂D of D was radiating with the intensity $h(t,x)$, we set $s\equiv 0$ above, we are left with (4) and are ready for.

4.5.2. *From the Markov property to the Huygens construction*

Assume we have a radiating surface ∂D producing a wave with the intensity $h(t,x)$. Then the field inside the unbounded domain is given by the first term in (4.37) above, i.e.,

$$\tilde{u}(\alpha,x)=E^x\left[e^{-\beta T}\tilde{h}(\alpha,B(T));T<\infty\right]. \tag{4.38}$$

Assume you know that the Green function $G_D^\beta(x,y)$ is the Laplace transform of a time dependent Green function $G_D(t;x,y)$ so that

$$u(t,x)=\int_0^t\int_{\partial D}-\frac{\partial G_D}{\partial n}(t,x,y)h(t-s,y)\,d\sigma(y). \tag{4.39}$$

Consider now some domain $D' \subset D$ with a surface $\partial D'$ enclosing ∂D or $D^c \subset (D')^c$, and denote by T' the first hitting time of $(D')^c$. From the continuity of the trajectories of the Brownian motion it is clear that $T' < T$ and that for x in D', $T = T' + T \circ \theta(T')$. Now invoking the strong Markov property, we have that (4.37) equals

$$\tilde{u}(\alpha, x) = E^x\big[\mathrm{e}^{-\beta T'} E^{B(T')}[\mathrm{e}^{-\beta T}\tilde{h}(\alpha, B(T)); T < \infty]; T' < \infty\big]$$

since $B(T') \in \partial D' \subset D$, by (4.39) this can be rewritten as

$$\tilde{u}(\alpha, x) = E^x\big[\mathrm{e}^{-\beta T'}\tilde{u}(\alpha, B(T')); T' < \infty\big]. \tag{4.40}$$

If the same assumption going into (6) is brought in here, namely, that there exist $G_{D'}(t, x, y)$ whose Laplace transform $\widetilde{G}^{\alpha}_{D'}(x, y)$ inverts $(\beta - \frac{1}{2}\Delta)$, we can rewrite (4.39) as

$$u(t, x) = -\int_0^t \mathrm{d}s \int_{\partial G_D} \frac{\partial G_{D'}}{\partial n}(s, x, y)u(t - s, y)\,\mathrm{d}\sigma(y)$$

which states that the wave field at any (t, x), produced by a distribution of sources at some surface S_1, (∂D above) can be built up from the radiation field at some intermediate surface S_2 enclosing S_1, at times prior to t.

We shall add that the Chapman–Kolmogorov equation $P_{t+s} = P_t P_s$, satisfied by any probabilistic transition semigroup, is sometimes likened as an analogue to the Huygens construction. This asserts that a wave field regenerates itself at any surface. This is basically what (4.40) says for us. We have thus connected the strong Markov property to the Huygens construction.

Below we deal with the issue of producing a candidate for $G_D(t, x, y)$.

4.5.3. Sommerfeld's radiation condition

To answer the question of uniqueness of the solution to Dirichlet's problem in an unbounded domain, i.e., to

$$\left(\beta - \frac{1}{2}\Delta\right)u(\alpha, x) = \tilde{s}(\alpha, x), \quad x \in D, \alpha > 0, \tag{4.41}$$

$$u(\alpha, x) = \tilde{h}(\alpha, x), \quad x \in D, \alpha > 0 \tag{4.42}$$

it is important to know the value of $\tilde{u}(\alpha, x)$ as $|x| \to \infty$.

It is not difficult to verify that

$$\begin{aligned}\tilde{u}(\alpha, x) &= E^x\big[\mathrm{e}^{-\beta T}\tilde{h}(\alpha, B(T)), T < \infty\big] + AP^x(T \leqslant \infty)\\ &\quad + E^x\int_0^{\infty} \mathrm{e}^{-\beta t}\widetilde{S}(\alpha, B(t))\,\mathrm{d}t\end{aligned}$$

is the probabilistic representation of the unique solution to (4.41) such that $\tilde{u}(\alpha, x) \to A$ as $|x| \to \infty$, while it is implicit in (4.37) that $\tilde{u}(\alpha, x) \to 0$ as $|x| \to \infty$.

We show below that (for $\alpha > 0$) the strong Markov property combined with the hitting distribution of a sphere from the exterior allows us to obtain

$$\left(\frac{\partial \tilde{u}}{\partial r} + \frac{\alpha}{c}\tilde{u}\right) = \mathrm{o}\left(\frac{1}{r}\right)$$

for large r.

Note to begin with that

$$F(t,x) = \mathrm{e}^{-\beta t}\frac{\mathrm{e}^{-\frac{\alpha}{c}|x-x_0|}}{|x-x_0|}$$

satisfies

$$\frac{\partial F}{\partial t} + \frac{1}{2}\Delta F = 0, \quad (t,x) \in (0,\infty) \times (\mathbb{R}^3 - \{x_0\})$$

and consider the process $F(t, B(t))$. Let R be such that the ball $B(0, R)$ encloses D^c and $|x| > R$.

Since, for $d \geqslant 3$, the Brownian motion does not hit x_0, $F(t, B(t))$ is a martingale, an application of the Optional Sampling Theorem yields (with $S = \inf\{t > 0 : B(t) = R\}$),

$$E^x\left[\mathrm{e}^{-\beta S}\frac{\mathrm{e}^{-\frac{\alpha}{c|B(S)-x_0|}}}{|B(S)-x_0|}\right] = \frac{\mathrm{e}^{-\frac{\alpha}{c|x-x_0|}}}{|x-x_0|}.$$

Recalling the expansion

$$\frac{\mathrm{e}^{-\lambda\ell}}{\mathrm{e}} = \frac{2}{\pi}\sum_0^\infty(2n+1)\left\{\sqrt{\frac{\pi}{2\lambda r}}I_{n+1/2}(\lambda r)\right\}\left\{\sqrt{\frac{\pi}{2\lambda\rho}}K_{n+\frac{1}{2}}\right\}P_n(\cos\theta)$$

with $\ell = (r^2 + \rho^2 - 2r\rho\cos\theta)^{1/2}$, substituted above, and equating the coefficients of $I_{n+1/2}(\frac{X}{c}|x_0|)$, we obtain

$$\left(\frac{\alpha}{c}R\right)^{-1} E^x\left[\mathrm{e}^{-\beta s}P_n(\cos\theta_S)K_{n+\frac{1}{2}}\left(\frac{\alpha}{c}R\right)\right] = \left(\frac{\alpha}{c}|x|\right)^{-1} K_{n+\frac{1}{2}}\left(\frac{\alpha}{c}|x|\right)P_n(\cos\Gamma),$$

where Γ = angle $(x_0 0 x)$ and θ_t = angle $(B_t 0 x)$. Since x_0 is arbitrary, we chose it lying along x so that $\Gamma = 0$ and

$$E^x\left[\mathrm{e}^{-\beta S}P_n(\cos\theta_S)\right] = \left(\frac{R}{|x|}\right)^{1/2}\frac{K_{n+1/2}(\frac{\alpha}{c}|x|)}{K_{n+\frac{1}{2}}(\frac{\alpha}{c}R)}.$$

Let us now apply (4.40) with D' being the complement of the sphere $B(0, R)$ and T' replaced by S:

$$\tilde{u}(\alpha, x) = E^x\left[e^{-\beta s}\tilde{u}(\alpha, B(S)); S < \infty\right]$$

and clearly it does not matter what happens on $\{S = \infty\}$, unless $\alpha = 0$.

Denote by $\tilde{\varphi}(\alpha, \xi)$ the restriction of $\tilde{u}(\alpha, \xi)$ to the sphere $\partial B(0, R)$. Expanding it in terms of the spherical harmonics, we obtain

$$\tilde{\varphi}(\alpha, \xi) = \sum_0^\infty \left\{ a_{n,0} P_n(\cos\theta) + \sum_{n=1}^n a_{nn} e^{i\varphi h} P_{n,n}(\cos\theta) \right\},$$

where the coefficients $a_{n,h}$ depend on α and R. Due to the symmetry with respect to rotations around the $\overline{0x}$ axis, the second term in the summation does not contribute to the expectation above and we obtain

$$\begin{aligned}\tilde{u}(\alpha, x) &= \sum_{n=0}^\infty a_{n,0} E^x\left[e^{-\beta S} P_n(\cos\theta(S))\right] \\ &= \left(\frac{R}{|x|}\right)^{1/2} \sum_{n=0}^\infty a_{n,0} \frac{K_{n+1/2}(\frac{\alpha}{c}|x|)}{K_{n+\frac{1}{2}}(\frac{\alpha}{c}R)}.\end{aligned}$$

Since the Bessel functions involved here can be expanded as

$$K_{n+\frac{1}{2}}(\rho) = \sqrt{\frac{\pi}{2}\frac{e^{-\rho}}{\rho}} \sum_{k=0}^n \left(n + \frac{1}{2}, k\right)\left(\frac{1}{2\rho}\right)^k$$

with $(n + \frac{1}{2}, k) = \frac{(n+k)!}{k!(n-k)!}$, by substituting and regrouping terms we obtain

$$\tilde{u}(\alpha, x) = \sqrt{\frac{\pi}{2}}\sqrt{\frac{C_0}{2}} \frac{e^{-\frac{\alpha}{c}|x|}}{|x|} \sum_{n=0}^\infty b_n + o\left(\frac{1}{|x|}\right)$$

with $b_n = a_{n,0} R^{1/2} / K_{n+1/2}(\frac{\alpha}{c}R)$.

Setting $|x| = r$, we see that $\tilde{u}(\alpha, r) = o(\frac{1}{r})$ and that

$$\frac{\partial \tilde{u}}{\partial r} + \left(\frac{\alpha}{c}\right)\tilde{u} = o\left(\frac{1}{r}\right).$$

Thus, the probabilistic representation is the unique solution to the problem. For more about these matters the reader can check the references.

4.6. Extended Hadamard's construction

Originally Hadamard's construction allowed us to obtain an explicit solution for the wave propagation in $d = 2$ starting from the construction for $d = 3$.

We shall see that this is so for any two values of d. So let us consider, for the sake of definiteness, the problem

$$\frac{\partial^2 u}{\partial t} = c^2 \Delta u, \quad t > 0, \ x \in \mathbb{R}^d$$

with prescribed $u(0+, x) \equiv 0$ and $u_t(0+, x) = g(x)$.

Assume now that the initial data depend only on the first d_1 components of x. Write $x = (x_1, x_2)$ with $x_1 \in \mathbb{R}^{d_1}$ and $x_2 \in \mathbb{R}^{d_2}$, $d_2 = d - d_1$, and so on. According to (1.4), the Laplace transform of the solution is as follows, with $\beta = \alpha^2/2c^2$,

$$\tilde{u}(\alpha, x) = \int \mathrm{e}^{-\beta t} P_t g(x)\,\mathrm{d}x = \int \mathrm{e}^{-\beta t}\left(\int \frac{\mathrm{e}^{-(x-y)^2/2t}}{(2\pi t)^{d/2}} g(y)\,\mathrm{d}y\right)\mathrm{d}t.$$

Since

$$\int \frac{\mathrm{e}^{-(x-y)^2/2t}}{(2\pi t)^{d/2}}\,\mathrm{d}y = \frac{\mathrm{e}^{-(x_1-y_1)^2/2t}}{(2\pi t)^{d_1/2}},$$

it is clear that, if $g(y)$ is actually $g(y_1)$, then

$$\tilde{u}(\alpha, x) = \int \mathrm{e}^{-\beta t} \int \frac{\mathrm{e}^{-(x_1-y_1)^2/2t}}{(2\pi t)^{d_1/2}} g(y_1)\,\mathrm{d}y_1 = \tilde{u}(\alpha, x_1).$$

Apart from the fact that we constantly sin in using the same symbols for different mathematical objects, the conclusion is clear: If the initial data do not depend on a certain number of degrees of freedom, the solution is independent of the same variables.

In Hadamard's original situation $d = 3$ and $d_1 = 2$, and recall that for $d = 3$

$$\begin{aligned} u(\alpha, x_1) &= \tilde{u}(\alpha, x) = \int \frac{\mathrm{e}^{-\frac{\alpha}{c}|x-y|}}{2\pi |x-y|} g(y)\,\mathrm{d}y \\ &= \int \frac{\mathrm{e}^{-\frac{\alpha}{c}((x_1-y_1)^2+\zeta^2)^{1/2}}}{((x_1+y_1)^2+\zeta^2)^{1/2}} g(y_1)\,\mathrm{d}y_1\,\mathrm{d}\zeta, \end{aligned}$$

where we used the change of variables $x^3 - y^3 \to \zeta$.

To integrate over $y_1 = (y^1, y^2)$, use polar coordinates on the plane to rewrite the identity above as follows

$$u(\alpha, x_1) = \int_{-\infty}^{\infty}\int_0^{\infty} \frac{\mathrm{e}^{-\frac{\alpha}{c}(r^2+\zeta^2)^{1/2}}}{2\pi (r^2+\zeta^2)^{1/2}} \langle g(x_1 + r(\cdot))\rangle 2r\,\mathrm{d}r\,\mathrm{d}\zeta,$$

where, obviously,

$$\langle g(x_1 + r(\cdot))\rangle_2 = \int_0^{2\pi} g(x + r(\cos\theta, \sin\theta))\,\mathrm{d}\theta.$$

To continue, introduce $c\tau = (\rho^2 + \zeta^2)^{1/2}$ and thus $c\,\mathrm{d}\tau = r\,\mathrm{d}r/(\rho^2 + \zeta^2)^{1/2}$, and we obtain

$$\tilde{u}(\alpha, x_1) = \frac{c}{2\pi}\int_{-\infty}^{\infty} \mathrm{d}\zeta \int_0^{\infty} \mathrm{e}^{-\alpha\tau}\big\langle g(x_1 + \sqrt{(c\tau)^2 - \zeta^2}(\cdot))\big\rangle_2 \theta(c\tau - \zeta^2)\,\mathrm{d}\tau,$$

where the Heaviside function shows up because we must keep $\rho > 0$ in the change of variables. We thus obtain

$$\widetilde{u}(\alpha, x_1) = \int_0^{\infty} \mathrm{e}^{-\alpha\tau} \int_{-c\tau}^{c\tau} \frac{c}{2\pi}\big\langle g(x_1 + \sqrt{(c\tau)^2 - \zeta^2}(\cdot))\big\rangle_2\,\mathrm{d}\zeta$$

or

$$u(t, x) = u(t, x_1) = \frac{c}{2\pi}\int_{-ct}^{ct} \big\langle g(x_1 + \sqrt{(c\tau)^2 - \zeta^2}(\cdot))\big\rangle_2\,\mathrm{d}\zeta,$$

and the reader can verify that this is the usual result.

4.7. From resolvents to propagators

There is a pending issue from Section 4, namely, to produce a propagator (or fundamental solution to the wave equation) $G_D(t, x, y)$ whose Laplace transform $G_D(\alpha, x, y)$ satisfies

$$E^x \int_0^T \mathrm{e}^{-}\beta t\tilde{s}(\alpha, B(t))\,\mathrm{d}t = \int_D G_D(\alpha, x, y)\tilde{s}(\alpha, y)m(\mathrm{d}y) \tag{4.43}$$

as well as

$$E^x\big[\mathrm{e}^{-\beta T}\tilde{h}(\alpha, B(T)\big] = -\int_{\partial D} \frac{\partial G}{\partial n} D(\alpha, x, y)\widetilde{m}(\mathrm{d}y), \tag{4.44}$$

where, recall $\frac{\partial G_D}{\partial n}(\alpha, x, y)$ denotes the derivative of $G_D(\alpha, x, y)$ along the inner normal to ∂D at y, and $\widetilde{m}(\mathrm{d}y)$ denotes the surface measure on ∂D induced by the Lebesque measure $m(\mathrm{d}y)$ on $\mathbb{R}^d$.

There are two issues to take care of: First, if there exists a Green function $G_D(\alpha, x, y)$ for every unbounded domain D. The second one is, whether, for each fixed (x, y) $G_D(\alpha, x, y)$, there is the Laplace transform of some measure on $[0, \infty)$.

To begin with, note that the map

$$f(x) \to E^x \int_0^T \mathrm{e}^{-\beta s} f(B(s))\,\mathrm{d}s$$

defined on $C_K(D)$ to $C_0(D)$ (the class of continuous functions vanishing at ∂D) is linear, continuous, positive and absolutely continuous with respect to the Lebesque measure, i.e., if $m(A)=0$

$$E^x\int_0^T e^{-\beta t} I_A(X_t)\,dt \leqslant E^x\int_0^\infty e^{-\beta t} I_A(X_t)\,dt = 0.$$

Thus, the existence of a function $G_D(\alpha,x,y)$ such that

$$E^x\int_0^T e^{-\beta t} f(B(t)\,dt = \int_D G_D(\alpha,x,y)f(y)m(dy) \tag{4.45}$$

is easy to establish. It is considerably more elaborate to verify that

$$\left(\beta-\frac{1}{2}\Delta\right)G_D(\alpha,x,y)=\delta(x-y), \quad x,y\in D.$$

Below we indicate how to verify it. Note that a simple application of the strong Markov property yields

$$\begin{aligned} E^x&\int_0^T e^{-\beta t} f(B(t))\,dt \\ &= E^x\int_0^\infty e^{-\beta t} f(B(t))\,dt - E^x\left[e^{-\beta T}E^{X(T)}\left\{\int_0^\infty e^{-\beta t} f(B(t))\,dt\right\}\right] \end{aligned} \tag{4.46}$$

and recall that

$$E^x\int_0^\infty e^{-\beta t} f(B(t))\,dt = \int \frac{e^{-\sqrt{2\beta}|x-y|}}{4\pi|x-y|}f(y)\,dy$$

which offers the following representation for $G_D(\alpha,x,y)$:

$$G_D(\alpha,x,y)=g(\alpha,x-y)-E^x\big[e^{-\beta T}g(\alpha,B(T)-y)\big], \tag{4.47}$$

where we introduced $g(\alpha,\xi)=e^{-\alpha|\xi|}/4\pi|\xi|$. Note that $(\beta-\frac{1}{2}\Delta)g(\alpha,x-y)=\delta(x-y)$, and we only have to know that $R(\alpha,x,y)=E^x[e^{\beta T}g(\alpha,\beta(T)-y)]$ satisfies $(\beta-\frac{1}{2}\Delta)R(\alpha,x,y)=0$ and that as $x\in D\to x_0\in\partial D$, $R(\alpha,x,y)\to g(\alpha,x_0-y)$. These two facts imply that $(\beta-\frac{1}{2}\Delta)G(\alpha,x,y)=\delta(x-y)$ and $G(\alpha,x,y)\to 0$ as $x\in D\to x_0\in\partial D$.

To verify that $G(\alpha, x, y)$ satisfies the equation, we proceed as follows. Let now D' be the complement of $B(0, R)$ with R large enough so that $D' \subset D$ (or so that $D^c \subset B(0, R)$), and let T_1 be the hitting time of $\partial D'$ by $B(t)$. Writing

$$\begin{aligned} & V^\beta \tilde{s}(\alpha, x) \\ &\quad = E^x \int_0^T \mathrm{e}^{-\beta t} \tilde{s}(\alpha, B(t))\, \mathrm{d}t \\ &\quad = E^x \int_0^{T_1} \mathrm{e}^{-\beta t} \tilde{s}(\alpha, B(t))\, \mathrm{d}t + E^x \left[\mathrm{e}^{-\beta T_1} E^{X(T_1)} \int_0^T \mathrm{e}^{-\beta t} \tilde{s}(\alpha_1(B(t)) \right] \mathrm{d}t \\ &\quad = U^\beta \tilde{s}(\alpha, x) + E^x \big[\mathrm{e}^{-\beta T_1} V^\beta \tilde{s}(\alpha_1 B(T_1)) \big], \end{aligned} \tag{4.48}$$

it is not hard to verify that $U^\beta \tilde{s}(\alpha, x)$ is the Laplace transform of the solution to

$$\begin{aligned} &\partial_t^2 u = c^2 \Delta u + s(t, x), \quad t > 0, x \in D', \\ &u(t, x) = 0, \quad t > 0, x \in \partial D, \end{aligned}$$

which we have (essentially) done above. Equation (4.48) can be solved by iteration on a (slightly) smaller set $B(0, R + \varepsilon)^c$: Let $K^\beta h$ be defined on $L_\infty(D')$ by

$$h(x) \to K^\beta h(x) := E^\beta \big[\mathrm{e}^{-\beta T} h(B(T)); T < \infty \big],$$

where, again, the presence of the exponential makes restriction of the integration to $\{T < \infty\}$ unnecessary.

Note now that, for $x \in B(0, R + \varepsilon)^c = D''$,

$$\|K^\beta h\|_\infty \leqslant \sup_{x \in D''} E^x [\mathrm{e}^{-\beta T}] \|h\|_\infty = M \|h\|_\infty,$$

where $M < 1$ (had we considered D' instead of D'', we would have obtained $M = 1$). Anyway, iterating (4.48), we would obtain

$$V^\beta \tilde{s}(\alpha, x) = \sum_{n=0}^\infty (K^\beta)^n U^\beta \widetilde{s}(\alpha, x).$$

Since $\widetilde{s}(\alpha, x)$ is a Laplace transform, so is $K^\beta U^\beta \widetilde{s}(\alpha, x)$, and a fortiori so is $V^\beta \widetilde{s}(\alpha, x)$, and we are done.

4.8. Reciprocity: A probabilistic approach

The usual physical interpretation of reciprocity is the following: producing a signal at a point x_0 and observing it at x is the same as producing a signal at x and observing it at x_0.

We saw in Chapter 2 that this is described mathematically as follows: consider, for given x_0,

$$\frac{\partial^2 u}{\partial t^2} = c^2 \Delta u + f(t)\delta(x - x_0) \tag{4.49}$$

for $x \in \mathbb{R}^3$ homogeneous initial conditions. Also, assume that $f(t) = 0$ for $t < 0$ and is otherwise bounded. Denote by $u(t, x, x_0)$ the solution to (4.49). Reciprocity means $u(t, x, x_0) = u(t, x_0, x)$.

Of course, when $c(x)$ is constant, this is rather easy to establish. We know that when we take Laplace transforms of (4.49), we obtain the equation

$$\frac{1}{2}\left(\left(\frac{\alpha}{c}\right)^2 - \Delta\right)\tilde{u}(\alpha, x, x_0) = \frac{\hat{f}(\alpha)}{2c^2}\delta(x - x_0)$$

for $\tilde{u}(\alpha, x, x_0)$. This equation has the solution

$$\tilde{u}(\alpha, x, x_0) = \frac{\tilde{f}(\alpha)}{2c^2} g(\alpha_1 |x - x_0|) = \frac{\tilde{f}(\alpha)}{2c^2} \frac{e^{-\frac{\alpha}{c}|x - x_0|}}{4\pi |x - x_0|}$$

which is symmetric in (x, x_0). From $\tilde{u}(\alpha, x, x_0) = \tilde{u}(\alpha, x, x_0)$ the desired result follows by inverting the Laplace transforms. Consider now a problem similar to (4.49), but in a region $D \subset \mathbb{R}^3$, bounded or not, but such that $V = \text{int}(D^c)$ is open. That is, consider

$$\frac{\partial^2 u}{\partial t^2} = c^2 \Delta^2 u + f(t)\delta(x - x_0), \quad t > 0,\ x, x_0 \in D \tag{4.50}$$

and assume that ∂D is soft, that is,

$$u(t, x) = 0, \quad t > 0, x \in \partial D. \tag{4.51}$$

We want to prove that $u(t, x, x_0) = u(t, x_0, x)$. For this we note that, taking Laplace transforms, the problem becomes

$$\begin{aligned} &\left[\frac{1}{2}\left(\frac{\alpha}{c}^2 - \frac{1}{2}\Delta\right)\right]\tilde{u}(\alpha, , x, x_0) = \frac{\tilde{f}(\alpha)}{2c^2}\delta(x - x_0), \\ &\tilde{u}(\alpha, x, x_0) = 0, \quad x_0 \in D, x \in \partial D. \end{aligned} \tag{4.52}$$

Again, except for a factor $\tilde{f}(\alpha)/2c^2$, the solution to (4.52) is just the Green function for $(\beta - \frac{1}{2}\Delta)$ in D, when Dirichlet boundary conditions are imposed. Again, we set $\beta = \frac{1}{2}(\frac{\alpha}{c})^2$. To obtain the symmetry $\tilde{u}(\alpha, x, x_0) = \tilde{u}(\alpha, x_0, x)$ by a probabilistic procedure, we retake a theme started in Chapter 1.

Consider again the imposed motion $B(t)$ on $\mathbb{R}^3$, and define the killed semigroup

$$q_D(t; x, A) = P^x(B(t) \in A; T_D > t),$$

where $T_D = \inf\{t > 0 \mid B(t) \in D^c\} = \inf\{t > 0 \mid B(t) \in \partial D\}$.

Now, let $t > 0, x \in \mathbb{R}^3, A \in B(\mathbb{R}^3)$. Then

$$P^x(B(t) \in A) = P^x(X(t) \in A; T_D > t) + P^x(B(t) \in A; T_D \leqslant t)$$

which can be rewritten, using $\rho_t(x, y) = \frac{1}{(2\pi)^{3/2}} e^{-(x-y)^2/2t}$, as

$$\begin{aligned} q_d(t; x; A) &= P_t(x, A) - E^x\left[\int \rho_{t-T_D}(X(T_D), y); T_D \leqslant t\right] \\ &= \int_A \{\rho_t(x, y) - E^x[\rho_{t-T}(X(T_D)y); T_D \leqslant t]\} \, dy, \end{aligned}$$

after an obvious conditioning with respect to $\mathcal{L}_{T(D)}$ and using the strong Markov property. Now define

$$q_D(t; x, y) = \rho_t(x, y) - E^x[\rho_{t-T(D)}(X(T_D), y); T_D \leqslant t]. \tag{4.53}$$

After obtaining some regularity conditions, the following theorem is proved in Port and Stone [1-8].

THEOREM 4.1. *Let D be as above, and define $q_D(t; x, y)$ by (4.53), for $(t, x, y) \in (0, \infty) \times \mathbb{R}^3 \times \mathbb{R}^3$. Then for $s_T t > 0$*

$$q_D(tq\breve{s}t; x, y) = \int q_d(t, x, z) q_D(s, z, y) \, dz. \tag{4.54}$$

Also, $q_D(t, x, y)$ is symmetric in x, y and vanishes if either x or y are in $V^r = \{\xi \in \mathbb{R}^3 \mid P^\xi(T_D = D) = 1\}$, $q_D(t; x, y)$ is upper semicontinuous on $(0, \infty) \times (\mathbb{R}^3/\partial D) \times (\mathbb{R}^3/\partial D)$. Furthermore, $q_D(t, x, y) \to 0$ whenever x and y are in the same component of D and $q_D(t; x, y) = 0$ if x, y are in different components of D.

The proof exploits the symmetry and regularity properties of $\rho_t(x, y)$ plus some probabilistic arguments and is rather lengthy. Important for us is the fact that

$$g_D^B(x, y) = \int_0^\infty e^{-\beta t} q_D(t, x, y) \, dt$$

the Green function for $(\beta - \frac{1}{2}\Delta)$ on D with homogeneous boundary conditions, that is, $g_0^\beta(x, y)$ satisfies

$$\left(\beta - \frac{1}{2}\Delta\right) g_D^\beta(x, y) = \delta(x - y), \quad x, y \in D,$$

$$g_D^\beta(x, y) = 0, \quad x \in D, x \in D\partial D.$$

And for us, $g_D^\beta(x,y) = g_D^\beta(y,x)$ for all $x, y \in D$. The final step consists in realizing that $\tilde{u}(\alpha, x, x_0) = \frac{f(\alpha)}{2c^2} g_d^\beta(x, x_0)$.

With all this we came back to a theme started in Chapter 2. If we define the β-harmonic measure

$$h_D^\beta(x, A) := E^x\big[\mathrm{e}^{-\beta T(0)}; X(T_D) \in A\big]$$

for $\beta > 0$, or if you prefer, for bounded or positive $f : \partial D \to \mathbb{R}$,

$$h_D^\beta f(x) = E^x\big[f(x(T_D))\mathrm{e}^{-\beta T(D)}\big].$$

It is an easy verification, and we have done it above, noting that

$$\int_0^\infty \mathrm{e}^{-\beta t}\big[\rho_{t-T(D)}(X(T_D), y); T_D \leqslant t\big]\,\mathrm{d}t$$
$$= E^x\left[\frac{\mathrm{e}^{-\sqrt{2\beta}|X(T(D))-y|}}{4\pi|X(T(D)) - y|}\mathrm{e}^{\beta T}\right],$$

where we used the fact that $\int \mathrm{e}^{-\beta t}\rho_t(x, y) = \frac{\mathrm{e}^{-\sqrt{2\beta}|x-y|}}{4\pi|x-y|}$.

Therefore, multiplying (4.53) by $\mathrm{e}^{-\beta t}$, integrating with respect to t and using the last identity we found, we arrive at

$$E^x\left[\mathrm{e}^{-\beta T(D)}\frac{\mathrm{e}^{-\frac{\alpha}{c}|X(T(D))-x_0|}}{|X(T(D)) - x_0|}\right] = E^{x_0}\left[\mathrm{e}^{-\beta t(D)}\frac{\mathrm{e}^{-\frac{\alpha}{c}|X(T_D)-x|}}{|X(T_D) - x|}\right].$$

This will be used in Chapter 9 below.

References

There was not much about waves in this chapter, or was there? Most of the computations are standard, but you could check with [1-6], [2-1] or [1-7] in case something looks too funny. More connections between potential theory and probability theory are exposed in

1. Chung, K. L. and Zhao, Z.: *From Brownian Motion to Schrödinger's Equation*, Springer-Verlag, Berlin, 1995.

A short table of Laplace transforms appears in

2. Abramowitz, M. and Stegun, I.: *Handbook of Mathematical Functions*, Dover, New York, 1965.

And a longish table and general identities appear in

3. Ditkine, V. and Proudnikov, A.: *Transformations Integrales et Calcul Operationel*, Mir, Moscow, 1982.

The material for Sommerfeld's radiation condition appears in references [2-5]–[2-6], and the connection between the Markov property and Huygens construction was first sketched in

4. Gzyl, H. and Jiménez, R.: Diffusions and waves, *J. Math. Anal. Appl.* **174** (1993), 498–517.

CHAPTER

FIVE

WAVES AND DIFFUSIONS

The bulk of this chapter is devoted to waves in heterogeneous but isotropic media: that is the equations we consider describe media in which the velocity of propagation depends on position but not on direction. As in the previous chapter, we aim at obtaining the appropriate Green functions as path integrals over diffusions or over appropriately modified imposed motions.

5.1. Waves in full space

As above, we begin with

$$\partial_t^2 u(t,x) = c^2(x)\Delta u(t,x) + s(t,x), \quad t > 0,\ x \in \mathbb{R}^d \tag{5.1}$$

and assume that the initial data is part of the source term $s(t,x)$ and set $u(0+,x) = 0$, $u_t(0+,x) = 0$. Assuming that both $s(t,x)$ and $u(t,x)$ do not grow too fast as t becomes large, we put

$$\tilde{u}(\alpha,x) = \int_0^\infty \mathrm{e}^{-\alpha t} u(t,x)\,\mathrm{d}t \tag{5.2}$$

etc., for $\alpha > 0$ and transform (1) into

$$\frac{1}{2}(\alpha^2 - c^2(x)\Delta)\tilde{u}(\alpha,x) = \tilde{s}(\alpha,x)/2. \tag{5.3}$$

Then, multiplying by $1/2c^2(x) \equiv q(x)/2$ and setting $\widetilde{S}(\alpha, x) = q(x)\tilde{s}(\alpha, x)/2$, we obtain

$$\frac{1}{2}(\alpha^2 q(x) - \Delta)\tilde{u}(\alpha, x) = \widetilde{S}(\alpha, x). \tag{5.4}$$

There are two ways of writing down $\tilde{u}(\alpha, x)$ depending on whether we choose to consider (5.3) or (5.4).

If we consider (5.4), then, according to the results in Section 1 of Chapter 2, we can write the solution to (5.4) as

$$\tilde{u}(\alpha, x) = E^x \int_0^\infty \mathrm{e}^{-\frac{\alpha^2}{2}A(t)} \widetilde{S}(\alpha, B(t))\, \mathrm{d}t, \tag{5.5}$$

where $A(t) = \int_0^t q(B_s)\, \mathrm{d}s$. Here we see that (5.5) makes sense as long as $q(x)$ is positive and $\tilde{s}(\alpha, x)$ is at least bounded or positive in x, for each α.

One can verify that $\tilde{u}(\alpha, x)$ is actually (a path integral representation of) a weak solution to (5.4). To obtain a classical solution, we would have to throw in some smoothness assumptions on $q(x)$ and $\tilde{s}(\alpha, x)$.

A similar representation is obtained in terms of the diffusion $x(t)$ obtained by solving

$$\mathrm{d}X(t) = c(X_t)\, \mathrm{d}B(t), \qquad X_0 = x$$

which has the generator $G = c(x)^2\Delta/2$. It is given by

$$\tilde{u}(\alpha, x) = \frac{1}{2} E^x \int_0^\infty \mathrm{e}^{-\frac{\alpha^2 t}{2}} \tilde{s}(\alpha, X(t))\, \mathrm{d}t, \tag{5.6}$$

where $X(t)$ and the corresponding E^x are described in Section 3 of Chapter 2. This looks pretty much like the stuff in the previous chapter, except that this time $E^x[\tilde{s}(\alpha, X(t))] = \int \rho(k, x, y)\tilde{s}(\alpha, y)\, \mathrm{d}y$, where the transition density satisfies

$$\frac{\partial \rho}{\partial t}(t, x, y) = \frac{\Delta}{2}(c^2(y)\rho(t, x, y)).$$

And still another possibility comes from thinking of $\frac{c^2(x)}{2}\Delta$ as the infinitesimal generator of a time changed imposed motion. For that, suppose that $c_{\min} \leqslant c(x) \leqslant c_{\max}$, which is an a fortiori hypothesis from the physical point of view.This makes $A(t)$ strictly increasing, bounded above and below by $q_{\max}t$ and $q_{\min}t$, respectively. If τ_t denotes the inverse of $A(t)$, i.e., $A(\tau(t)) = t$, $\tau(A(t)) = t$, a simple change of variables transforms (5.5) into

$$\begin{aligned}
\tilde{u}(\alpha, x) &= \frac{1}{2} E^x \int_0^\infty \mathrm{e}^{-\frac{\alpha^2}{2}t} \widetilde{S}(\alpha, B(\tau(t)))\, \mathrm{d}t / q(B(\tau_t)) \\
&= \frac{1}{2} E^x \int_0^\infty \mathrm{e}^{-\frac{\alpha^2}{2}t} \tilde{s}(\alpha, \widehat{B}(t))\, \mathrm{d}t,
\end{aligned} \tag{5.7}$$

where we brought in the fact that $d\tau(t) = dt/q(B(\tau(t)))$ as follows from the chain rule applied to $A(\tau(t)) = t$. Both in (5.5) and (5.7), $E^x(\cdot)$ refers to integration with respect to the Wiener measure on a path space. Also, $\widehat{B}(t) = B(\tau(t))$ was discussed in Section 7 of Chapter 1. The similarity of (5.7) and (5.6) is more than coincidental: the distributions of $B(\tau(A))$ under the Wiener measure and that of $X(t)$ in (5.6) are the same.

We designated Chapters 1 and 2 to set up the notation that would allow us to obtain the path integral representation of the Laplace transform $\tilde{u}(\alpha, x)$ of the solutions to

$$\frac{\partial^2 u}{\partial t}(t,x) = c^2(x)\Delta u(t,x) + b(x)\cdot\nabla u(t,x) - \gamma(x)u(t,x) + s(t,x) \tag{5.8}$$

plus vanishing initial conditions. The two terms $b(x)\cdot\nabla u(t,x)$ and $-\gamma(x)u(t,x)$ describe, respectively, the effect of the drift of the medium and the absorption in the medium. The corresponding equation for $\tilde{u}(\alpha, x)$ is

$$\left(\frac{\alpha^2}{2} + \frac{\gamma(x)}{2} - \frac{b(x)}{2}\nabla - \frac{c^2(x)}{2}\Delta\right)\tilde{u}(\alpha,x) = \frac{1}{2}\tilde{s}(\alpha,x). \tag{5.9}$$

If now we denote by $X(t)$ the solution to the stochastic differential equation

$$dX(t) = c(X(t))\,dB(t) + b(X(t))\,dt, \qquad X(0) = x, \tag{5.10}$$

where $B(t)$ is the Brownian motion issued from 0, and we denote by E^x the measure induced by the solutions to (5.10) in the path space, then the solution to (5.9) can be represented by

$$\tilde{u}(\alpha,x) = \frac{1}{2}E^x\int_0^\infty M(t)\tilde{s}(\alpha, X(t))\,dt, \tag{5.11}$$

where $M(t) = \exp -\frac{1}{2}(\alpha^2 t + \int_0^t \gamma(X_s)\,ds)$. Again, unless more regularity assumptions are placed upon $q(x)$, $\gamma(x)$, $b(x)$ and $\tilde{s}(\alpha,x)$, (5.11) denotes the weak solution to (5.9). To obtain a different representation for $\tilde{u}(\alpha,x)$, in (5.9), divide by $c(x)^2$ and α to obtain

$$\left(\frac{\alpha}{2}q(x) + \frac{k(x)}{2\alpha} - \frac{\beta(x)}{2\alpha}\nabla - \frac{1}{2\alpha}\Delta\right)\tilde{u}(\alpha,x) = \frac{1}{\alpha}\tilde{s}(\alpha,x),$$

where $k(x) = \gamma(x)q(x)$, $\beta(x) = b(x)q(x)$, $q(x) = 1/c^2(x)$. Instead of (5.11), we may write

$$\tilde{u}(\alpha,x) = E^x\int_0^\infty M_1(t)Z(t)\widetilde{S}(\alpha, B_\alpha(t))\,dt,$$

where $B_\alpha(t) = B(0) + \frac{1}{\sqrt{\alpha}}(B(t) - B(0))$, and now $E^x(\cdot)$ in (5.12) denotes the usual Wiener integral. Also,

$$M_1(t) = \exp -\frac{1}{2}\left\{\alpha\int_0^t q(B_\alpha(s))\,ds + \frac{1}{\alpha}\int_0^t k(B_\alpha(s))\,ds\right\}$$

and

$$Z(t) = \exp\left\{\int_0^t \beta(B_\alpha(s))\,\mathrm{d}B_\alpha(s) - \frac{1}{2\alpha}\int_0^t \|B_\alpha(s)\|^2\,\mathrm{d}s\right\}.$$

It is this last factor that relates the Wiener measure to the measure induced on the path space by the solutions to

$$\mathrm{d}X_\alpha(t) = \mathrm{d}B_\alpha(t) + \beta(X_\alpha(t))\,\mathrm{d}t.$$

5.2. Existence of solutions to the wave equations

Throughout this section we shall restrict ourselves to the case $d \geqslant 3$, and indicate where it is essential. Consider

$$\frac{\partial^2 u}{\partial t^2} = c^2(x)\Delta u + b(x)\cdot\nabla u - \gamma(x)u + s(t,x) \tag{5.12}$$

for $t > 0$, $x \in \mathbb{R}^d$ with vanishing initial conditions. We assume that $\gamma(x)$ is continuous and positive, $b(x)$ is continuous and satisfies the assumptions in Section 3 of Chapter 2, and that $c(x)$ is continuous, bounded above and below away from zero, i.e., $c_1 c(x) \leqslant c_2$. Also, we assume that $s(t,x)$ is bounded in both variables and that $\int_0^\infty s(t,x)\,\mathrm{d}t$ exists and has a compact support in $\mathbb{R}^d$.

If $u(t,x)$ is a solution to (5.12) and $\tilde{u}(\alpha,x)$ denotes its Laplace transform, then

$$\frac{1}{2}\left\{\frac{\alpha}{2}q(x) - \left(\frac{k(x)}{2\alpha} + \frac{\beta(x)\cdot\nabla}{2\alpha} + \frac{1}{2\alpha}\Delta\right)\right\}\tilde{u}(\alpha,x) = \widetilde{S}(\alpha,x)/\alpha \tag{5.13}$$

with $\widetilde{S}(\alpha,x) = q(x)\tilde{s}(\alpha,x)/2$ as usual. Assuming the existence and uniqueness of elliptic problem (5.13), and if we prove that $\tilde{u}(\alpha,x)$ is the Laplace transform of a function $\tilde{u}(t,x)$, the same arguments used in Section 4 of Chapter 4 would imply that $u(t,x)$ satisfy (5.12). So, let us prove that $\tilde{u}(\alpha,x)$ is the Laplace transform of a measure on $[0,\infty)$.

There is no problem in exchanging integrals in (5.11) to obtain

$$\tilde{u}(\alpha,x) = \int_0^\infty \tilde{v}(\alpha,t,x)\,\mathrm{d}t \tag{5.14}$$

and recall that the physical interpretation of t in (5.14) has nothing to do with time variable in (5.12). Let us prove that there exists a measure $v(\mathrm{d}s,t,x)$ on $(0,\infty)$ such that

$$\tilde{v}(\alpha,t,x) = \int \mathrm{e}^{-\alpha s} v(\mathrm{d}s,t,x).$$

From the boundedness and continuity assumptions it is clear that

$$\tilde{v}(\alpha,t,x) = \lim E^x\left\{\widetilde{S}(\alpha, B_\alpha(t))Z^{(n)}(t)\exp\left\{-\left[\frac{\alpha t}{n}\sum_0^{n-1} q\left(B_\alpha\left(\frac{kt}{n}\right)\right)\right.\right.\right.$$
$$\left.\left.\left.+\frac{t}{\alpha n}\sum k\left(B_\alpha\left(\frac{kt}{n}\right)\right)\right]\right\}\right\}\Big/\alpha,$$

where we have set

$$Z^{(n)}(t) = \exp\left\{\sum_{k=0}^{n-1}\beta\left(B_\alpha\left(\frac{kt}{n}\right)\right)\cdot\left(B_\alpha\left(\frac{k+1}{n}t\right)-B_\alpha\left(\frac{kt}{n}\right)\right)\right.$$
$$\left.-\frac{1}{2\alpha}\frac{t}{n}\sum_{k=0}^{n-1}\left\|\beta\left(B_\alpha\left(\frac{tk}{n}\right)\right)\right\|^2\right\}.$$

Write $\tilde{v}_n(\alpha,t,x)$ for the expected values introduced above and note that, as a consequence of the simple Markov property,

$$\tilde{v}_n(\alpha,t,x) = \int\cdots\int I_n(\alpha,t,x_0,x_1,\ldots,x_n)\,\mathrm{d}x_1\cdots\mathrm{d}x_n,$$

where $x_0 = x$, all integrations are over $\mathbb{R}^d$ and $I_n(\alpha,t,x_0,\ldots,x_n)$ is a product of the following factors

$$\widetilde{S}(\alpha,x_n);\qquad \prod_{k=1}^{n} p_t^\alpha(x_{k-1},x_k);\qquad \exp\left\{-\frac{\alpha t}{2n}\sum_{k=0}^{n-1} g(x_k)\right\};$$

$$\exp\left\{-\frac{t}{2\alpha n}\sum_{k=0}^{n-1} k(x_k)\right\},\qquad \exp\left\{\sum_{k=0}^{n-1}\beta(x_k)\cdot(x_{k+1}-x_k)-\frac{t}{2\alpha n}\|\beta(x_k)\|^2\right\},$$

where, of course, $p_t^\alpha(x,y) = (\frac{\alpha}{2\pi t})^{d/2}\exp\{-(\frac{\alpha}{2t}\|x-y\|^2)\}$.

To verify that each of the factors of $I_n(\alpha,t,x_0,\ldots,x_n)$ is a Laplace transform, we only need a few basic facts. First note that $g(s) = (\frac{s}{a})^{\frac{v-1}{2}}J_{v-1}(2\sqrt{sa})$ has the Laplace transform $\tilde{g}(\alpha) = (\frac{1}{\alpha})^v\exp\{-a/\alpha\}$, with $\alpha, a > 0$. If $f(s)$ has the Laplace transform $f(\alpha)$ and $h(s) = \int_0^\infty J_0(2\sqrt{st})f(t)\,\mathrm{d}t$, then $\tilde{h}(\alpha) = \frac{1}{\alpha}\tilde{f}(\frac{1}{\alpha})$, which, applied to $g(s)$ defined above, implies that $\tilde{h}(\alpha) = \alpha^v\exp\{-\alpha x\}$ is a Laplace transform.

Since $\frac{1}{\alpha}\exp\{-a/\alpha\}$ is the Laplace transform of $J_0(2\sqrt{as})$, $\exp\{-a/\alpha\}$ is a Laplace transform.

Therefore, each $I_n(\alpha,t,x_0,\ldots,x_n)$ is a Laplace transform, and so is $\tilde{v}(\alpha,t,x)$. Since $\tilde{v}_n(\alpha,t,x)\to\tilde{v}(\alpha,t,x)$, by the extended continuity theorem, $\tilde{v}(\alpha,t,x)$ is the Laplace transform of a measure $v(\mathrm{d}s,t,x)$ on $(0,\infty)$, i.e.,

$$\tilde{v}(\alpha,t,x) = \int_0^\infty \mathrm{e}^{-\alpha s}v(\mathrm{d}s,t,x)$$

and, if we invoke Fubini's to exchange the integrals again, we obtain

$$\tilde{u}(\alpha, x) = \int_0^\infty e^{-\alpha s} \int_0^\infty v(ds, t, x)\, dt = \int_0^\infty e^{\alpha s} U(ds, x).$$

Proceeding as in Section 4 of Chapter 4, we obtain $U(dt, x) = u(t, x)\, dt$, and that $u(t, x)$ satisfies (5.12) is obtained similarly.

5.3. An evaluation of some path integrals

The following procedure, proposed by J. Bertrand and B. Gaveau allows us to "compute" the path integral

$$E^x\left[\exp\left\{-\frac{\alpha^2}{2} A(t)\widetilde{S}(\alpha, B(t))\right\}\right] \tag{5.15}$$

in terms of an auxiliary Poisson process. For this we have to assume that we can recover $\widetilde{S}(\alpha, x)$ from its Fourier transform

$$\widetilde{S}(\alpha, x) = \int e^{-ik\cdot x} \widehat{S}(\alpha, k) \frac{dk}{(2\pi)^d}$$

and that $q(x)$ is such that

$$q(x) = \int e^{i(x\cdot y + \phi(y))} \mu(dy)$$

for some finite measure $\mu(dy)$ on $\mathbb{R}^d$ and some appropriate $\phi : \mathbb{R}^d \to \mathbb{R}$. Let $m = \int \mu(dy)$.

Let $(W, \mathcal{G}, \mathcal{P})$ be an auxiliary probability space on which we have defined a random measure $\nu(ds, dk)$ by

$$\mathcal{E} \exp\left\{i \int_0^t \int F(s, k)\nu(ds, dk)\right\} = \exp\left\{\int_0^t ds \int \left(e^{iF(s,k)} - 1\right) d\mu(k)\right\} \tag{5.16}$$

for any bounded, complex-valued $F(s, k)$.

Or, if you prefer, for $A \times B \in B([0, \infty)) \otimes B(\mathbb{R}^d)$, $\nu(A, B)$ obeys a compound Poisson law whose characteristic function is described by (5.16).

To this random measure ν we associate a compound Poisson process $N(t)$ by

$$N(t) = \int_0^t \int k\nu(ds, dk)$$

with values in $\mathbb{R}^d$. To refer everything to the Brownian motion issued from 0, write (5.16) as

$$E^0\left[\exp\left\{-\frac{\alpha^2}{2}\left\{\int_0^t q(x + B(s))\, ds\right\}\widetilde{S}(\alpha, x + B(t))\right\}\right].$$

Now, proceed as follows

$$\exp\left\{-\frac{\alpha^2}{2}\int_0^t q(x+B(s))\,\mathrm{d}s\right\}$$
$$=\mathrm{e}^{mt}\exp\left\{\int_0^t \mathrm{d}s\int\left[\mathrm{e}^{i(x+B(s)\cdot y}\mathrm{e}^{i(\phi(y)-i\beta)}-1\right]\mu(\mathrm{d}y)\right\},$$

where $\beta=\lg(\frac{\alpha^2}{2})+i\pi$. From (5.16) the last integral equals

$$\mathrm{e}^{mt}\mathcal{E}\left\{\exp\left\{i\int_0^t\int(x+B(s))\cdot y+\phi(y)-i\beta\nu(\mathrm{d}s,\mathrm{d}y)\right\}\right\}$$
$$=\mathrm{e}^{mt}\mathcal{E}\left\{\mathrm{e}^{i\int_0^t(x+B(s))\cdot\mathrm{d}N(s)}\mathrm{e}^{i\int_0^t\int(\phi(y)-i\beta)\nu(\mathrm{d}s,\mathrm{d}y)}\right\}.$$

Substituting and exchanging integrations, we obtain that (5.15) equals

$$E^x\left\{\widetilde{S}(\alpha,B(t))\exp\left\{-\frac{\alpha^2}{2}A(t)\right\}\right\}$$
$$=\mathrm{e}^{mt}\int\widetilde{S}(\alpha,k)\mathcal{E}\left\{\mathrm{e}^{i\int_0^t\int(\phi(y)-i\beta)\nu(\mathrm{d}s,\mathrm{d}y)}\right.$$
$$\left.\times E^0\left[\mathrm{e}^{i\int_0^t(x+B(s))\cdot\mathrm{d}N(s)+ik\cdot(x+B(s))}\right]\right\}\frac{\mathrm{d}k}{(2\pi)^d}.$$

To compute the expectation with respect to $\mathrm{d}P^0$, observe that

$$\int B(s)\cdot\mathrm{d}N(s)+k\cdot B(t)=-\int_0^t(N(t)-N(s)-k)\cdot\mathrm{d}B(s),$$

and if now we make use of

$$E^0\left[\exp\left\{i\int_0^t\Gamma(s)\cdot\mathrm{d}B(s)\right\}\right]=\exp\left\{-\frac{1}{2}\int_0^t\|\Gamma(s)\|^2\,\mathrm{d}s\right\}$$

whenever $\Gamma(s)$ is independent of $B(t)$. We apply this to compute the inner integral with $\Gamma(s)=-(N(t)-N(s)-k)$ to finally obtain

$$E^x\left\{\widetilde{S}(\alpha,B(t))\mathrm{e}^{-\frac{\alpha^2}{2}A(t)}\right\}$$
$$=\mathrm{e}^{mt}\int\widetilde{S}(\alpha,k)\left\{\mathrm{e}^{i\int_0^t\int(\phi(y)\cdot i\beta)u(\mathrm{d}s,\mathrm{d}y)}\mathrm{e}^{ix\cdot(N_t+kt)}\mathrm{e}^{-\frac{1}{2}\int_0^t\|N_t-N_s+k\|^2\,\mathrm{d}s}\right\}\frac{\mathrm{d}k}{(2n)^d}.$$

5.4. Waves in stratified media

We consider a situation in which the speed of propagation varies only in one direction. Assume $d=3$ and denote the coordinate along which the velocity varies by z.

We consider first the propagation in a one-dimensional space, afterwards in a one-dimensional subspace with reflecting boundary conditions, and then proceed to add on the rest of the coordinates.

5.4.1. *Waves in a one-dimensional stratified medium*

This time our old friend looks like this:

$$\frac{\partial^2 u}{\partial t^2} = c^2(z)\frac{\partial^2 u}{\partial z^2} + s(t, z), \quad t > 0,\ z \in \mathbb{R} \tag{5.17}$$

with $u(0+, z) = 0$, $u_t(0+, z) = 0$. Again, $0 < c_1 < c(z) < c_2$ for appropriate c_1 and c_2.

We know that the Laplace transform $\tilde{u}(\alpha, z)$ of $u(t, z)$ can be represented as

$$\tilde{u}(\alpha, z) = E^z \int_0^\infty \mathrm{e}^{-\frac{\alpha^2}{2}A(t)}\widetilde{S}(\alpha, B(t))\,\mathrm{d}t, \tag{5.18}$$

where this time $B(t)$ is the one-dimensional Brownian motion and $E^z(\cdot)$ denotes integration with respect to the Wiener measure on the continuous trajectories issued from z.

Let us explore a way of evaluating (5.19) in terms of the local times L_t^a of the one-dimensional Brownian motion.

Recall from Section 7 of Chapter 1 that L_t^a is a continuous, increasing additive functional such that

(i) $L_t^a = \int_0^t I_{\{a\}}(B_s)\,\mathrm{d}L_s^a$,

(ii) $\int_0^t F(B_s)\,\mathrm{d}s = \int_{-\infty}^{\infty} F(a) L_t^a\,\mathrm{d}a$

for positive or bounded $F(z)$. So we see that L_t^a is some sort of occupation density of B_t at a.

A simple exchange of orders of integration, and an application of (ii) above should convince us that

$$\int_0^t F(t, B(t))\,\mathrm{d}A(t) = \int_{-\infty}^{\infty} q(a)\left(\int_0^\infty F(t, a)\,\mathrm{d}L_t^a\right)\mathrm{d}a, \tag{5.19}$$

where again $F(t, z)$ is positive and $A(t) = \int_0^t q(B_s)\,\mathrm{d}s$, and the inner integral is, with respect to the Stieltjes measure, induced by the increasing function L_t^a.

On other elementary result to recall is the following: Since $L^a(t)$ is continuous and increases to infinity as $t \to \infty$, it can be inverted as follows: set $\tau_t^a = \inf\{s > 0 : L_t^a > t\}$, and verify that, since $L^a(\tau_t^a) = t$,

$$\int_0^\infty \phi(t)\,\mathrm{d}L_t^a = \int_0^\infty \phi(\tau_t^a)\,\mathrm{d}t. \tag{5.20}$$

Comment: The proof that $L_t^a \to \infty$ as $t \to \infty$ relies on the recurrence of the one-dimensional imposed motion.

Recall that $\widetilde{S}(\alpha,z)=\frac{1}{2}q(z)\tilde{s}(\alpha,z)$, and that $\mathrm{d}A(t)=q(B_t)\,\mathrm{d}t$. From (5.19)–(5.20), we can rewrite (5.18) as

$$\begin{aligned}\tilde{u}(\alpha,z)&=\frac{1}{2}\int_{-\infty}^{\infty}\widetilde{S}(\alpha,a)q(a)\left(E^z\int_0^{\infty}\mathrm{e}^{-\frac{\alpha^2}{2}A(t)}\,\mathrm{d}L_t^a\right)\mathrm{d}a\\&=\frac{1}{2}\int_{-\infty}^{\infty}\widetilde{S}(\alpha,a)q(a)\left(E^z\int_0^{\infty}\mathrm{e}^{-\frac{\alpha^2}{2}A(\tau_t^a)}\,\mathrm{d}t\right)\mathrm{d}a\\&=\frac{1}{2}\int_{-\infty}^{\infty}\widetilde{S}(\alpha,a)q(a)\left(\int_0^{\infty}E^z\big[\mathrm{e}^{-\frac{\alpha^2}{2}A(\tau_t^a)}\big]\,\mathrm{d}t\right)\mathrm{d}a.\end{aligned}$$

Since the Brownian motion does not stop at zero, just passes through it, the closed random set $\{t:B_t=0\}$ has the Lebesque measure zero (but L_t^a grows on it!), its complement is the union of a collection of open sets corresponding to the excursions of $B(t)$ away from zero. These excursions correspond to the flat spots in the graph of L_t^a or to jumps of τ_t^a; i.e., the excursion set is the union $\bigcup(\tau_{s-}^a,\tau_s^a)$ with s in $\mathbb{R}$.

We can thus write

$$A(\tau_t^a)=A(\tau_0^a)+\sum_{s\leqslant t}A(\tau_s^a)-A(\tau_{s-}^a),\tag{5.21}$$

since $A(t)$ does not start growing until τ_0^a: the hitting time of a by $B(t)$.

Also, since L_t^a does not grow until $t>\tau_0^a$, we have

$$\begin{aligned}\tau_t^a&=\inf\{s>\tau_0^a:L_s^a>t\}=\tau_0^a+\inf\{s>0:L_{s+\tau_0^a}^a>t\}\\&=\tau_0^a+\tau_t^a\cdot\theta(\tau_0^a)\end{aligned}$$

since $L^a(s+\tau_0^a)=L^a(\tau_0^a)+L_s^a\cdot\theta(\tau_0^a)=L_s^a\cdot\theta(\tau_0^a)$ due to the additivity of L_t^a and $L^a(\tau_0^a)=0$. Thus, the strong Markov property and (5.21) yield

$$\begin{aligned}E^z\big[\mathrm{e}^{-\frac{\alpha^2}{2}A(t)}\big]&=E^z\big[\mathrm{e}^{-\frac{\alpha^2}{2}A(\tau_0^a)}\big]E^a\big[\mathrm{e}^{-\frac{\alpha^2}{2}A(\tau_t^a)}\big]\\&=E^z\big[\mathrm{e}^{-\frac{\alpha^2}{2}A(\tau_0^a)}\big]E^a\big[\mathrm{e}^{-\frac{\alpha^2}{2}\sum_{s\leqslant t}\{A(\tau_s^a)-A(\tau_{s-}^a)\}}\big].\end{aligned}$$

It turns out that the behavior of the imposed motion during the (disjoint) excursion intervals is statistically independent, and then the increments $A(\tau_s^a)-A(\tau_{s-}^a)$ are independent, with respect to P^a, and therefore $A(t)$ has independent increments. Thus,

$$E^a\big[\mathrm{e}^{-\frac{\alpha^2}{2}A(t)}\big]=E^a\left[\exp\left\{-\frac{\alpha}{2}\sum_{s\leqslant t}\big(A(\tau_s^a)-A(\tau_{s-}^a)\big)\right\}\right]=\exp\{-tJ(\alpha,a)\},$$

where $J(\alpha,a)=[E^a\int_0^\infty\exp\{-\frac{\alpha^2}{2}A(\tau_t^a)\}\,\mathrm{d}t]^{-1}$. To find $J(\alpha,a)$, just integrate both sides of the previous identity to obtain

$$\frac{1}{J(\alpha,a)}=\int_0^{\infty}\mathrm{e}^{-tJ(\alpha,A)}\,\mathrm{d}t=E^a\int_0^{\infty}\mathrm{e}^{-\frac{\alpha^2}{2}A(\tau_t^a)}\,\mathrm{d}t=E^a\int_0^{\infty}\mathrm{e}^{-\frac{\alpha^2}{2}A(t)}\,\mathrm{d}L_t^a.$$

To pin $J(\alpha, a)$ down any further, consider

LEMMA 5.1. *Let* $W_\alpha^>(a, z)$, $W_\alpha^<(a, z)$ *be equal to* $E^z[e^{-\frac{\alpha^2}{2}A(\tau_0^a)}] = W_\alpha(\alpha, z)$ *for* $z > a$ *and* $z < a$, *respectively. Then*

$$\frac{d^2}{dz^2}W^>(a, z) - \alpha^2 q(2) W_\alpha^>(a, z) = 0, \quad z > a, \tag{5.22}$$

$$\lim_{z\downarrow a} W_\alpha^>(a, z) = 1, \qquad \lim_{z\to\infty} W_\alpha^>(a, z) = 0 \tag{5.23}$$

and similar identities are satisfied by $W_\alpha^<(a, z)$.

PROOF. The first assertion is left for the reader. To obtain the second one, note that

$$E^z\left[\exp\left\{-\frac{\alpha^2}{2}A(\tau_0^a)\right\}\right] \leqslant E^z\left[\exp\left\{-\frac{\alpha^2}{2}q_2\tau_0^a\right\}\right],$$

where $q_2 = 1/c_2^2$. Let us now vary z. Consider $a < z_1 < z_2$, and use T_a, instead of τ_0^a, starting from z_2, it is clear that $T_a = T_{z_1} + T_a \cdot \theta(T_{z_1})$ and the strong Markov property yields

$$E^{z_1}\left[\exp\left\{-\frac{\alpha^2}{2}q_2 T_a\right\}\right] = E^{z_2}\left[\exp\left\{-\frac{\alpha^2}{2}q_2 T_{z_1}\right\}\right]E^{z_1}\left[\exp\left\{-\frac{\alpha^2}{2}q_2 T_a\right\}\right].$$

In other words, $E^z[e^{-\beta T_a}] = e^{-k(z-a)}$. Since this function has to satisfy (5.22) with $q(z)$ replaced by q_2, it follows that $k = \alpha\sqrt{q_2}$, and therefore (5.23) holds. □

LEMMA 5.2. *With the notation introduced above, if* $q(z)$ *and* $\tilde{s}(\alpha, z)$ *are continuous at* z, *then*

$$J(\alpha, z) = \frac{1}{2}\left\{\left(\frac{dW_\alpha^>}{dz}\right)(z, z) - \left(\frac{d}{dz}W_\alpha^<\right)(z, z)\right\},$$

where

$$\left(\frac{dW_\alpha^<}{dz}\right)(z, z) = \lim_{a\downarrow z}\frac{dW_\alpha^<}{dz}(a, z).$$

PROOF. We already know that

$$\begin{aligned}\tilde{u}(\alpha, z) &= \frac{1}{2}\int_{-\infty}^{\infty}\tilde{s}(\alpha, a)q(a)\frac{W_\alpha(a, z)}{J(\alpha, a)}da\\ &= \frac{1}{2}\left\{\int_{-\infty}^{z}\tilde{s}(\alpha, a)q(a)\frac{W_\alpha^>(a, z)}{J(\alpha, a)} + \int_z^{\infty}\tilde{s}(\alpha, a)q(a)\frac{W_\alpha^<(a, z)}{J(\alpha, a)}da\right\}\end{aligned}$$

which, after differentiation with respect to z, yields

$$\frac{d\tilde{u}}{dz}(\alpha, z) = \frac{1}{2}\left[\frac{\tilde{s}(\alpha, z)q(z)}{J(\alpha, z)} - \frac{\tilde{s}(\alpha, z)q(z)}{J(\alpha, z)}\right] + \frac{1}{2}\left\{\int_{-\infty}^{z} \frac{\tilde{s}(\alpha, a)q(a)}{J(\alpha, a)}\frac{dW_\alpha^>}{dz}(a, z)\, dq + \int_{z}^{\infty} \frac{\tilde{s}(\alpha, a)}{J(\alpha, z)}q(a)\frac{dW_\alpha^<}{dz}(a, z)\, da\right\}.$$

Differentiating once more we obtain, making use of Lemma 5.11, that

$$\frac{d^2\tilde{u}}{dz^2}(\alpha, z) = \frac{1}{2}\left\{\frac{\tilde{s}(\alpha, z)q(z)}{J(\alpha, z)}\frac{dW_\alpha^>}{dz}(z, z) - \frac{\tilde{s}(\alpha, z)q(z)}{J(\alpha, z)}\frac{dW_\alpha^<}{dz}\alpha(z, z)\right\} + \int_{-\infty}^{z} \frac{\tilde{s}(\alpha, a)q(a)}{J(\alpha, a)}\alpha^2 q(z)\, da + \int_{z}^{\infty} \frac{\tilde{s}(\alpha, a)q(a)}{J(\alpha, a)}\alpha^2 q(z)\, da,$$

$$\frac{d^2\tilde{u}}{dz^2}(\alpha, z) = \frac{1}{2}\frac{\tilde{s}(\alpha, z)q(z)}{J(\alpha, z)}\left(\frac{dW_\alpha^>}{dz}(z, 2) - \frac{dW_\alpha^<}{dz}(z, z)\right) + \alpha^2\tilde{u}(\alpha, z)$$

and since

$$\frac{d^2\tilde{u}}{dz^2}(\alpha, z) - \alpha^2 q(z)\tilde{u}(\alpha, z) = -\widetilde{S}(\alpha, z)q(z),$$

we obtain

$$J(\alpha, z) = \frac{1}{2}\left\{\frac{dW_\alpha^<}{dz}(z, z) - \frac{dW_\alpha^>}{dz}(z, z)\right\}.$$ □

5.4.2. Waves in a stratified half-space

In a variety of problems in geophysics, the earth is modeled as a half-space $D = \{x : z > 0\}$, and it is assumed that in a first approximation the speed of sound depends on depth only, yielding the problem of considering the solutions of

$$\frac{\partial^2\tilde{u}}{\partial t^2}(t, x) = c^2(z)\Delta u(t, x) + s(t, x), \quad t > 0,\ x \in D$$

with zero initial conditions plus the free boundary condition

$$\frac{\partial u}{\partial z}(t, x) = 0, \quad x \in D.$$

We shall consider first the one-dimensional version of this problem, and in Section 4.4 below we throw in the rest of the coordinates. For this we consider problem (1), i.e.,

$$\frac{\partial u}{\partial t}(t, z) = c^2\frac{\partial^2 u}{\partial z^2} + s(t, z), \quad z > 0,\ t > 0 \tag{5.24}$$

with vanishing initial conditions plus the free boundary term

$$\frac{\partial u}{\partial z}(t, 0+) = 0, \quad t > 0. \tag{5.25}$$

Taking the Laplace transform of (5.24), we obtain

$$\frac{1}{2}\left(\alpha^2 q(z) - \frac{\mathrm{d}^2}{\mathrm{d}z^2}\right)\tilde{u}(\alpha, z) = \widetilde{S}(\alpha, z) \tag{5.26}$$

plus the free boundary condition

$$\frac{\mathrm{d}\tilde{u}}{\mathrm{d}z}(\alpha, z) = 0, \quad \alpha > 0. \tag{5.27}$$

The process $X(t) = |B(t)|$ is a candidate for a reflected imposed motion on the line. In terms of this process the solution to (5.27) is given by

$$\tilde{u}(\alpha, z) = E^z \int_0^\infty \mathrm{e}^{-\frac{\alpha^2}{2}A(t)}\widetilde{S}(\alpha, X(t))\,\mathrm{d}t, \tag{5.28}$$

where this time $A(t) = \int_0^t q(X_s)\,\mathrm{d}s$. We can evaluate (5.29) in terms of $B(t)$ by considering the even extensions $\widetilde{S}_1(\alpha, z)$ and $q_1(z)$ to $\mathbb{R}$. With this we can write (5.28) as

$$\begin{aligned}\tilde{u}(\alpha, z) &= E^2 \int_0^\infty \mathrm{e}^{-\frac{\alpha^2}{2}A_1(t)}\widetilde{S}_1(\alpha, B(t))\,\mathrm{d}t \\ &= \int_{-\infty}^\infty \widetilde{S}(\alpha, a)q(a)\left\{\frac{W_\alpha(a, z)}{J(\alpha, a)} + \frac{W_a(-a, z)}{J(\alpha, -a)}\right\}\mathrm{d}a.\end{aligned} \tag{5.29}$$

When $q(z) = 1/c^2$ is a constant, $W_\alpha(a, z) = \mathrm{e}^{-\frac{\alpha}{c}|z-a|}$ and $J(\alpha, a) = 1/\alpha$, the identity above yields

$$\tilde{u}(\alpha, z) = \int_0^\infty \frac{\widetilde{S}(\alpha, a)}{c^2}\frac{1}{\alpha}\left(\mathrm{e}^{-\frac{\alpha}{c}(z+a)} + \mathrm{e}^{-\frac{\alpha}{c}|a-z|}\right)\mathrm{d}a$$

and we leave for the reader to obtain $u(t, z)$ from this. Only note that $\frac{\partial \tilde{u}}{\partial z}(\alpha, 0) = 0$.

5.4.3. A change of spatial scales

A different way of associating a diffusion to wave equation (5.17) is the following: note that

$$c^2(z)\frac{\mathrm{d}^2(\cdot)}{\mathrm{d}z^2} = c(z)\frac{\mathrm{d}}{\mathrm{d}z}\left(c(z)\frac{\mathrm{d}}{\mathrm{d}z}(\cdot)\right) - \left(\frac{\mathrm{d}c}{\mathrm{d}z}\right)c(z)\frac{\mathrm{d}}{\mathrm{d}z}(\cdot)$$

and if we introduce the change of variables

$$\zeta(z)=\int_0^z \frac{\mathrm{d}z^1}{c(z^1)} \quad \text{or} \quad \int_z^0 \frac{\mathrm{d}z^1}{c(z^1)}$$

depending on whether, $z>0$ or $z<0$, respectively. This is well defined since $0<c_1\leqslant c(z)\leqslant c_2$. Defining $\tilde{w}(\alpha,\zeta)=\tilde{u}(\alpha,z(\zeta))$ and $2b(\zeta)=(\frac{\partial c}{\partial z})(z(\zeta))$, we obtain

$$c^2(z)\frac{\mathrm{d}^2\tilde{u}}{\mathrm{d}z^2}=\left(\frac{\mathrm{d}^2}{\mathrm{d}\zeta^2}-2b(\zeta)\frac{\mathrm{d}}{\mathrm{d}\zeta}\right)\tilde{w}$$

and, setting $\mathcal{L}=\frac{1}{2}\frac{\mathrm{d}^2}{\mathrm{d}\zeta 2}-b(\zeta)\frac{\mathrm{d}}{\mathrm{d}\zeta}$, we can write the path integral solutions to

$$\left(\frac{\alpha^2}{2}-\mathcal{L}\right)\tilde{w}(\alpha,\zeta)=\frac{1}{2}\widetilde{S}(\alpha,\zeta)$$

in terms of the diffusion having $\mathcal{L}$ as an infinitesimal generator.

There are two possibilities, depending on whether we consider the problem of solving (5.17) for $-\infty<z<\infty$ or for $0\leqslant z<\infty$.

In the first case, we have to solve

$$\mathrm{d}Z(t)=\mathrm{d}B(t)-b(Z(t))\,\mathrm{d}t \tag{5.30}$$

and in the second case, we have to solve

$$\mathrm{d}Z(t)=\mathrm{d}B(t)-b(Z(t)+I_{\{0\}}(Z(t))\,\mathrm{d}L_t. \tag{5.31}$$

For both cases we need the standard assumptions on $c(z)$ to insure that $b(z)$ is such that $Z(t)$ exists and is uniquely determined.

In terms of $Z(t)$ we can write the solution to $(\frac{\alpha^2}{2}-\mathcal{L})\tilde{w}(\alpha,\zeta)=\widetilde{S}(\alpha,\zeta)/2$ as the following path integral

$$\tilde{w}(\alpha,\zeta)=\frac{1}{2}\widehat{E}^{\zeta}\int_0^{\infty}\mathrm{e}^{-\frac{\alpha^2}{2}t}\tilde{s}(\alpha,Z(t))\,\mathrm{d}t. \tag{5.32}$$

To transform this further, invoke the Cameron–Martin–Girsanov theorem to obtain

$$\tilde{w}(\alpha,\zeta)=\frac{1}{2}E^{\zeta}\int_0^{\infty}\mathrm{e}^{-\int_0^t b(B(s))\cdot\mathrm{d}B(s)-\frac{1}{2}\int_0^t(b(B(s)))^2\mathrm{d}s}\tilde{s}(\alpha,B(t))\,\mathrm{d}t. \tag{5.33}$$

In (5.32) $\widehat{E}^{\zeta}$ denotes the measure on the class $(\Omega,\mathcal{F})$ of continuous paths obtained from the Wiener measure P^{ζ} on $(\Omega,\mathcal{F})$ as indicated in Section 4 of Chapter 2.

To deal with the reflected case, i.e., to solve $(\frac{\alpha^2}{2}-\mathcal{L})\tilde{w}(\alpha,\zeta)=\widetilde{S}(\alpha,\zeta)/2$ subject to the condition $\frac{\partial\tilde{w}}{\partial\zeta}(\alpha,0+)=0$, we have to consider (5.31).

Before writing down a path integral representation for $\tilde{w}(\alpha,\zeta)$, note that we can write (5.31) as

$$\mathrm{d}Y(t) = \mathrm{d}W(t) + I_{\{0\}}(Y(t))\,\mathrm{d}L_t^0, \tag{5.34}$$

where now $W(t) = B(t) - \int_0^t b(Y(s))\,\mathrm{d}s$ is a Brownian motion with respect to the measure $\widetilde{P}^\zeta$ with the density $\mathrm{d}\widetilde{P}^\zeta/\mathrm{d}P^\zeta = \exp\{\int_0^t b(Y(s))\cdot\mathrm{d}B(s) - \frac{1}{2}\int_0^t b(Y(s))^2\,\mathrm{d}s\}$ when restricted to $\mathcal{F}_t$. Again, P^ζ is the Wiener measure on the space of continuous paths (issued from ζ).

What is interesting about (5.34) is that, in this presentation, $Y(t)$ is equivalent in law to $|W(t)|$ and we can write

$$\tilde{w}(\alpha, y) = \frac{1}{2}\widetilde{E}^\zeta\left\{\int_0^\infty \mathrm{e}^{-\frac{\alpha^2}{2}t}\widetilde{S}(\alpha, |W(t)|)\,\mathrm{d}t\right\}. \tag{5.35}$$

5.4.4. *Waves in multidimensional stratified media*

We examine now the problem consisting in obtaining path integral representations for the solutions to

$$\frac{\partial^2 u}{\partial t^2}(t,x) = c^2(z)\Delta u + s(t,x), \quad t > 0,\ x = (x^1, x^2, z) \in D \tag{5.36}$$

with vanishing initial conditions, $D = \mathbb{R}^3$ or $D = \mathbb{R}^3_t = \{(x^1, x^2, z) : z \leqslant 0\}$. In this last case, we supplement (5.37) with

$$\frac{\partial u}{\partial z}(t, x^1, x^2, 0) = 0. \tag{5.37}$$

There are several ways of dealing with this problem. The first, and similar to what we did in the first sections, consists in first taking Fourier transforms of $u(t,x)$ with respect to (x^1, x^2) and then the Laplace transform with respect to t, to transform (15) into

$$\left(\frac{1}{2}\left(\frac{\alpha^2}{q}(z) + k^2\right) - \frac{1}{2}\frac{\mathrm{d}}{\mathrm{d}z^2}\right)\tilde{u}(\alpha, k, z) = \widetilde{S}(\alpha, k, z)$$

supplemented with

$$\frac{\partial \tilde{u}}{\partial z}(\alpha, k, 0) = 0$$

if need be. The solution to this equation can be written as either of the path integrals

$$\tilde{u}(\alpha, k, z) = E^z \int_0^\infty \mathrm{e}^{-F(t)}\widetilde{S}(\alpha, k, B(t))\,\mathrm{d}t, \tag{5.38}$$

$$\tilde{u}(\alpha, k, z) = E^z \int_0^\infty \mathrm{e}^{-F(t)}\widetilde{S}(\alpha, k, |B(t)|)\,\mathrm{d}t, \tag{5.39}$$

where

$$F(t) = \frac{1}{2}\left(\alpha^2 \int_0^t q(B(s))\,\mathrm{d}s + tk^2\right).$$

A different approach, involving the processes $Z(t)$ introduced in the previous section, consists in changing variables to obtain

$$\frac{\partial^2 w}{\partial t^2} = \frac{\partial^2 w}{\partial \zeta^2} - b(\zeta)\frac{\partial w}{\partial \zeta} + c^2(\zeta)\Delta^1 w + s(t, \bar{x}, \zeta), \tag{5.40}$$

where $\bar{x} = (x^1, x^2)$, $\Delta^1 = \partial_{x_1}^2 + \partial_{x_2}^2$. At this point we can do two things. We can now transform Fourier and Laplace to obtain

$$\left\{\frac{\alpha^2}{2} - \left(\mathcal{L} - \frac{k^2}{2}c^2(\zeta)\right)\right\}\tilde{w}(\alpha, k, \zeta) = \frac{1}{2}\tilde{s}(\alpha, k, \zeta)$$

which has the path integral representation in terms of the process $Z(t)$ that solves either of (5.30) or (5.31)

$$\tilde{w}(\alpha, k, \zeta) = \frac{1}{2}\widehat{E}^{\zeta}\int_0^\infty \mathrm{e}^{-K(t)}\tilde{s}(\alpha, K, Z(t))\,\mathrm{d}t, \tag{5.41}$$

where now $K(t) = \frac{\alpha^2}{2}t + \frac{k^2}{2}\int_0^t c^2(Z(s))\,\mathrm{d}s$.

The other approach, that does not involve taking Fourier transforms, consists in taking $\frac{c(Z)^2}{2}\Delta^1$ as the infinitesimal generator of the two-dimensional, time changed Brownian motion $B(\tau(t))$ where $\tau(t) = \int_0^t c^2(Z(s))\,\mathrm{d}s$ is independent of $B(t) = (B^1(t), B^2(t))$.

We can then represent the solution to the Laplace transform $\tilde{w}(\alpha, \bar{x}, \zeta)$ as

$$\tilde{w}(\alpha, \hat{x}, z) = \frac{1}{2}\widehat{E}^{(\bar{x},\zeta)}\int_0^\infty \mathrm{e}^{-\frac{\alpha^2}{2}t}\tilde{s}(\alpha, Z(t), B(\tau(t)))\,\mathrm{d}t, \tag{5.42}$$

where $\widehat{E}^{(x,\zeta)}$ denotes integration with respect to $\mathrm{d}P^{\bar{x}}\mathrm{d}\widehat{P}^{\zeta}$. If we were to write $\tilde{s}(\alpha, \bar{x}, \zeta) = \int \mathrm{e}^{-ik\cdot\bar{x}}\tilde{s}(\alpha, k, \zeta)\frac{\mathrm{d}k}{(2\pi)^2}$ and to use $E^{\bar{x}}[\mathrm{e}^{-ik\cdot B(\tau(t))}] = \mathrm{e}^{ik\cdot x}\mathrm{e}^{-\frac{k^2}{2}\tau(t)}$, (5.42) would become

$$\tilde{w}(\alpha, \bar{x}, \zeta) = \int_{\mathbb{R}^2} \mathrm{e}^{-ik\cdot\bar{x}}\widehat{E}^{\zeta}\int_0^\infty \mathrm{e}^{-\frac{\alpha^2}{2}}\mathrm{e}^{-\frac{k^2}{2}\tau(t)}\widetilde{S}(\alpha, k, z(t))\,\mathrm{d}t$$

which is (5.41).

5.4.5. Examples

Example 1: To check whether everything reduces to the known results, the first example consists in examining the trivially stratified medium; $c(z) = c =$ constant. In this case, as we saw above, in the course of the proof of Lemma 6, $W_\alpha(a, z) = \exp\{-\frac{\alpha}{c}|x - y|\}$ and thus $J(\alpha, z) = 1/\alpha$. Now (5.1) and everything else reduces to what we found in Chapter 4.

Example 2: Consider now, for $z \geqslant 0$, $c(z) = c_1 + c_0 z/\ell = c_1 + \beta z$, with ℓ having units of length. In the notation of Subsection 4.2 $b(z) = \beta/2$ and $\zeta(z) = \frac{1}{\beta}\ln(1+\gamma z)$ with $\gamma = \beta/c_1$.

If we introduce $W(t) = B(t) - bt$, then $Z(t) = |B(t) - bt| = W(t)$ is a reflected imposed motion with constant drift, and it is easy to verify that

$$E^{\zeta}\left[e^{ikZ(t)}\right] = \int e^{ik\zeta^1}\left(\frac{1}{\sqrt{2\pi t}}\right)\{e^{-(\zeta+\zeta^1-bt)^2/2t} + e^{-(\zeta-\zeta^1+bt)^2/2t}\}\,d\zeta^1$$

and therefore $\rho_\gamma(t,\zeta,\bar{\zeta}^1) = \frac{1}{\sqrt{2\pi t}}\{e^{-(\zeta+\bar{\zeta}^1-bt)^2/2t} + e^{-(\zeta-\bar{\zeta}^1+bt)^2/2t}\}$ satisfies

$$\frac{\partial \rho_r}{\partial t}(t,\zeta,\bar{\zeta}^1) = \frac{1}{2}\frac{\partial^2}{\partial\bar{\zeta}^2}\rho_r + b\frac{\partial}{\partial\bar{\zeta}}\rho_r$$

with

$$\left.\frac{\partial\rho_r}{\partial\zeta}\right|_{\zeta=0} = 0, \qquad \rho_r(0+,\zeta,\zeta^1) = \delta(\zeta-\zeta^1).$$

The corresponding equations in the z-variables are somewhat messier to write down, only because we have to take adjoints with respect to the scalar product

$$(f,g) = \int_0^\infty f(z)g(z)\,dz/c(z).$$

So let us complete the computation in the ζ-variables.

According to (5.36), we know that

$$\begin{aligned}
\tilde{w}(\alpha,\zeta) &= \frac{1}{2}\widehat{E}^{\zeta}\int_0^\infty e^{-\frac{\alpha^2 t}{2}}\tilde{s}(\alpha,Z(t))\,dt \\
&= \frac{1}{2}\int_0^\infty e^{-\frac{\alpha^2 t}{2}}\left\{\int_0^\infty \rho_r(t,\zeta,\zeta^1)\tilde{s}(\alpha,\zeta^1)\,d\zeta^1\right\} \\
&= \frac{1}{2}\int_0^\infty \tilde{s}(\alpha,\zeta^1)\frac{1}{\alpha}\{e^{-\alpha|\zeta+\zeta^1-bt|} + e^{-\alpha|\zeta+\zeta^1-bt|}\}\,d\zeta^1
\end{aligned}$$

from which $w(t,\zeta)$ can be obtained as in Chapter 4, and therefore $u(t,z) = u(t,\zeta(z))$.

Let us now consider some examples in which $w_\alpha(a,z)$ can be found explicitly, basically because a change of variables brings the equations satisfied by $w_\alpha(a,z)$ into hypergeometric form.

Example 3: Consider a medium in which $c(z)$ is such that

$$q(z) = 1/c^2(z) = k_1^2 - \frac{k_1^2 - k_2^2}{1+e^{\beta z}}, \quad k_1 > k_2 > 0,\ \beta > 0.$$

Write now $\zeta(z) = -\mathrm{e}^{-\beta z}$ and put $s = \alpha k_1/\beta$. Let $W(z)$ be any function satisfying

$$\frac{\mathrm{d}^2 W}{\mathrm{d}z^2} - \alpha^2 q(z) W = 0,$$

and write $W(z) = -(\zeta(z))^s V(\zeta(z))$. Then $V(\zeta)$ satisfies

$$(1-\zeta)\frac{\mathrm{d}^2 V}{\mathrm{d}\zeta^2} + (2s+1)(1-s)\frac{\mathrm{d}V}{\mathrm{d}\zeta} - \left(k_1^2 - k_2^2\right)V = 0 \tag{5.43}$$

the solution to which can be expressed in terms of the hypergeometric function

$$F(p,q,r,\zeta) = \sum_{n\geqslant 0} \frac{(p)_n (q)_n}{(r)_n} \frac{\zeta^n}{n!},$$

where $(p)_n = p(p-1)\cdots(p-1+n)$. In our case, the solution to (5.43), decaying to zero as $n \to +\infty$, is

$$w(z) = \mathrm{e}^{-\beta s z} F\left(\frac{\alpha}{\beta}(k_1 - k_2), \frac{\alpha}{\beta}(k_1 + k_2), 2k_1\frac{\alpha}{\beta} + 1, -\mathrm{e}^{\beta z}\right).$$

To obtain $W_\alpha^>(a,z)$ and $W_\alpha^<(a,z)$, we change signs appropriately and normalize so that $W_\alpha^>(a,a) = W_\alpha^<(a,a) = 1$. Thus,

$$W_\alpha^>(a,z) = \mathrm{e}^{-\beta s(z-a)} F\left(\rho, q, r, \mathrm{e}^{-\beta(z)}\right) / F\left(\rho, q, r, \mathrm{e}^{-2a}\right)$$

for $a < z < +\infty$, and also

$$W_\alpha^<(a,z) = \mathrm{e}^{\beta s(z-a)} F\left(p, q, r, \mathrm{e}^{-2\beta}\right) / F\left(p, q, r, -\mathrm{e}^{2a}\right)$$

for $-\infty < z < a$. Here we set $p = \frac{\alpha}{p}(k_1 - k_2)$, $q = \frac{\alpha}{\beta}(k_1 + k_2)$, $\gamma = 2k_1\frac{\alpha}{\beta} + 1$ shortened script.

To find $J(\alpha, a)$, it does help to use the identity

$$\frac{\mathrm{d}F}{\mathrm{d}\zeta}(p,q,r,\zeta) = \frac{pq}{r} F(p+1, q+1, r+1, \zeta)$$

and we leave details for the (valiant) interested reader.

Example 4: Another velocity function leading to an integrable equation for $w_\alpha(a,z)$ is given by

$$q(z) = 1/c^2(z) = k_1^2 + k_2^2/\cosh\beta^2, \quad \beta > 0.$$

This time the proposed change of variables $\zeta(z) = \operatorname{tgh} \beta z$ and $w(z) = (1-\zeta)^{\varepsilon/z} V(\zeta)$ yields

$$(1-\zeta^2)^2 \frac{\mathrm{d}V}{\mathrm{d}\zeta^2} - 2\zeta(1-\zeta^2)(\varepsilon+1)\frac{\mathrm{d}V}{\mathrm{d}\zeta} + \left[\varepsilon^2\zeta^2 - \left(\frac{\alpha k_1}{\beta}\right)^2 - \frac{\alpha k_1^2}{\beta}(1-\zeta^2)\varepsilon\right] V = 0.$$

Take ε to be the positive root of $\varepsilon^2 + \varepsilon + (\frac{\alpha}{\beta}k_1)^2$ and put $2y = \zeta + 1$, to obtain, for $R(y) = V(\zeta(y))$, the equation

$$(1-y)yR'' + (\varepsilon - 1)(1-2y)R' - \left[\varepsilon^2 - \left(\frac{\alpha}{\beta}k_2\right)^2\right] R = 0$$

which can be solved in terms of $F(1, p, r, y)$ of $r = \varepsilon + 1$, $p + q = 2\varepsilon + 1$, $pq = \varepsilon^2 - (\frac{\alpha}{\beta}k_2)^2$.

Since $q(z) = q(-z)$, once we have $W_\alpha^>(a, z)$, we obtain $W_\alpha^<(a, z) = W_\alpha^>(-a, -z)$ for $z < 0$. From this we obtain $J(\alpha, a)$ which, when inserted in

$$\tilde{u}(\alpha, z) = \frac{1}{2}\int q(a)\tilde{s}(\alpha, z)\frac{W_\alpha(a, z)}{J(\alpha, a)}\,da,$$

can be used to obtain $u(t, z)$.

Note that in two last the examples, $q(x) \to$ constant $= q_0$ as $\beta \to 0$.

Finding what the corresponding limit is from the involved hypergeometric functions is a bit cumbersome, while

$$E^z\left[\exp\left\{-\frac{\alpha^2}{2}A(\tau_a^0)\right\}\right] \to \exp\{-\alpha/c_0|z-a|\}$$

can be verified easily. We set $q_0 = 1/c_0$.

5.5. Maxentropic equivalent linearization and approximate solutions to the wave equations

The procedure referred to in the title consists in producing a measure $\widetilde{P}$ on the space $(\Omega, \mathcal{F})$ of continuous trajectories, which is equivalent to the Wiener measure P^0 on the space of continuous trajectories issued from the origin, with respect to which the solutions to the (nonlinear) stochastic differential equation,

$$\mathrm{d}X_t = \sigma\,\mathrm{d}B_t + b(X_t)\,\mathrm{d}t, \qquad X(0) = x, \tag{5.44}$$

where $b(x)$ is continuous and bounded, is transformed into the solutions of a linear stochastic differential equation like

$$\mathrm{d}Y_t = \left\{\alpha_t + \beta_t(Y_t - m_t)\right\}\mathrm{d}t + \sigma\,\mathrm{d}B_t, \qquad Y(0) = x. \tag{5.45}$$

The time dependence of the coefficients in (b) can be thought of as a local linearization of (5.44) along the trajectory. In (5.45) $\alpha(t)$, $\beta(t)$, $m(t)$ are non-random functions determined in a self-consistent way (a bootstrapping procedure), indeed, from

$$\begin{aligned} \alpha(t) &= \int b(y)q_t(y)\,\mathrm{d}y, \\ \beta(t) &= \int b(y)(y-m(t))q_t(y)\,\mathrm{d}y \Big/ \int (y-m(t))^2 q_t(y)\,\mathrm{d}y, \end{aligned} \tag{5.46}$$

where $q_t(y)$ is a Gaussian density having the form

$$q_t(y) = \frac{1}{(2\pi V(t))^{1/2}} \exp\left(-\frac{(y-m(t))^2}{2V(t)}\right)$$

and $m(t)$, $V(t)$ are related to $\alpha(\beta)$, $\beta(t)$ by

$$\dot{m}(t) = \alpha(t), \qquad \dot{V}(t) = 2\beta(t)V(t) + \sigma^2. \tag{5.47}$$

To obtain the measure $\widetilde{E}$, with respect to which $Y(t)$ is distributed according to $q_t(Y)$, we proceed as follows. Recall from Section 2.2 of Chapter 3, that the solution to (5.45) induces a measure $\widehat{P}$ on $(\Omega, \mathcal{F})$, a measure equivalent to P^0, such that, restricted to $\mathcal{F}_t$, has the density

$$N(t) = \exp \int_0^t b(X(s))\,\mathrm{d}B(s) - \frac{1}{2}\int_0^t b^2(X(s))\,\mathrm{d}s,$$

where the generic $X(t)$ to denote the coordinate map $X(t)(w) = w(t)$.

It is actually a simple computation to verify that the measure $\widetilde{P}$ on $(\Omega, \mathcal{F})$ having the density

$$\left.\frac{\mathrm{d}\widetilde{P}}{\mathrm{d}\widehat{P}}\right|_{\mathcal{F}_t} = \exp \int_0^t (\tilde{b}(X_s)) - b(X_s))\,\mathrm{d}B(s) - \frac{1}{2}\int_0^t (\tilde{b}(X_s) - b(X_s))\,\mathrm{d}s \tag{5.48}$$

makes the coordinate functions $X(t)$ be a diffusion on $\mathbb{R}$ with the infinitesimal generator $\frac{1}{2}\frac{\mathrm{d}^2}{\mathrm{d}x^2} + \tilde{b}(x)\frac{\mathrm{d}}{\mathrm{d}x}$, that is, they can be thought of as the solutions to (5.45) with $\tilde{b}(x)$ replacing $b(x)$. To linearize, take $\tilde{b}(x) = \alpha + \beta(x-m)$ and define the concave entropy functional

$$S(\widetilde{P};\widehat{P}) = -\int \frac{\mathrm{d}\widetilde{P}}{\mathrm{d}\widehat{P}} \ln \frac{\mathrm{d}\widetilde{P}}{\mathrm{d}\widehat{P}}\,\mathrm{d}\widehat{P}. \tag{5.49}$$

Actually, we do not need $S(\widetilde{P}, \widehat{P})$ to be defined on the class

$$P = \left\{\widetilde{P} \ll \widehat{P} : \int \left|\ln \frac{\mathrm{d}\widetilde{P}}{\mathrm{d}\widehat{P}}\right| \mathrm{d}\widetilde{P} < \infty\right\}.$$

What we care for is $\tilde{b}(x)$ yielding the largest value for the entropy (5.49). If $S(\widetilde{P}, \widehat{P})$ is computed for $\widehat{P}$ such that (5.49) holds, we have, for each $T > 0$,

$$2c^2 S(\widetilde{P}, \widehat{P}) = \int_0^t \widetilde{E}\big[b(X_t) - (\alpha + \beta(y-m))^2\big]\,\mathrm{d}t.$$

If we assume α, β, and m to be deterministic and differentiate the identity above with respect to α and β (or rather, take Frechet derivatives), we obtain (5.46) easily if we choose $m = \widetilde{E}y(t)$. Since the solution $y(t)$ turns out to be a Gaussian process, its distribution is like $q_t(y)$ given above.

To obtain (5.47), we have to use (5.45) written as

$$Y(t) = Y(0) + \int_0^t \{\alpha(s) + \beta(s) - m(s)\}\,\mathrm{d}s + \sigma B(t).$$

Taking the expected values with respect to $\widetilde{P}$, we obtain

$$m(t) = \widetilde{E}(Y(t)) = \int_0^t \alpha(s)\,\mathrm{d}s,$$

and Ito's formula yields

$$\begin{aligned}(Y(t) - m(t))^2 &= (Y(0) - m(0))^2 + 2\int_0^t (Y(s) - m(s))\,\mathrm{d}(Y(s) - m(s)) \\ &\quad + \int_0^t \sigma^2\,\mathrm{d}s \\ &= 2\int_0^t (y(s) - m(s))^2 \beta(s)\,\mathrm{d}s + 2c\int_0^t (y(s) - m(s))\,\mathrm{d}B(s) + \sigma^2.\end{aligned}$$

And taking expectations, we obtain

$$V(t) = 2\int_0^t \beta(s)V(s)\,\mathrm{d}s + \sigma^2 t.$$

If $b(x)$ is such that $\alpha(m, v)$ and $\partial\alpha/\partial m$ have continuous bounded derivatives with respect to m, and v, we can devise a successive approximation scheme to solve (4) as follows: Pick m_0 and v_0 and define recursively

$$\begin{aligned}\dot{m}_{n+1}(t) &= \alpha(m_n(t), V_n(t)), \\ \dot{V}_{n+1} &= 2V_n(t)m_n(t)\left(\frac{\partial\alpha}{\partial m}\right)(\alpha_n(t), v_n(t)) + \sigma^2,\end{aligned} \tag{5.50}$$

where we used the fact that $\beta = \partial\alpha/\partial m$. For example, if we took $v_0 = 0$, then $\alpha(m_0, v_0) = b(m_0)$ and thus

$$m_1(t) = m(0) + b(m_0)t,$$

$$v_1(t) = t\sigma^2 + v_1(0),$$

etc. The two identities we used above to derive (5.48) suggest that we put $m(0) = 0$ and $v(0) = 0$ as initial conditions at every step. The difficulty, from the computational point of view, lies in being able to carry out the integral in definition (5.48) of α, or, equivalently,

$$\alpha(m, v) = \int \hat{b}(k) \mathrm{e}^{-ikm(t)} \mathrm{e}^{-\frac{k^2}{2}v(t)} \, \mathrm{d}k/2\pi.$$

This approach can be coupled with the change of variables, introduced in Section 4.3, to obtain approximate solutions to (5.31). We leave for the interested reader to carry out such a program.

References

For material in Section 3 check with.

1. Bertrand, J. and Gaveau, B.: Transformation canonique et renormalization pour certaines equations d'evolution, *J. Funct. Anal.* **50** (1983), 81–99.

For most of this chapter, the stuff in Chapters 1 and 2 is a must. Exact integration of the Dirac equation plus the example in Section 5 appears in

2. Bagrov, V. G. and Gitman, D. M.: *Exact Solutions of Relativistic Wave Equations*, Kluwer Acad. Publ., Dordrecht, 1990.

For the maxentropic linearization procedure check with

3. Ito, H.: An entropy-minimum aspect of the equivalent linearization of a non-linear stochastic differential equation, *Prog. Theoret. Phys.* **66** (1981), 337–339.

CHAPTER

SIX

ASYMPTOTIC EXPANSIONS

6.1. Digressive introduction

We have already seen that the probabilistic approach makes explicit the connection between the Markov property (for Markov processes) and the Huygens construction (for waves).

In this chapter, in order to carry out the probabilistic analogue of the high frequency asymptotic expansion detailed in Chapter 3, we make explicit the connection between geometric optics, Hamilton–Jacobi equations and the (probabilistic version of) classical limit in quantum mechanics.

We want to examine the behavior of the Laplace transform $\tilde{u}(\alpha, x)$ of the solution $u(t, x)$ to the wave equation for large values of α. The connection being that the change $\alpha \to -iw$ transforms $\tilde{u}(\alpha, x)$ into $\hat{u}(w, x)$: the Fourier component of $u(t, x)$ at the frequency w.

Even though we could be more general, for the sake of concreteness, in the applications we consider throughout the case $x \in \mathbb{R}^3$.

We begin again with our old friend

$$\partial_t^2 U(t, x) = c^2(x)\Delta U + S(t, x), \quad t > 0,\ x \in \mathbb{R}^3$$

and assume $U(0, x) = 0, \partial_t u(t, x) = 0$. We write $c(x) = c_0/n(x)$ where c_0 is some appropriate reference speed and $0 < n_1 \leqslant n(x) \leqslant n_2 < \infty$ describes the refractive index of the medium. We shall introduce a new variable $x_0 = c_0 t$ (the units of which are light years or sound seconds) and our equation will be

$$\frac{\partial^2}{\partial x_0^2} u(x_0, x) = \frac{1}{n(x)^2} \Delta u(x_0, x) + \frac{S(t, x)}{c_0^2} x > 0, \quad x \in \mathbb{R}^3.$$

After taking Laplace transforms in x_0, this becomes

$$\left(\frac{\alpha}{2} - \frac{1}{2\alpha n(x)^2}\Delta\right)\tilde{u}(\alpha, x) = \widetilde{S}(\alpha, x)/2\alpha c_0^2 \equiv \widetilde{S}'(\alpha, x), \tag{6.1}$$

where now α is measured in units of 1/length. Now, in terms of the diffusion $X^\alpha(t)$ with the infinitesimal generator $(1/2\alpha n^2(x))\Delta$, which solves the stochastic equation

$$\mathrm{d}Z^\alpha(t) = \frac{1}{n(Z^\alpha)\sqrt{\alpha}}\mathrm{d}B(t),$$

we can represent $\tilde{u}(\alpha, x)$ as

$$\tilde{u}(\alpha, x) = E^x \int_0^\infty \mathrm{e}^{-\frac{\alpha}{2}t}\widetilde{S}'(\alpha, Z^\alpha(t))\,\mathrm{d}t. \tag{6.2}$$

Keep bearing in mind that the "time" variables appearing in (6.2) also have units of 1/length but have nothing to do with the x_0 introduced above. Rewriting (6.2) as

$$\tilde{u}(\alpha, x) = \int_{\mathbb{R}^3} g_\alpha(x, y)\widetilde{S}'(\alpha, y)\,\mathrm{d}y,$$

we end up thinking of (6.2) as the path integral representation for the Green function $g_\alpha(x, y)$ of (6.1). In order to further analyze (6.2), we rewrite it as

$$\tilde{u}(\alpha, x) = \int_0^\infty \mathrm{e}^{-\frac{\alpha}{2}t}E^x\big[\widetilde{S}'_\alpha(\alpha, Z^\alpha(t))\big]\,\mathrm{d}t = \int_0^\infty \mathrm{e}^{-\frac{\alpha}{2}t}V_\alpha(t, x)\,\mathrm{d}t, \tag{6.3}$$

where $V_\alpha(t, x) = E^x[\widetilde{S}'(\alpha, Z^\alpha(t))]$ satisfies on $t > 0, x \in \mathbb{R}^3$:

$$\frac{\partial V_\alpha}{\partial t}(t, x) = \frac{1}{2\alpha n(x)^2}\Delta V_\alpha(t, x), \qquad V_\alpha(0, x) = \widetilde{S}'(\alpha, x). \tag{6.4}$$

We can propose the standard expansion in α either for $V_\alpha(t, x)$ or for $\widetilde{U}(\alpha, x)$. Let us examine what is involved in relating both of them.

If we substitute

$$V_\alpha(t, x) = \mathrm{e}^{-\alpha\Sigma(t,x)}\sum_{k\geqslant 0}^{\infty}\varphi_k(t, x)/\alpha^k \tag{6.5}$$

in (6.4) and identify equal powers of α, we obtain the recursive set

$$\frac{\partial\Sigma}{\partial t} + \frac{1}{2n(x)^2}(\nabla\Sigma)^2 = 0, \tag{6.6}$$

$$\frac{\partial\varphi_0}{\partial t} + \frac{1}{2n^2}(2\nabla\varphi_0\cdot\nabla\Sigma + \varphi_0\Delta\Sigma) = 0, \tag{6.7}$$

$$\frac{\partial \varphi_k}{\partial t} + \frac{2\nabla\varphi_k \cdot \nabla\Sigma + \varphi_k \Delta\Sigma}{2n(x)^2} = \frac{\Delta\varphi_{k-1}}{2n(x)}, \quad k \geqslant 1. \tag{6.8}$$

In order to establish initial conditions for this recursive family, we must place assumptions on $S(t,x)$. If we admit that $S(t,x) = \delta(t - \tau(x))h(x)$, then

$$V_\alpha(0,x) = \widetilde{S}'(\alpha,x) = \frac{e^{-\alpha\tau(x)}h(x)}{2\alpha c_0^2} H(\tau(x)) = e^{-\alpha\Sigma(0,x)} \sum_{k\geqslant 0} \frac{\varphi_k(0,x)}{\alpha^k}$$

implies that

$$\Sigma(0,x) = \tau(x), \qquad \varphi_0(0,x) = 0, \qquad \varphi_1(0,x) = h(x)H(\tau(x))/2c_0^2, \tag{6.9}$$

where $H(x)$ is the usual Heaviside function.

If we assume that $\Sigma(t,x) = \tau(x) - t/2$ and substitute (6.5) in (6.3) we obtain

$$\tilde{u}(\alpha,x) = e^{-\alpha\tau(x)} \sum_{k\geqslant 0}^{\infty} \varphi_k(x)/\alpha^k, \tag{6.10}$$

where $\varphi_k(x) = \int_0^\infty \varphi(t,x)\,dt$. We have to take for granted the existence of these integrals at this point. Integrating each of the equations in the family (6.7)–(6.8) with respect to time, we obtain

$$(\nabla\tau(x))^2 = n^2(x), \tag{6.11}$$

$$2\nabla\varphi_0 \cdot \nabla\tau(x) + \varphi_0\Delta\tau(x) = 0, \tag{6.12}$$

$$2\nabla\varphi_1 \cdot \nabla\tau + \varphi_1\Delta\tau = n^2(x)\frac{h(x)}{c^2} + \Delta\varphi_0, \tag{6.13}$$

$$2\nabla\varphi_k \cdot \nabla\tau + \varphi_k\Delta\tau = \Delta\varphi_{k-1}, \quad k \geqslant 2. \tag{6.14}$$

We leave for the reader to compare with Section 3 of Chapter 3 and verify that, if we set $f_0(u) = \delta(u)$ in (3.52), we obtain the corresponding results of this section if we identify $\tau(x)$ there with the $\tau(x)$ introduced here.

All this is very similar to what is done in elementary quantum mechanics to obtain classical mechanics as Planck's constant h tends to zero. To see how some of this comes about (we shall need some of it below anyway), consider the flow in $\mathbb{R}^3$ defined by

$$\frac{dx}{dt}(t) = \frac{\nabla\tau(x(t))}{n^2(x(t))}, \qquad x(0) = x. \tag{6.15}$$

This equation will have a solution for all values of t when, for example, its right-hand side has bounded derivatives with respect to all variables.

Note also that on account of (6.11), the right-hand side of (6.15) is a vector normal to the surfaces $\{x : \tau(x) = \tau_0\}$, $-\infty < \tau_0 < +\infty$. Not only that, it is easy to verify that

$$\tau(x(t)) = \tau(x) - t. \tag{6.16}$$

To integrate the transport equations (6.12)–(6.14), consider the flow

$$\phi : \mathbb{R} \times \mathbb{R}^3 \to \mathbb{R}^3,$$

where $\phi(t, x) = x(t)$ is the solution to (6.7), which on account of uniqueness satisfies

$$\phi(t, \phi(s, x)) = \phi(t + s, x), \quad t, s \in \mathbb{R}, x \in \mathbb{R}^3.$$

Also consider the surfaces

$$D_a = \{x : \tau(x) = a\}$$

and note that $\phi(t, \cdot)$ maps $D_0 = \{x : \tau(x) = 0\}$ on to D_t. Also, from (6.16), if $x \in D_t$, $\phi(-\tau(x), x) \in D_0$ (this is true for both positive and negative values of t). The generic transport equation looks like

$$\nabla\psi \cdot \frac{\nabla\tau}{\eta^2} + \psi g = f \tag{6.17}$$

which, after computing along $x(t)$ and multiplying both sides by $\exp\int_0^t g(x(s))\,\mathrm{d}s$, becomes

$$\frac{\mathrm{d}}{\mathrm{d}t}\left(\psi(x(t))\exp\int_0^t g(x(s))\,\mathrm{d}s\right) = f(x(t))\exp\int_0^t g(x(s))\,\mathrm{d}s$$

which, after integration and solving for $\psi(x(t))$, yields

$$\psi(\phi(t, x)) = \psi(x)\exp\left\{-\left(\int_0^t g(\phi(s, x))\,\mathrm{d}s\right)\right\} + \int_0^t f(\phi(s, x))\,\mathrm{e}^{\int_s^t g(\phi(t', x))\,\mathrm{d}t'}.$$

To get rid of the running time t, we proceed as follows: First replace x by $\phi(-t, x)$ to obtain

$$\begin{aligned}
\psi(x) &= \psi(\phi(-t, x))\exp\left\{-\int_0^t g(\phi(s - t, x))\,\mathrm{d}s\right\} \\
&\quad + \int_0^t \phi(s - t, x)\exp\left\{-\int_0^t g(\phi(t' - t, x))\,\mathrm{d}t'\right\}\mathrm{d}s \\
&= \psi(\phi(-t, x))\exp\left\{-\int_0^t g(\phi(-s, x))\,\mathrm{d}s\right\} \\
&\quad + \int_0^t f(\phi(-s, x))\exp\left\{-\int_0^t g(\phi(-t', x))\,\mathrm{d}t'\right\}\mathrm{d}s.
\end{aligned}$$

Now refer to a fixed surface D_0. For that replace t by $\tau(x)$ to obtain

$$\psi(x) = \psi_0(x)\exp\left\{-\int_0^{\tau(x)} g(\phi(-s,x))\,\mathrm{d}s\right\} + \int_0^{\tau(x)} f(\phi(-s,x))\exp\left\{-\int_0^{s} g(\phi(-t',x))\,\mathrm{d}t'\,\mathrm{d}s\right\}, \tag{6.18}$$

where $\psi_0(x) = \psi(\phi(-s(x),x))$ depends on the values of ψ on D_0 only. Also note that, if x varies on D_t, then $\phi(-\tau(x),x)$ varies on D_0, and if x varies along a ray, $\psi_0(x)$ remains constant. With this it is really easy to verify that (6.18) is indeed a solution to (6.17)

Comment: Implied in all these computations is the fact that $\tau(x) \geqslant 0$. Some words about this bellow. The application of the previous results to the set (6.11), (6.14) is contained in

LEMMA 6.1. *Assume that there exsists a twice continuously differentiable function $\tau(x)$ satisfying* (6.14) *and $\nabla\tau(x)/n^2(x)$ is a continuously differentiable vector field, having bounded derivatives, so that the flow $\phi(t,x)$ described by* (6.15) *exists for all t. Assume that the values of $n(x)$ are known on $T_0 = \{x : \tau(x) = 0\}$, in particular, $\varphi_0(x) = 0$, according to* (6.9). *Then the solutions to the recursive set* (6.12)–(6.14) *are given by* (6.18) *with $g(x) = \nabla\varphi_{n-1}(x)/2n^2(x)$ and $f(x) = h(x)H(\tau(x))/2c_1^2$ or $f(x) = \nabla\varphi_{n-1}(x)/2n^2(x)$.*

Comment: Note that, if $\{x : \tau(x) < 0\} = D_<$ is not empty and $\varphi_k(x) = 0$ on $D_0 = \{x : \tau(x) = 0\}$, then $\tilde{u}(\alpha,x) = 0$ on D. The nicest interpretation occurs when T_0 reduces to a single point x_0.

To close down this section, let us recall some very basic identities about Hamiltonian mechanics, just to have a different way of thinking about equations (3.59)–(3.61) in Section 4 of Chapter 3, which describe the method of characteristics to find a solution to (6.11). To further understand (6.15), first note that it implies $n^2\dot{x}^2 = 1$ when (6.11) is taken into account. Consider

$$\frac{\mathrm{d}}{\mathrm{d}t}\left(n^2(x)\dot{x}\right) = \dot{x}\cdot\nabla\frac{\partial\tau}{\partial x_i}, \quad i = 1,2,3. \tag{6.19}$$

To evaluate the right-hand side, differentiate $(\nabla\tau)^2 = n^2$ to obtain $2\frac{\partial\tau}{\partial x_k}\frac{\partial}{\partial x_k}(\frac{\partial\tau}{\partial x_i}) = \frac{\partial}{\partial x_i}n^2$ and, using (6.15) and (6.11) again, we obtain

$$\dot{x}\cdot\nabla\frac{\partial\tau}{\partial x_i} = \frac{\dot{x}^2}{2}\frac{\partial}{\partial x_i}n^2.$$

If we introduce the (Lagrangian) function $L(x,\dot{x}) : \mathbb{R}^3 \times \mathbb{R}^3 \to \mathbb{R}$ by

$$L(x,\dot{x}) = \frac{1}{2}n^2(x)\dot{x}^2$$

in which x and $\dot{x}$ are thought to be independent variables, equations (6.11)–(6.14) can be written as

$$\frac{\mathrm{d}}{\mathrm{d}t}(\nabla_{\dot{x}}\mathcal{L}) = \nabla_x\mathcal{L}, \tag{6.20}$$

which are the Euler–Lagrange equations of some variational problem. Another way of obtaining (6.15) and (6.19) consists in looking for the curve $x(t)$ which makes

$$I_1(\gamma) = \int_0^t L(\gamma, \dot{\gamma})\,\mathrm{d}s - \tau(\gamma(t))$$

a minimum. $I(\gamma)$ is defined over the class of continuously differentiable curves $\gamma(t)$ such that $\gamma(0) = x$. The value of γ at t, and t itself are left to vary.

The curve that minimizes $I(\gamma)$ satisfies

$$\frac{\mathrm{d}}{\mathrm{d}t}(\nabla_{\dot{x}}\mathcal{L}) = \nabla_x\mathcal{L}$$

subject to

$$x(0) = x, \qquad \dot{x}(t) = \frac{\nabla\tau}{n^2(x)}(x).$$

Had we considered

$$I_2(\gamma) = \int_0^t L(\gamma, \dot{\gamma})\,\mathrm{d}s,$$

we would have obtained, for the minimizing curve $x_0(t)$, the set of equations

$$\frac{\mathrm{d}}{\mathrm{d}s}\left(n^2\dot{x}_0\right) = \frac{\dot{x}_0^2}{2}\nabla n^2(x_0)$$

subject to

$$x_0(0) = x, \qquad \dot{x}_0(t) = -\frac{\nabla\tau(x_0)}{n^2(x_0)}.$$

This equation can be obtained from the previous one by time reversal. The problem now is to provide a sensible interpretation of $I_1(\gamma)$ and $I_2(\gamma)$.

Had we wanted to obtain equation (6.11) from the same variational principle, we should have considered the functionals

$$I(\gamma) = \int_0^t L(\gamma, \dot{\gamma})\,\mathrm{d}s \mp \tau(\gamma(t)) + \frac{t}{2}$$

and t to be free.

6.2. Probabilistic approach to geometrical optics

Even though more is required to obtain the same results of Section 1, the probabilistic approach will reveal its use in the next section. There we treat the case of propagation in full space, and our starting point will be (6.3), namely,

$$u(\alpha, x) = \int_0^\infty e^{-\frac{\alpha}{2}t} E^x[\widetilde{S}'(\alpha, z^\alpha H)]\, dt.$$

When we consider a source term like above, i.e., $S(t, x) = \delta(t - \tau(x))h(x)$, we have

$$\widetilde{S}(\alpha, x) = e^{-\alpha\tau(x)} h(x) H(\tau(x))/2\alpha c_0^2.$$

If we denote $T_0 = \inf\{t > 0 : Z^\alpha(t) \in D_<\} = \inf\{t > 0 : Z^\alpha(t) \in D_0\}$ (due to the continuity of the trajectories of $Z^\alpha(t)$ and the fact that $\partial D_< = D_0$), we can rewrite $\tilde{u}(\alpha, x)$ as

$$\tilde{u}(\alpha, x) = E^x \left\{ \int_0^{T_0} e^{\frac{-\alpha}{2}t} - e^{\alpha\tau(Z^{\alpha(t)})} h(z^\alpha(t))\, dt \right\} / 2\alpha c_0^2. \tag{6.21}$$

Note, now, that

$$\frac{dt}{2} + d\tau(Z^\alpha(t)) = \frac{1}{2}\left(\frac{\nabla\tau}{n}\right)^2 (Z^\alpha(t))\, dt + \nabla\tau(Z^\alpha(t))\, dZ^\alpha(t) + \frac{1}{2}\alpha\left(\frac{\nabla\tau}{n^2}\right)(Z^\alpha(t))\, dt$$

which allows us to rewrite $\tilde{u}(\alpha, x)$ as

$$\tilde{u}(\alpha, x) = e^{-\alpha\tau(x)} E^x \left\{ \int_0^{T_0} M(t) e^{-\frac{1}{2}\int_0^t g(Z^\alpha(s))\, ds} h(Z^\alpha(t))\, dt \right\} / 2\alpha c_0^2$$

with $M(t) = \exp\{-\int_0^t \sqrt{\alpha}\frac{\nabla\tau}{n} \cdot dB^\alpha(s) + \frac{\alpha}{2}\int_0^t (\frac{\nabla\tau}{n})^2\, dt\}$ and $g(x) = \nabla\tau(x)/n^2(x)$. Introducing the process $X^\alpha(t)$ by the solution of the stochastic differential equation

$$\begin{aligned} dX^\alpha(t) &= -\frac{\nabla\tau}{n^2}(X^\alpha(t))\, dt + dZ^\alpha(t) \\ &= -\frac{\nabla\tau}{n^2}(X^\alpha(t)) + \frac{1}{\sqrt{\alpha}n(X^\alpha(t))^2}\, dB(t). \end{aligned} \tag{6.22}$$

Invoke Cameron–Martin–Girsanov–Maruyama to rewrite $\widetilde{u}(\alpha, x)$ as

$$\widetilde{u}(\alpha, x) = e^{\alpha\tau(x)} E^x \int_0^{T_0^\alpha(0)} h(X^\alpha(t)) e^{-\frac{1}{2}\int_0^t g(X^\alpha(s))\, ds}\, dt / 2\alpha c_0^2, \tag{6.23}$$

where $\widetilde{P}^x$ denotes the measure induced by $X^\alpha(t)$.

If we denote $\nabla\tau(x)/n^2(x)$ by $b(x)$ and $1/n^2(x)$ by $\sigma(x)$, it is a well known fact that, if the components $b_i(x)$ and $\sigma(x)$ have continuous bounded derivatives with respect to all variables up to order $k+1$, then the solution to (6.22) can be expanded as

$$X^\alpha(t) = X^0(t) + \frac{1}{\sqrt{\alpha}}X'(t) + \cdots + \left(\frac{1}{\sqrt{\alpha}}\right)^k X^{(k)}(t) + R^\varepsilon_{k+1}(t), \tag{6.24}$$

where for any interval $[0, t_0]$

$$\sup_{0\leqslant t\leqslant t_0} \left(E^x[|R^\varepsilon_{k+1}(t)|]^2\right)^{1/2} \leqslant c\left(\frac{1}{\sqrt{\alpha}}\right)^{k+1}$$

and $X^{(0)}, X', \ldots$ are obtained solving the recursive set

$$\mathrm{d}X^0(t) = -b(X^0(t))\,\mathrm{d}t, \qquad X^0(0) = x, \tag{6.25}$$

$$\mathrm{d}X^1(t) = -b(X^0(t))X^1(t)\,\mathrm{d}t + \sigma(X^0(t))\,\mathrm{d}B(t), \qquad X^1(0) = 0, \tag{6.26}$$

$$\begin{aligned}\mathrm{d}X^{(2)} + b(X^0(t))x^2(t) = {} & \frac{1}{2}\big(X^{(1)}(t), bt\text{'}t\text{'}(X^0(t))X^{(1)}(t)\big)\,\mathrm{d}t \\ & + \sigma'(x^0(t))X^{(1)}(t)\,\mathrm{d}B(t), \qquad X^{(2)}(0) = 0,\end{aligned} \tag{6.27}$$

etc. Here $(y, bt\text{'}t\text{'}(x)y)$ denotes $\sum y_k y_e \partial^2_{ke} b_i(x)$ and $\sigma'(x)y\,\mathrm{d}B = y_k\partial_k\sigma_{ij}\,\mathrm{d}B_j$. Anyway, to obtain this recursive set, one just substitutes (6.24) into (6.22), expands and collects coefficients of some powers of $1/\sqrt{\alpha}$.

The following preliminaries will allow us to state the general result from which the asymptotic behavior of (6.3) drops out as α becomes large.

Consider the following elliptic boundary problem

$$\frac{1}{2\alpha n^2(x)}\Delta w(\alpha, x) - b(x)\cdot\nabla w(\alpha, x) - q(x)w(\alpha, x) = -q(x) \tag{6.28}$$

for $x \in D$, with the boundary condition

$$w(\alpha, x) = \psi(x), \quad x \in \partial D. \tag{6.29}$$

We know that, in terms of $X^\alpha(t)$ satisfying (6.22), the solution to (6.28) can be written explicitly as

$$\begin{aligned} w(\alpha, x) = {} & E^x\big[\mathrm{e}^{-\int_0^{T(D)} g(X^\alpha(s))\,\mathrm{d}s}\psi[X^\alpha(T(D))]\big] \\ & + E^x\left[\int_0^{T(D)} S(X^\alpha(t))\mathrm{e}^{-\int_0^t g(X^\alpha(s))\,\mathrm{d}s}\,\mathrm{d}t\right],\end{aligned}$$

where $T(D) = \inf\{t > 0 : X(t) \in D^e\}$. We assume that $S(x)$ is bounded in D, $\psi(x)$ continuous and bounded ∂D, is every point in ∂D is regular for D^c, and, of course, that $g(x) \geqslant 0$. When $q = 0$ in D, we assume that $P^x(T(D) < \infty) = 1$.

Also, we assume that, if the solution $X^0(t)$ of (6.24) ever leaves D, it does it regularly, i.e., $T_0(x) = \inf\{t : X^0(t) \notin D\} < \infty$ and $X^0(T_0(x) + \delta) \notin \overline{D}$ for sufficiently small $\delta > 0$. It is known that

THEOREM 6.1. *Under the assumptions stated above, if the trajectory $X^0(t)$ never leaves D, then*

$$w(\infty, x) = \int_0^\infty s(X^0(t))\mathrm{e}^{-\int_0^t q(X^0(s))\,\mathrm{d}s}\,\mathrm{d}t.$$

If the trajectory of $X^0(t)$ leaves D regularly, then

$$\begin{aligned}\lim_{\alpha\to\infty} w(\alpha, x) = w(\infty, x) &= \psi\big(X^0(T_0(x))\big)\mathrm{e}^{-\int_0^{T_0(x)} q(X^0(s))\,\mathrm{d}s}\\ &\quad + \int_0^{T_0(x)} S(X^0(t))\mathrm{e}^{-\int_0^t q(X^0(s))\,\mathrm{d}s}\,\mathrm{d}t.\end{aligned}$$

For all of this the reader should consult the book by Freidlin.

To adapt Theorem 6.1 to our situation, D will be what we should call

$$D_> = \{x : \tau(x) > 0\},$$

and $T_0(x)$ will correspond to our old friend $\tau(x)$. Thus, for very large α

$$\begin{aligned}\tilde{u}(\alpha, x) &\approx \frac{\mathrm{e}^{-\alpha\tau(x)}}{2\alpha c_0^2}\lim_{\alpha\to\infty} E^x\left\{\int_0^{T(0)} h(X^\alpha(t))\mathrm{e}^{-\int_0^t q(X^\alpha(s))\,\mathrm{d}s}\right\}\\ &= \frac{\mathrm{e}^{-\alpha\tau(x)}}{2\alpha c_0^2}\int_0^{\tau(x)} h(X^0(t))\mathrm{e}^{-\int_0^t q(X^0(s))\,\mathrm{d}s}\end{aligned} \tag{6.30}$$

as expected from Section 1, if we set $\varphi_0(x) = 0$ on D_0.

6.3. Geometrical optics and the Dirichlet problem

We want to study the "high frequency" behavior of waves in a region D smaller than the whole space. It could be a bounded or unbounded region, but such that $T_D = \inf\{t > 0 : X(t) \in D^c\}$ satisfies $P^x(T_D < \infty) = 1$ for every x in D.

This time we are led to consider equation (6.1) again, i.e.,

$$\left(\frac{\alpha}{2} - \frac{1}{2\alpha n(x)^2}\Delta\right)\tilde{u}(\alpha, x) = \widetilde{S}'(\alpha, x) \tag{6.31}$$

supplemented with the boundary conditions

$$\tilde{u}(\alpha, x) = \tilde{g}(\alpha, x), \quad x \in \partial D.$$

Recall that, if ∂D is regular and $\tilde{g}(\alpha, x)$ is continuous on ∂D, then $\tilde{u}(\alpha, x)$ is continuous up to ∂D, and the meaning of the boundary condition is $\lim \tilde{u}(\alpha, y) = g(\alpha, x)$ as $y \in D \to \in \partial D$. In most applications ∂D is smooth (has a continuously differentiable normal) which makes it "quite regular".

Without further ado, the solution to (6.1) can be represented by the following path integrals

$$\tilde{u}(\alpha, x) = E^x\big[\mathrm{e}^{-\frac{\alpha}{2}T(0)}\tilde{g}(Z^\alpha, (T_\alpha))\big] + E^x\bigg[\int_0^{T_\alpha} \mathrm{e}^{-\frac{\alpha}{2}t}\widetilde{S}'(\alpha, Z^\alpha(t))\,\mathrm{d}t\bigg], \tag{6.32}$$

where now E^x denotes the expectation with respect to P^x induced on the space $(\Omega, \mathcal{F})$ by the diffusion $Z^\alpha(t)$, introduced in Section 1. In (6.32) $T_\alpha = \inf\{t >: Z^\alpha(t) \in D\}$.

So above, if we want $\tilde{u}(\alpha, x)$ to be of type $\mathrm{e}^{-\alpha\tau(z)}\Phi(\alpha, x)/\alpha$, with $\Phi(\alpha, x)$ a series in $(1/\alpha)$, we have to consider a boundary term $g(t, x)$ of the type

$$g(t, x) = \delta(t - \tau(x))g(x)H(\tau(x))$$

which yields

$$\tilde{g}(\alpha, x) = \mathrm{e}^{-\alpha\tau(x)}g(x)H(\tau(x)).$$

Again, and as above, the presence of the Heaviside function $H(\tau(x))$ reflects the fact that we are measuring time starting at $t = 0$. The routine to compute each of the terms in (6.30) is pretty much as above. To begin with the first term, consider

$$\begin{aligned}\tau(x^\alpha(t)) + \frac{1}{2}T_\alpha &= \tau(z^\alpha(0)) + \int_0^{T_\alpha} (\nabla\tau)(Z^\alpha(s)) \cdot \mathrm{d}Z_s^\alpha \\ &\quad + \frac{1}{2}\int_0^{T_\alpha} \left(\frac{\nabla\tau}{n}\right)^2 (Z_s^\alpha)\,\mathrm{d}s + \frac{1}{2^\alpha}\int_0^{T_\alpha} \left(\frac{\Delta\tau}{n^2}\right)(Z_s^\alpha)\,\mathrm{d}s\end{aligned}$$

and therefore, bringing in the Cameron–Martin change of variables once more

$$\begin{aligned}&E^x\big[\mathrm{e}^{-\frac{\alpha}{2}T_\alpha}\tilde{g}(\alpha, Z^\alpha(T_\alpha))\big] \\ &\quad = \mathrm{e}^{-\alpha\tau(x)}\widetilde{E}\big[\mathrm{e}^{-\frac{1}{2}\int_0^{T_\alpha} q(X^\alpha(s))\,\mathrm{d}s}g(\widetilde{X}^\alpha(T_\alpha))H(\tau(X^\alpha(T_\alpha)))\big],\end{aligned} \tag{6.33}$$

the notation being as in Section 2.

Now, as $\alpha \to \infty$, $X^\alpha(t) \to X^0(t)$ and $T_\alpha \to T_0$ as explained above, and we have to add the assumption that $X^0(t)$ leaves D regularly. Since $H(\tau(X^0(T^0))) = 1$ if and only if $X^0(T^0) \in D_> = \{x : \tau(x) > 0\}$, for large α the dominant term in (3) is

$$\mathrm{e}^{-\alpha\tau(x)}\mathrm{e}^{-\frac{1}{2}\int_0^{T_0} q(X^0(s))\,\mathrm{d}s}g(X^0(T^0))I_{D_>}(X^0(T^0)). \tag{6.34}$$

We leave for the reader to proceed as in Section 2 and verify that

$$E^x\left[\int_0^{T_\alpha} \mathrm{e}^{-\frac{\alpha}{2}T_\alpha}\widetilde{S}(\alpha, X_s)\,\mathrm{d}s\right] \sim \frac{\mathrm{e}^{-\alpha\tau(x)}}{2\alpha c_0^2}\int_0^{T_0\wedge\tau(x)} \mathrm{e}^{-\int_0^t q(X_s^0)\,\mathrm{d}s} h(X^0(t))\,\mathrm{d}t. \tag{6.35}$$

Putting (6.33) and (6.34) together, we obtain the first term in the large α ("high frequency") asymptotic expansion of (6.32).

6.4. Two variations on a theme

This section is concerned with rewriting the results of Section 2 in a way that allows an easy approach to the problem of obtaining the "high frequency" approximation to waves in a half-space with a reflecting boundary, and hopefully, we end up understanding further the way probability enters this game.

6.4.1. The first variation on the theme

Instead of considering (6.1), we consider now

$$\left(\frac{\alpha^2}{2}n^2(\alpha) - \frac{1}{2\alpha}\Delta\right)\tilde{u}(\alpha, x) = \widetilde{S}'(\alpha, x), \tag{6.36}$$

where this time (and until further notice) $\widetilde{S}(\alpha, x) = n^2(x)\tilde{s}(\alpha, x)/2\alpha c_0^2$, with $s(\alpha, x)$ as in Sections 1 and 2. Instead of the process $Z_\alpha(t)$ we consider

$$B_\alpha(t) = B(0) - \frac{1}{\sqrt{\alpha}}(B(t) - B(0)),$$

in terms of which the solution to (6.34) can be written as

$$\tilde{u}(\alpha, x) = E^x\int_0^\infty \mathrm{e}^{-\frac{\alpha}{2}A(t)}\widetilde{S}'(\alpha, B_\alpha(t))\,\mathrm{d}t, \tag{6.37}$$

where $A(t) = \int_0^t n^2(B_\alpha(s))\,\mathrm{d}s$. To rewrite the exponent, consider

$$\begin{aligned}
&\tau(B_\alpha(t)) + \frac{1}{2}\int_0^t n^2(B_\alpha(s))\,\mathrm{d}s \\
&\quad = \tau(B(0)) + \int_0^t \nabla\tau(B_\alpha(s))\cdot \mathrm{d}B_\alpha(s) \\
&\qquad + \frac{1}{2}\int_0^t (\nabla\tau)^2(B_\alpha(s))\,\mathrm{d}s + \frac{1}{2\alpha}\int_0 \Delta\tau(B_\alpha(s))\,\mathrm{d}s.
\end{aligned}$$

If we now invoke the Cameron–Martin formula of changing variables, and skip one step, (6.36) can be rewritten as

$$u(\alpha,x)=\frac{\mathrm{e}^{-\alpha\tau(x)}}{2\alpha c_0^2}E^x\int_0^\infty \mathrm{e}^{-\int_0^t q(Y_s^\alpha)\,\mathrm{d}s}g\big(Y_s^\alpha\big)H\big(\tau(Y_s^\alpha)\big)\,\mathrm{d}s, \tag{6.38}$$

where again $q(x)=\Delta\tau(x)/2$ and Y_s^α is the diffusion that satisfies

$$\mathrm{d}Y^\alpha(t)=-(\nabla\tau)(Y^\alpha(s))\,\mathrm{d}t+\frac{1}{\sqrt{\alpha}}\mathrm{d}B(t),\qquad Y^\alpha(0)=x, \tag{6.39}$$

which looks simpler than (2.2). Now, set

$$T^\alpha(D)=\inf\{t>0:Y^\alpha(t)\in D_<\}=\inf\{t>0:Y^\alpha(t)\in\partial D_<=D_0\}$$

and $T_0(x)$ will be obtained by solving

$$\tau(x)=\int_0^T n^2(Y_0(s))\,\mathrm{d}s, \tag{6.40}$$

where $\mathrm{d}Y_0(s)=-(\nabla\tau)(Y_0^\alpha(s))\,\mathrm{d}s$. Thus, to be consistent, we should have written $T_0(x)$ as $T^\infty(D)$ and pointed out that $T^\alpha(D)\to T_0(x)=T^\infty(D)$ in probability as $\alpha\to\infty$. Note that, since $\tau(Y_0(s))=\tau(x)-\int_0^t n^2(Y_0(s))\,\mathrm{d}s$, it is clear that $T_0(x)$ is the hitting time of D_0 by Y_0. Thus, as α becomes large, (6.37) becomes

$$\tilde{u}(\alpha,x)=\frac{\mathrm{e}^{-\alpha\tau(x)}}{2\alpha c_0^2}\int_0^{T_o(x)} n^2(Y_0(t))h(y_0(t))\mathrm{e}^{-\int_0^t q(Y_0(s))\,\mathrm{d}s}\,\mathrm{d}t. \tag{6.41}$$

To relate to (6.29), we perform the following time change along the trajectories of $Y_0(t)$. Define

$$\sigma(t)=\int_0^t n_2(Y_0(s))\,\mathrm{d}s \tag{6.42}$$

and note that, since $n(x)$ is strictly positive, $\sigma(t)$ can be inverted to obtain $t=t(\sigma)$ such that $\mathrm{d}t/\mathrm{d}\sigma=1/n^2(Y_0t(\sigma))$. If we now set $X_0(\sigma)=Y_0(t(\sigma))$, then

$$\frac{\mathrm{d}}{\mathrm{d}\sigma}X_0(\sigma)=-\frac{(\nabla\tau)(X_0(\sigma))}{n^2(X_0(\sigma))}$$

as well as

$$\tau(X_0(\sigma))=\tau(x)-\sigma.$$

Therefore

$$\sigma(T_0(x))=\tau(x) \tag{6.43}$$

and we leave it as homework for the reader to use (6.41) and (6.42) to transform (6.40) into (6.29).

6.4.2. The second variation on the theme

Assume that as $|x| \to \infty$, $c(x) \to c_0$ our reference speed. It means that $n(x) \to 1$ as $|x| \to \infty$. We write $n^2(x) = 1 + V(x)$ and $V(x)$ will be bounded and tend to zero as $|x| \to \infty$. Now rewrite (6.34) as

$$\left(\frac{\alpha}{2} - \left(\frac{1}{2\alpha}\Delta - \frac{\alpha}{2}V(x)\right)\right)\tilde{u}(\alpha, x) = \widetilde{S}'(\alpha, x). \tag{6.44}$$

Denote by Q_t^x the semigroup (go back to Chapter 1 if need be)

$$Q_t^\alpha f(x) = E^x\left\{e^{-\frac{\alpha}{2}\int_0^t V(B_\alpha(s))\,ds} f(B_\alpha(t))\right\}$$

for bounded $f(x)$. Q_t^α will be a transition semigroup when $V(x) \geqslant 0$. Its infinitesimal generator is $\Delta/2\alpha - \alpha V(x)/2$. We can write the solution to (6.43) as

$$\tilde{u}(\alpha, x) = \int_0^\infty e^{\frac{-\alpha t}{2}} Q_t^\alpha \widetilde{S}'(\alpha, x)\,dt = \int_0^\infty E^x\left[e^{-\alpha F(t)}\widetilde{S}'(\alpha, B_\alpha(t))\right] dt, \tag{6.45}$$

where $F(t) = \frac{t}{2} + \frac{1}{2}\int_0^t V(B_\alpha(s))\,ds = \frac{1}{2}\int_0^t n^2(B_\alpha(s))\,ds$.

Instead of the usual procedure, at this point we refer the reader to Varadhan who proves that performing

$$E^x\left[e^{-\alpha F(t)} f(B_\alpha(t))\right] \approx f(\hat{X}_0(t))e^{-I(\hat{X}_0(t))}, \tag{6.46}$$

where $\hat{X}_0(t)$ is the curve minimizing the functional

$$I(\gamma) = \int_0^t \left\{\frac{\dot{\gamma}}{2} + \frac{V(\gamma)}{2}\right\} ds + \tau(\gamma(t)) + \frac{t}{2}$$

or $+\infty$ if $\dot{\gamma}(t)$ does not exist. Note that

$$I(\gamma) = \int_0^t \left\{\frac{\dot{\gamma}^2}{2} + \frac{n^2(\gamma)}{2}\right\} ds + \tau(\gamma(t)) \tag{6.47}$$

and now the results in Section 2 can be seen in the light of large deviation theory. As α becomes large, of all possible curves contributing to (6.45), are those that lie close to the curve minimizing (6.46).

6.5. Geometrical optics and the Neumann problem

Now we want to study the "high frequency" behavior of the solutions to

$$\frac{\partial^2 u}{\partial t^2}(t,x) = c^2(x)\Delta u(t,x) + s(t,x), \quad t>0, x\in D$$

supplemented with the boundary condition

$$\frac{\partial u}{\partial x_3}(t,x_1,x_2,0) = 0, \quad t>0.$$

Here $D=\{x : x_3>0\}$. After taking Laplace transforms, we obtain

$$\left(\frac{1}{2}n^2(x) - \frac{1}{2\alpha}\Delta\right)\tilde{u}(\alpha,x) = \widetilde{S}'(\alpha,x)$$

supplemented with

$$\frac{\partial \tilde{u}}{\partial x_3}(\alpha,x_1,x_2,0) = 0,$$

where, as in the previous section, $\widetilde{S}'(\alpha,x) = n^2(x)\tilde{s}(\alpha,x)/2\alpha c_0^2$.

We already know several different ways of representing the solution to this problem in terms of path integrals in several ways. In particular, if we introduce the reflected Brownian motion $B^r(t) = (B^1_\alpha(t), B^2_\alpha(t), |B^3_\alpha(t)|)$, we can write

$$\tilde{u}(\alpha,x) = E^x\int_0^\infty e^{-\frac{\alpha}{2}A(t)}\widetilde{S}'(\alpha, B^r_\alpha(t))\,dt. \tag{6.48}$$

Now, if we extend $n(x)$ and $\tilde{s}(\alpha,x)$ to $\mathbb{R}^3$ by reflection on ∂D, and denote the even extensions by $n_e(x)$ and $\tilde{s}_e(\alpha,x)$, respectively, we rewrite (6.46) as

$$\tilde{u}(\alpha,x) = E^x\int_0^\infty e^{-\frac{\alpha}{2}A_e(t)}\widetilde{S}'_e(\alpha, B_\alpha(t))\,dt. \tag{6.49}$$

By extending to functions which are invariant under reflection, we can consider $B_\alpha(t)$ instead of $B_\alpha(t)$.

We denote by $\tau_e(x)$ the even solution to $(\nabla\tau_e)^2 = n_e^2$, and assume its existence.

The behavior of (6.49), as α becomes large, is contained in the little.

THEOREM 6.2. *With the same notation, and under the same assumptions as in Section* 4, *the asymptotic behavior of* (2), *as α becomes large, is given by*

$$\tilde{u}(\alpha,x) = e^{-\alpha\tau(x)}\int_0^{T_0(x)} e^{-\int_0^t q(x_0(s))\,ds} g(x_0(t)) H\left(x_0^3(t)\right) dt$$

$$+\,\mathrm{e}^{-\alpha\tau(Rx)}\int_0^{T_0(Rx)} \mathrm{e}^{-\int_0^t q(x_0^R(s))} g\big(x_0^R(t)\big) H(-x_0(t))\,\mathrm{d}t. \tag{6.50}$$

PROOF (Sketch). Denote by $R:\mathbb{R}^3\to\mathbb{R}^3$ the reflection $x\to Rx=(x_1,x_2,-x_3)$, and by R as well the induced map in the space of continuous trajectories $(R\omega)(t)=R(\omega(t))$. It is easy to verify that, for $H\in b\mathcal{L}$ and $x\in\mathbb{R}^3$,

$$E^x[H\circ R]=E^{Rx}[H]$$

(this is the invariance of imposed motion under reflection). To verify it, use induction on H of the type $H=\prod_1^n f_i\,(B_\alpha(t_i))$, $0\leqslant t<t_2<\cdots$, etc. Now rewrite (6.48) as

$$\begin{aligned}\tilde{u}(\alpha,x) &= E^x\int_0^\infty \mathrm{e}^{-\frac{\alpha}{2}A_e(t)}\widetilde{S}'_e(\alpha,B_\alpha(t))H\big(B_\alpha^3(t)\big)\,\mathrm{d}t\\ &\quad+E^x\int_0^\infty \mathrm{e}^{-\frac{\alpha}{2}A(t)}\widetilde{S}'_e(\alpha,B_\alpha(t))H\big(-B_\alpha^3(t)\big)\,\mathrm{d}t.\end{aligned}$$

Next, invoke the previous comment to rewrite the second term on the right-hand side as

$$\begin{aligned}&E^x\int_0^\infty \big\{\mathrm{e}^{-\frac{\alpha}{2}A_e(t)}\widetilde{S}'_e(\alpha,\cdot)H(\cdot)\big\}\cdot R\,\mathrm{d}t\\ &=E^{Rx}\int_0^\infty \mathrm{e}^{\frac{1}{2}A_e(t)}S'_e(\alpha,B(t))H\big(B_t^3\big)\,\mathrm{d}t.\end{aligned}$$

To complete the preliminaries, note that, if $X_0(t)$ minimizes,

$$\int_0^t \frac{1}{2}\big\{\dot{\gamma}(t)^2+n^2(\gamma(s))\big\}\,\mathrm{d}s+\tau_e(\gamma(t)) \tag{6.51}$$

and since for any vector field $v(x)$ on $\mathbb{R}^3$, $V(Rx)=R(V(x))$, it follows from the invariance of (6.48) that $R(X_0(t))=X_0^R(t)$ minimizes (6.51) as well, and satisfies $X_0^R(t)=-\nabla\tau_e(X_0^R(t))$!

Invoke now the results of Section 4 to obtain (6.50).

6.6. Example

To begin with, we obtain the asymptotic behavior of $\tilde{u}(\alpha,x)$ in dimensions 1 and 3 when the speed of propagation is constant and equals 1.

We know that, in this case,

$$\begin{aligned}\tilde{u}(\alpha,x) &= \int_{-\infty}^\infty \frac{\mathrm{e}^{-\alpha|x-y|}}{\alpha}\widetilde{S}(\alpha,y)\,\mathrm{d}y\\ &=\frac{1}{2\alpha}\int_{-\infty}^\infty \mathrm{e}^{-\alpha\{|x-y|+|y-x_0|\}}g(y)\,\mathrm{d}y\end{aligned}$$

when $\widetilde{S}(\alpha, x) = \mathrm{e}^{-\alpha\tau(x)} g(x)$ and $\tau(x) = |x - x_0|$. We assume that $x < x_0$ and split the integral into 3 pieces

$$\begin{aligned}\tilde{u}(\alpha, x) &= \frac{1}{2\alpha}\left\{\int_{-\infty}^{x} \mathrm{e}^{-\alpha(x+x_0-2y)} g(y)\,\mathrm{d}y + \mathrm{e}^{-\alpha(x_0-x)} \int_{x}^{x_0} g(y)\,\mathrm{d}y\right. \\ &\quad \left. + \int_{x_0}^{\infty} \mathrm{e}^{-\alpha(2y-(x+x_0))} g(y)\,\mathrm{d}y\right\}.\end{aligned}$$

A simple change of variables allows us to rewrite the first and last terms as

$$\frac{\mathrm{e}^{-\alpha(x_0-x)}}{4\alpha^2}\int_0^{\infty} \mathrm{e} - yg\left(x + \frac{y}{\alpha}\right)\mathrm{d}y \sim \frac{\mathrm{e}^{-\alpha(x_0-x)}}{4\alpha^2} g(x)$$

from which we obtain that

$$\tilde{u}(\alpha, x) \sim \frac{\mathrm{e}^{-\alpha|x-x_0|}}{2\alpha}\int_0^{|x-x_0|} g(x+s)\,\mathrm{d}s \tag{6.52}$$

plus the terms $\mathrm{O}(\mathrm{e}^{-\alpha|x-x_0|}/\alpha^2)$. This is (6.30) if you fill in the details.

The three-dimensional case: This time the starting point is

$$\begin{aligned}\tilde{u}(\alpha, x) &= \frac{1}{2}\int \frac{\mathrm{e}^{-\alpha|x-y|}}{2\pi|x-y|}\widetilde{S}(\alpha, y)\,\mathrm{d}y \\ &= \frac{1}{4\pi}\int \frac{\mathrm{e}^{-\alpha\{|x-y|+|x_0-y|\}}}{|x-y|} g(y) H(\tau(y))\,\mathrm{d}y,\end{aligned}$$

where now $\tau(x) = |x - x_0|$. We locate the origin at the point $(x_0 + x)/2$ and orient the axes so that the 1-axis is along $(x_0 - x)$ and choose the other two coordinate in a plane through $(x_0 + x)/2$ perpendicular to $(x_0 - x)$.

Now $|x_0 - x| + |x_0 - y|$ becomes

$$F(y_1, \eta) = \left|(y_1 + a)\frac{(x_0, z)}{|x_0, x|} + \eta\right| + \left|(y_1 - a)\frac{(x_0 - x)}{(x_0 - x)} + \eta\right|$$

with $\eta \perp (x_0 - x)$. For α large, we are interested in the quadratic approximation to $F(y, \eta)$

$$F(y_1, \eta) = |y_1 + a| + |y_1 - a| + \frac{1}{2\sigma}\eta^2,$$

where $a = |x - x_0|/2$ and $1/\sigma = |y_1 - a|^{-1} + |y_1 + a|^{-1}$.

With all this, and putting

$$g_1(\eta) = g\left(\frac{x_0 + x}{2} + \frac{(x_0 - x)}{|x_0 - x|} y_1 + \eta\right) \Big/ \left|(y_1 + a)\frac{(x_0 - x)}{|x_0 - x|} + \eta\right|$$

we rewrite the last integral as

$$\begin{aligned}\tilde{u}(\alpha,x) &= \frac{1}{4\pi}\int \mathrm{e}^{-\alpha\{|y_1+a|+|y_1-a|\}}\left(\int \mathrm{e}^{-\frac{\alpha}{2\sigma}\eta^2} g_1(\eta)\,\mathrm{d}\eta\right)\mathrm{d}y_1\\ &= \frac{1}{2}\int_{-\infty}^{\infty} \mathrm{e}^{-\alpha\{|y_1+a|+|y_1-a|\}}\sigma \int \frac{\mathrm{e}^{-\eta^2/2}}{\mathrm{e}\pi} g_1\left(\sqrt{\frac{\sigma}{\alpha}}\eta\right)\mathrm{d}\eta\,\mathrm{d}y_1\end{aligned}$$

and, considering the limit of the inner integral as $\alpha \to \infty$, we obtain

$$\begin{aligned}\tilde{u}(\alpha,x) &= \frac{1}{2\alpha}\int_{-\infty}^{\infty} \mathrm{e}^{-\alpha\{|y_1+a|+|y_1-a|\}}\frac{|y_1+a|+|y_1-a|}{|y_1-a|}\\ &\quad\times g\left(\frac{x_0+x}{2}+\frac{(x_0-x)}{|x_0-x|}y_1\right)\mathrm{d}y_1.\end{aligned}$$

Now invoke the results for the one-dimensional case to obtain

$$\tilde{u}(\alpha,x) = \frac{\mathrm{e}^{-\alpha a}}{2\alpha}\int_{-a}^{a}\frac{|y_1-a|}{|y_1-a|+|y_1+a|}g\left(\frac{x_0+x}{2}+\frac{(x_0-x)}{|x_0-x|}y_1\right)\mathrm{d}y_1,$$

shift y_1+a to t and obtain

$$\tilde{u}(\alpha,x) = \frac{\mathrm{e}^{-\alpha|x-x_0|}}{2\alpha}\int_0^{|x-x_0|}\frac{|x_0-x|-s}{|x_0-x|}g\left(x+\frac{(x_0-x)}{|x_0-x|}s\right)\mathrm{d}s. \tag{6.53}$$

To compare with (6.29), note that $\tau(x) = |x-x_0|$ yields $\Delta\tau(x) = \frac{2}{|x-x_0|}$ and $\tau(X_0(s)) = |x-x_0|-s$ and that $X_0(t) = x+\frac{(x_0-x)}{|x_0-x|}t$ solves $X_0(t) = -\nabla\tau(X_0)$ with $X_0(0)=x$.

The following example will be incorporated as part of the one after it.

Assume that $\tau_p(x) = p\cdot x$ with $p^2=1$, $c(x)=1$, and that we are in $\mathbb{R}^3$, then

$$u(\alpha,x) = \frac{1}{2}\int\frac{\mathrm{e}^{-\alpha|x-y|}}{2\pi|x-y|}\mathrm{e}^{-\alpha 2\alpha n^2(x)p\cdot x}g(y)H(p\cdot y)\,\mathrm{d}y.$$

Write $x = (x\cdot p)p + x_1$ with $x_1 \perp p$, shift the origin to x and orient the axes so that the 1-axis lies along p, then rewrite $\tilde{u}(\alpha,x)$ as

$$\tilde{u}(\alpha,x) = \frac{\mathrm{e}^{-\alpha p\cdot x}}{4\pi}\int\frac{\mathrm{e}^{-\alpha(|ap+\eta|-a)}}{|ap+\eta|}g(x+ap+\eta)H(xp+a)\,\mathrm{d}a\,\mathrm{d}\eta,$$

where we set $y = ap+\eta$, $\eta\perp p$ in the new coordinate system. The integration with respect to η is over $\mathbb{R}^2$. Since we only care for large α's we expand $|ap+\eta|-a$ about $\eta=0$ to obtain $|ap+\eta|-a = |a|-a-\frac{1}{2|\alpha|}\eta^2$, and therefore

$$\tilde{u}(\alpha,x) = \frac{\mathrm{e}^{-\alpha p\cdot x}}{2}\int \mathrm{e}^{-(|a|-a)}\left(\int\frac{\mathrm{e}^{-\frac{\alpha\eta^2}{2|a|}}}{2\pi|ap+\eta|}g(x+ap+\eta)\,\mathrm{d}\eta\right)H(a+p\cdot x)\,da$$

$$\sim \frac{e^{-\alpha p\cdot x}}{2\alpha}\int e^{-\alpha(|a|-a)}g(x+ap)H(a+p\cdot x)\,da. \tag{6.54}$$

We leave it up to the interested reader to simplify this further.

Before we quit note the following: If $\widetilde{S}'(\alpha, x)$ appearing in equation (6.54) were to depend on some parameter p (as is the previous example), and this dependence is measurable and $\mu(dp)$ is some measure on the space of parameters, then

$$\tilde{u}(\alpha, x) = \int \tilde{u}_p(\alpha, x)\mu(dp)$$

also satisfies the appropriate Helmholtz equation, and the asymptotics is carried under the integral sign.

In our second example, we may consider x_0 to be the parameter and vary, say, on a surface Σ. Thus from (6.53) we obtain

$$\tilde{u}(\alpha, x) = \int_\Sigma \left\{ \frac{e^{-\alpha|x-x_0|}}{2\alpha}\int_0^{|x-x_0|}\frac{|x_0-x|-s}{|x_0-x|}g\left(x+\frac{x_0-x}{|x_0-x|}s\right)ds\right\}\mu(dx_0). \tag{6.55}$$

Or if we consider p as the interesting parameter distributed according to $\mu(dp)$ over some set Λ, and couple this with (6.54) to obtain

$$\tilde{u}(\alpha, x) = \int_\Lambda \frac{e^{-\alpha p\cdot x}}{2\alpha}\left\{\int_{-\infty}^{\infty} e^{-\alpha(|a|-a)}g(x+ap)H(a+p\cdot x)\,da\right\}\mu(dp).$$

6.7. Long time asymptotics

Note that

$$\alpha\tilde{u}(\alpha, x) = \alpha\int_0^\infty e^{-\alpha t}u(t, x)\,dt = \int_0^\infty e^{-t}u(t/\alpha, x)\,dt$$

from which it is clear that the behavior of $u(t, x)$ for short times (respectively, for long times) and the behavior of $\alpha u(\alpha, x)$ for large (respectively, small) valves of α are related. The mathematical results relating the two come under the heading of Abelian and Tauberian theorems.

It is clear that, since waves move away from the sources, in order to have a stationary situation, one must have either sources that radiate forever or bounded regions with reflecting walls.

Consider a wave in a region D, with a smooth boundary ∂D, described by

$$\frac{\partial^2 u}{\partial t^2} = \frac{1}{2}\sum_{i,j} a_{ij}(x)\frac{\partial^2}{\partial x_i \partial x_j}u + \sum_i b_i(x)\frac{\partial u}{\partial x_i} - k(x)u + s(t, x)$$

for $t > 0, x \in D$, with initial conditions

$$u(0,x) = f(x), \qquad u_t(0,x) = g(x), \quad x \in D$$

and reflecting boundary conditions

$$\langle n(x), \operatorname{grad} u(t,x)\rangle + h(t,x) = 0, \quad t > 0, x \in \partial D.$$

Taking Laplace transforms, we obtain

$$\begin{aligned}&(\alpha^2 - (G - k(x)))\tilde{u}(\alpha,x) = \tilde{s}(\alpha,x), \quad x \in D, \alpha > 0,\\ &\langle n(x), \operatorname{grad}\tilde{u}(\alpha,x)\rangle + \hat{h}(\alpha,x) = 0, \quad x \in \partial D, \alpha > 0.\end{aligned} \tag{6.56}$$

We have seen in Section 4 of Chapter 2, that under certain hypothesis ensuring the symmetry of the infinitesimal generator $G = \frac{1}{2}\sum a_{ij}\partial_{ij}^2 + \sum b_i\partial_i$, the solution to (6.54) can be represented as a path integral over the trajectories of the diffusion $X(t)$ that has G as an infinitesimal generator, as

$$\begin{aligned}\tilde{u}(\alpha,x) = {}& E^x\int_0^\infty e^{-\alpha^2 t}e^{-\int_0^t k(X(s))\,ds}\{\tilde{s}(\alpha,X(t)) + \alpha f(X(t))H + g(X(t))\}\,dt\\ &+ E^x\int_0^\infty e^{-\int_0^t (X(s))\,ds}e^{-\int_0^\infty (X(s))\,ds}e^{-\alpha^2 t}\tilde{h}(\alpha,X(t))\,dL_t,\end{aligned} \tag{6.57}$$

where, recall, L_t is the local time at the boundary.

To take the limit as $\alpha\tilde{u}(\alpha,x)$ tends to zero, we need to assume that $\alpha\tilde{s}(\alpha,x) \to \bar{s}(x)$ and $\alpha\tilde{h}(\alpha,x) \to \bar{h}(x)$ as $\alpha \to 0$ for some appropriate $\bar{s}(x)$ and $\bar{h}(x)$. The hypothesis to place upon $\bar{s}(x)$ and $\bar{h}(x)$ depends on whether $k(x) > 0$ or $k(x) \equiv 0$.

When $k(x) \geqslant k_0 > 0$, it suffices that $\bar{s}(x)$ and $\bar{h}(x)$ be bounded and that for some M, $|\alpha\tilde{s}(\alpha,x) - \bar{s}(x)| \leqslant \mu$, $|\alpha\tilde{h}(\alpha,x) - \bar{h}(x)| \leqslant \mu$.

When $h(x) \equiv 0$, it would suffice that for each α

$$\begin{aligned}&E^x\int_0^\infty |\tilde{s}(\alpha,X(t))|\,dt < \infty, \qquad E^x\int_0^\infty |\tilde{h}(\alpha,X(t))|\,dt < \infty,\\ &E^x\int_0^\infty |\bar{s}(X(t))|\,dt < \infty, \qquad E^x\int_0^\infty |\bar{h}(X(t))|\,dt < \infty\end{aligned}$$

as well as that $\alpha\tilde{s}(\alpha,x) \to \bar{s}(x)$, $\alpha\tilde{h}(\alpha,x) \to \bar{h}(x)$ for every $x \in D$. From (2) it then follows that $\alpha\tilde{u}(\alpha,x)$ tends to $\bar{u}(x)$ where

$$\begin{aligned}\bar{u}(x) = {}& E^x\int_0^\infty e^{-\int_0^t k(X(s))\,ds}\bar{s}(X(t))\,dt\\ &+ E^x\int_0^\infty e^{-\int_0^t k(X(s))\,ds}e^{-\int_0^t n(X(s))\,ds}\bar{h}(X(t))\,dL_t\end{aligned}$$

which satisfies

$$(G - k(x))U(x) = -\bar{s}(x), \quad x \in D$$

with the reflecting boundary condition

$$\langle n(x), \operatorname{grad} \bar{u}(x) \rangle + \bar{h}(x) = 0, \quad x \in \partial D.$$

References

In references [3-3]–[3-6]–[3-9] topics related to the material in Sections 1 and 2 are discussed. The basics of Hamiltonian mechanics, its connections with geometrical optics are beautifully explained in

1. Arnold, V. I.: *Mathematical Methods in Classical Mechanics*, Springer-Verlag, Berlin.
2. Dubrovin *et al.*: *Modern Geometry: Methods and Applications*, Springer-Verlag, Berlin.

Different aspects of the probabilistic approach to geometrical optics are treated in

3. Freidlin, M. I. and Wentzell, A. D.: *Random Perturbations of Dynamical Systems*, Springer-Verlag, Berlin, 1984.
4. Freidlin, M. I.: *Functional Integration and partial Differential Equations*, Princeton Univ. Press, Princeton, 1985.

And take a look at the material on large deviations in [2-4] if need be. The two examples were cooked up on the spot.

CHAPTER

SEVEN

TRANSMUTATION OPERATIONS

7.1. Basic transmutations

In this section, we shall examine how to transform the following three problems:

$$\text{(P1)}\qquad \frac{\partial u}{\partial t}(t) = Gu(t) + s_1(t), \qquad u(0) = f,$$

$$\text{(P2-a)}\qquad \frac{\partial^2 v}{\partial t^2} = Gv + s_2(t), \qquad u(0) = f, \qquad u_t(0) = 0,$$

$$\text{(P2-b)}\qquad \frac{\partial^2 v}{\partial^2 t} = Gv + s_2(t), \qquad u(0) = 0, \qquad v_t(0) = f,$$

$$\text{(P3)}\qquad \frac{\partial^2 w}{\partial t^2} + Gw = s_3(t), \qquad u(0) = f$$

on to each other. For the applications we are interested in, G will be an operator independent of t, and usually it will just be a second order elliptic operator. For the computations that we carry out below, if suffices to assume that G operates on a closed domain $D(G)$ of a Banach space B, and that it generates a C_0 semigroup there.

We follow the work of Bragg and Detmann to connect (P1)–(P3) and ask the reader to consult the treatise by Carroll, where the full fledged mathematical issues are discussed.

To go from (P2) to (P1) is easy. For that consider the transmutation maps

$$(T_1 f)(t) = \int_0^\infty \frac{e^{-s^2/4t}}{(\pi t)^{1/2}} f(s)\, ds = \int_0^\infty \frac{e^{-s^2/4}}{\sqrt{\pi}} f\big(\sqrt{t}s\big)\, ds, \tag{7.1}$$

$$(T_2 f)(t) = \int_0^\infty s \frac{e^{-s^2/4t}}{(4\pi t^3)^{1/2}} f(s)\, ds = \int \frac{se^{-s^2/4}}{(4\pi t)^{1/2}} f\big(\sqrt{t}s\big)\, ds. \tag{7.2}$$

To verify that they are well defined, it suffices to assume that $s \to \|f(s)\|$ is bounded by a polynomial in s, where $\|\cdot\|$ denotes the norm in B.

To verify that the behavior at $t = 0$ is as described, it suffices to assume that $f(t) = f(0) + tf'(0) + t^2R(t)$, with $R(t)$ being a bounded curve in B.

Note that the kernel in (7.2) is obtained from the kernel in (7.1) differentiating with respect to s. Also, the kernel in (7.1) satisfies

$$\frac{\partial^2}{\partial s^2}\left(\frac{e^{-s^2/4t}}{(\pi t)^{1/2}}\right) = \frac{\partial}{\partial t}\left(\frac{e^{-s^2/4t}}{(\pi t)^{1/2}}\right),$$

and so does the kernel in (7.2).

Integration by parts yields the following result.

PROPOSITION 7.1. *Let $v(t)$ denote a solution to* (P2-a) (*resp., to* (P2-b)) *such that $v(t) = v(0) + tv'(0) + t^2 v_R(t)$ with $\sup_t \|v_R(t)\| < \infty$, and that $\sup_t \|Gv(t)\| < \infty$. Let $u(t)$ denote $(T_1 v)(t)$ (resp., $(T_2 v)(t)$) and $s_1(t)$ denote either $(T_1)(t)$ or $(T_2)(t)$. Then $u(t)$ satisfies* (P1), *i.e.,*

$$\frac{\partial u}{\partial t} = Gu + s_1 \quad \text{for } t > 0, \qquad u(0) = f.$$

Comment: When **G** denotes the spatial part of a wave operator, and $s_2(t)$ vanishes for large t, the solutions to (P2) will decay in time as $t \to \infty$, so the boundedness conditions imposed hold naturally.

To go backwards, denote $T_1 f$ and $T_2 f$ by $\tilde{f}_1$ and $\tilde{f}_2$, and observe that a simple change of variables and then the substitution $t \to 1/4t$ yields

$$\sqrt{\pi/4}\, t^{-1} \tilde{f}_1(1/4t) = \int_0^\infty e^{-st} \frac{f(\sqrt{s})}{\sqrt{s}}\, ds, \tag{7.3}$$

$$\sqrt{\pi/4}\, t^{-3/2} \tilde{f}_2(1/4t) = \int_0^\infty e^{-st} f(\sqrt{s})\, ds. \tag{7.4}$$

For given B-valued functions $\tilde{f}_1(t)$ and $\tilde{f}_2(t)$, if it were known that the left-hand sides of (7.3) and (7.4) are Laplace transforms, then $f(t)$ defined by

$$f(t) = \sqrt{\pi/4}\, t\, \mathcal{L}^{-1}\big(s^{-1/2} \tilde{f}_1(1/4s)\big)(t^2) \tag{7.5}$$

as well as

$$f(t) = \sqrt{\pi/4}\, \mathcal{L}^{-1}\big(s^{-3/2} \tilde{f}_2(1/4s)\big)(t^2) \tag{7.6}$$

satisfies (7.3) and (7.4). Here $\mathcal{L}$ denotes the Laplace transform operator, and $\mathcal{L}^{-1}(h(s))$ denotes the inverse Laplace transform operator. The implied notation suggests that s is the Laplace transformed variable and t the inverse Laplace transformed variable.

To obtain (P2) from (P1), we go as follows.

PROPOSITION 7.2. *Consider problem* (P1) *with* $s_1(t)$ *being either of* $(T_1 s_2)(t)$ *or* $(T_2 s_2)(t)$. *Assume that the solution* $u(t)$ *to such* (P1) *exists and that* $s^{-1/2}u(1/4s)$ *and* $s^3 u(1/4s)$ *are Laplace transforms. Then the solutions to* (P2) *are given by*

$$v(t,x) = \sqrt{\pi/4t}\,\mathcal{L}^{-1}\big(s^{-1/2}u(1/4s)\big)(t^2),$$

in the case of (P2-a) *as well as*

$$v(t,x) = \sqrt{\pi/4}\,\mathcal{L}^{-1}\big(s^{-3/2}u(1/4s)\big)(t^2)$$

for (P2-b).

The proof is easy and spelled out in the references. We will come back to this point below. Let us now consider the connection of (P1) to (P3).

If we denote by P_t the C_0 semigroup on B generated by G, then the solution to (P1) can be displayed as follows

$$u(t) = P_t f + \int_0^t P_{t-t'}s(t')\,\mathrm{d}(t'). \tag{7.7}$$

Introduce now the transmutation operator

$$(T_3 f)(t) = \int_0^\infty \frac{t\mathrm{e}^{-t^2/4s}}{4\pi s} f(s)\,\mathrm{d}s = 2\int_0^\infty \mathrm{e}^{-u} f(t^2/4u)\frac{\mathrm{d}u}{\sqrt{\pi u}} \tag{7.8}$$

acting on bounded curves $s \to f(s)$ in B. Note that it has the same kernel as T_2, but the integration is with respect to a different variable.

When applied to the solution $u(t)$ of (P1), we see that $v(t) \to f$ as $t \to 0$ weakly or in the norm of B according to whether $u(t) \to f$ weakly or in the norm of B.

The integration by parts yields (P3) from (P1), whenever we can exchange G with the integration over s. For this it suffices to assume that $GP_t f$ and $G\int_0^t P_{t-t'}s(t')\,\mathrm{d}t'$ stay bounded in t. We have

PROPOSITION 7.3. *Under the stated assumption, if* $u(t)$ *satisfies* (P1) (*and is given by* (7.7)), *then* $w(t) = (T_s v)(t)$, *given by* (7.8) *satisfies* (P3) *with* $s_3(t) = (T_3 s_1)(t)$.

Comment: When we played the time changing game, we saw that

$$Q_t = \int \frac{t\mathrm{e}^{-t^2/4s}}{(4\pi s^3)^{1/2}} P_s\,\mathrm{d}s$$

was a semigroup with the generator $\widetilde{G} = \sqrt{-G}$. This explains why $\frac{\partial^2 Q_t}{\partial t^2} = -GQ_t$!

If you combine transforms (T_2) and (T_3), you can go directly from (P2) to (P3). The resulting transform is

$$(T_4 f)(t) = \frac{2t}{\pi}\int \frac{f(s)}{t^2+s^2}\,\mathrm{d}s \tag{7.9}$$

which, when applied to a solution of (P2-a), looks like this

$$w(t) = (T_4 v)(t) = \frac{2t}{\pi} \int \frac{v(s)}{t^2 + s^2} \, \mathrm{d}s. \tag{7.10}$$

The basic result being

THEOREM 7.1. *Let $v(t)$ be a (twice continuously differentiable) solution of* (P2-a). *Assume that for large T and all $t \geqslant T$, $(\mu(v)) \leqslant k(\mu)t^{\alpha}$ and $|\mu(v_{tt})| \leqslant M(\mu)t^{\beta}$, where $0 < \alpha < 1, 0 < \beta < 1$, $\mu \in B^*$ is any continuous linear functional on B, and $k(\mu)$, $M(\mu)$ are bounded. Then $w(t)$ given by* (7.10) *is a solution to* (P3) *with $s_3(t) = (T_4 s_2)(t)$.*

There are two facts to be verified: First, that the equation is satisfied. This is just a computation, made plausible because (P2) is equivalent to (P1) and (P1) to (P3). To verify the boundary condition, write (7.10) as

$$w(t) = \frac{2}{\pi} \int_0^{\infty} \frac{F(ts)}{1 + s^2} \, \mathrm{d}s$$

and since $T_4 1 \equiv 1$, it is clear that the bounded convergence theorem yields what we want.

7.2. Probabilistic version of transmutation operations

In this section, we provide a probabilistic setup for the transmutation operations introduced above. What we do is a variation on a theme developed by Rosencranz.

To begin with, assume that there exists a continuous bounded curve $u(\xi) : \mathbb{R} \to D(G)$ which satisfies the differential equation

$$\frac{1}{2}\sigma^2(\xi)\frac{\partial^2 u}{\partial \xi^2} + b(\xi)\frac{\partial u}{\partial \xi} \equiv Gu(\xi) + s(\xi). \tag{7.11}$$

Assume that $\sigma(\xi)$ and $b(\xi)$ satisfy the conditions spelled out in Chapter 3, so that

$$\mathrm{d}X(t) = \sigma(X(t))\,\mathrm{d}B(t) + b(X(t))\,\mathrm{d}t, \qquad x(0) = \xi \tag{7.12}$$

has a global solution. Here $B(t)$ denotes the standard one-dimensional imposed motion started at 0. With $u(\xi)$ as in (7.11) and $X(t)$ as in (7.12), define

$$\hat{u}(t, \xi) = E^{\xi}(X(t)).$$

For the following to be true, it suffices that

$$\int_0^t E^{\xi}\left\{\|u_{\xi}(X_s)\|^2 \sigma^2(X_s)\right\} \mathrm{d}s < \infty \tag{7.13}$$

for each t.

PROPOSITION 7.4. *With the notation and assumptions introduced above*

$$\frac{\partial \hat{u}}{\partial t} = G\hat{u} + \hat{s}(t), \qquad \hat{u}(0) = u(\xi_0)$$

for $\hat{s}(t_1) = E^{\xi}[s(X(t))]$.

PROOF. Just invoke (extension of) Ito's formula (to the Banach space setup)

$$\begin{aligned} \mathrm{d}u(X(t)) &= \left\{ \hat{u}_{\xi}(X(t))b(X(t)) + \frac{\sigma^2(X(t))}{2}\hat{u}_{\xi\xi}(X(t)) \right\} \mathrm{d}t \\ &\quad + u_{\xi}(X(t))\sigma(X(t))\,\mathrm{d}B(t). \end{aligned}$$

Due to assumption (7.3) this makes sense and the last term provides us with a martingale (when integrated) with zero mean. Thus, making use of (7.11), we have

$$\hat{u}(t) = E[u(X_t)] = u(\xi_0) + E\int_0^t \left\{ Gu((\xi(t')) + s(X(t')) \right\} \mathrm{d}t'.$$

Since we assumed $u(\xi)$ to be continuous and bounded in $D(G)$, we get $E\{Gu(X(t))\} = GE[u(X(t))] = G\hat{u}(t)$, and therefore

$$\frac{\partial \hat{u}}{\partial t} = G\hat{u}(t) + \hat{s}(t), \qquad \hat{u}(0) = u(\xi_0).$$

This trick can be transported to fields $u(\xi) : S \subset \mathbb{R}^d \to B$, where S is a domain on $\mathbb{R}^d$, such that

$$G_0 u(\xi) = Gu(\xi) + s(\xi),$$

where G_0 is the infinitesimal generator of some Markov process $X(t)$ living in $\mathbb{R}^d$. If $T = \inf\{t > 0 : X(t) \notin S\}$ and we put

$$\hat{u}(t) = E^x[u(X(t)); t < T], \quad t \geqslant 0,$$

then

$$\frac{\partial \hat{u}}{\partial t} = G\hat{u} + \hat{s}, \qquad \hat{u}(0) = u(x) \quad \text{for } x \in S.$$

□

In order to completely relate with the operators, described in Section 1, we assume in what follows that $B(\tau)$ is a one-dimensional imposed motion with variance 2, i.e., its transition density is given by $\rho_t(\xi) = (4\pi t)^{-1/2}\exp(-(\xi^2/4t))$.

PROPOSITION 7.5 (The time change again). *Let $u(\tau)$ satisfy $\frac{\partial u}{\partial \tau} = Gu + s(\tau)$. If both $u(\tau)$ and $Gu(\tau)$ stay bounded in τ, and $T(t)$ is the first time that $B(\tau)$ hits the level t, then $\hat{u}(t) = E[u(T_t)]$ solves the problem*

$$\frac{\partial^2 \hat{u}}{\partial t^2} + G\hat{u}(t) + \hat{s}(t) = 0, \qquad \hat{u}(0) = 0, \qquad \lim_{t\to\infty} \hat{u}(t) = 0$$

with $\hat{s}(t) = E[s(T(t))]$.

PROOF. Just note that $P(T_t \in \mathrm{d}\tau) = \frac{t e^{-t^2/4\tau}}{(4\pi\tau^3)^{1/2}}\,\mathrm{d}\tau$. We have already done the rest. □

Comment: In other words, if $u(t)$ satisfies (P1), then $\hat{u}(t)$ satisfies (P3). Consider now $M(t) = \sup_{s\leqslant t} B(s)$, the maximum displacement of $B(t)$ to the rights up to time t. It is clear that $M(T_t) = t$ and it also happens that

$$P(M(t) \in \mathrm{d}\xi) = \frac{e^{-\xi^2/4t}}{(\pi t)^{1/2}}\,\mathrm{d}\xi$$

for $\xi > 0$.

PROPOSITION 7.6. *If $u(t)$ satisfies $\frac{\partial^2 u}{\partial t^2} = Gu + s(t)$, then*

$$\hat{u}(t) = E[u(M(t))]$$

satisfies

$$\frac{\partial \hat{u}}{\partial t} = G\hat{u} + \hat{s}, \qquad \hat{u}(0) = u(0).$$

Comment: In other words, if $u(t)$ satisfies (P2-a), then $\hat{u}(t) = E[u(M(t))]$ satisfies (P1).

If we consider two independent imposed motions $B^1(\tau)$ and $B^2(\tau)$, both of variance 2, and consider the time T_0 that it takes $B^2(\tau)$ to hit 0, and start $B^2(\tau)$ from $t > 0$ as well as $B^1(\tau)$ from 0, and put

$$\hat{u}(t) = E^{(0,t)}[u(B^1(T_0))] = \frac{2t}{\pi}\int_0^\infty \frac{u(\xi)\,\mathrm{d}\xi}{t^2+\xi^2},$$

then we have

PROPOSITION 7.7. *If $u(t)$ satisfies $\frac{\partial^2 u}{\partial \xi^2} = Gu + s(\xi)$ and $\hat{u}(t)$ is defined as right above, then $\hat{u}(\xi)$ satisfies*

$$\frac{\partial^2 \hat{u}}{\partial t^2} + G\hat{u} = \hat{s}(t).$$

Similarly, if $u(\xi)$ satisfies an equation as

$$\Delta u(\xi) = Gu(\xi) + S(\xi)$$

for ξ in $\mathbb{R}^d$, and if $B = (B^1(t), \ldots, B^d(t))$ is an $\mathbb{R}^d$ imposed motion issued from zero, all the components of which have variance 2, and $X(t)$ is a similar one-dimensional imposed motion, but independent of $B(t)$, and if $T = \inf\{t > 0 : X(t) = 0\}$, and if we set

$$\hat{u}(t) = E^{(t,0)}[u(B(T))] = \frac{\Gamma\left(\frac{d+1}{2}\right)t}{\pi^{(d+1)/2}} \int \frac{u(\xi)\,\mathrm{d}\xi}{(\xi^2 + t^2)^{(d+1)/2}} \tag{7.14}$$

then it happens to satisfy

$$\frac{\mathrm{d}^2\hat{u}}{\mathrm{d}t^2} + G\hat{u} = \hat{s}(t) \tag{7.15}$$

with $\hat{s}(t) = E^{(0,t)}[s(B(T))]$. Let us follow Dettman in the analysis of the conditions at $t = 0$ for (7.15).

PROPOSITION 7.8. *If $u(t\xi)$ and $\frac{u(t\xi)-u_0(\xi)}{t}$ converges boundedly to u_0 and v_0, then $\hat{u}(t)$, given by* (7.14), *is continuously differentiable at $t = 0$ and, $\hat{u}(0) = u_0$, $\mathrm{d}\hat{u}/\mathrm{d}t(0) = 0$.*

PROOF. Note (and it is obvious from 1) that, if $u(0) = u_0$, then $\hat{u}_0 = u_0$. Changing variables

$$\hat{u}(t) = \frac{\Gamma\left(\frac{d+1}{2}\right)}{\pi^{(d+3)/2}} \int \frac{u(t\xi)\,\mathrm{d}\xi}{(\xi^2 + 1)^{(d+1)/2}}$$

from which the convergence of $\hat{u}(t)$ to u_0 follows.

Clearly, setting $u \equiv 1$ in (7.14) and differentiating with respect to t, we obtain

$$\frac{\Gamma\left(\frac{d+1}{2}\right)}{\pi^2} \int \frac{\xi^2 - t^2 d}{(\xi^2 + t^2)^{(d+1)/2}} u\,\mathrm{d}\xi = 0.$$

Also, from the differentiability of $\hat{u}(t)$ at $t > 0$

$$\frac{\partial\hat{u}}{\partial t} = \frac{\Gamma\left(\frac{d+1}{2}\right)}{\pi^{(d+1)/2}} \int \frac{\xi^2 - t^2 d}{(\xi^2 + t^2)^{(d+1)/2}} u(\xi)\,\mathrm{d}\xi$$

from which we obtain

$$\frac{\partial\hat{u}}{\partial t}(t) = \frac{\Gamma\left(\frac{d+1}{2}\right)}{\pi^{(d+1)/2}} \int \frac{\xi^2 - t^2 d}{(\xi^2 + t^2)^{(d+3)/2}} [u(\xi) - u_0]\,\mathrm{d}\xi$$

and making a change of variables $\xi \to \xi/t$, we obtain

$$\frac{\partial \hat{u}}{\partial t} = \frac{\Gamma\left(\frac{d+1}{2}\right)}{\pi^{(d+1)/2}} \int \frac{\xi^2 - d}{(\xi^2+1)^{(d+3)/2}} \frac{\{u(t\xi) - u_0\}}{t} \,\mathrm{d}\xi \to 0$$

as $t \to 0$ which is due to the fact that

$$\int \frac{\Gamma\left(\frac{d+1}{2}\right)}{\pi^{(d+1)/2}} \int \frac{(\xi^2 - d)}{\xi^2+1}^{(d+3)/2} \mathrm{d}\xi = 0$$

as we have already seen. □

7.3. Examples

Below we present two ways of integrating the wave equation

$$\frac{\partial^2 \bar{v}}{\partial t^2} + 2\beta \frac{\partial \bar{v}}{\partial t} + m^2 \bar{v} = c^2 \Delta \bar{v} + s_2, \qquad t > 0, x \in \mathbb{R}^n$$

with initial data $\bar{v}(0,x) = 0$, $\bar{v}_t(0,x) = g(x)$. We take $\beta > 0$ and $m > 0$. If we set $v(t,x) = \mathrm{e}^{-\beta t}\bar{v}(t,x)$, we obtain

$$\frac{\partial^2 v}{\partial t^2} = c^2 \Delta v + (\beta^2 - m^2)v + s_2, \quad t > 0, \ x \in \mathbb{R}^n \tag{7.16}$$

again with initial conditions $v(0,x) = 0$, $v_t(0,x) = g(x)$.

We assume that $c = 1$, which amounts to $t \to ct$, $\beta^2 - m^2 \to (\beta^2 - m^2)/c^2$ and $s_2 \to s_2/c^2$. Denote, for short, $a^2 = (\beta^2 - m^2)/c^2$, and consider $a^2 > 0$ or $a^2 < 0$.

In Section 1, we transmuted this problem into

$$\frac{\partial u}{\partial t} = \Delta u + a^2 u + s_1 \tag{7.17}$$

with $u(0,x) = g(x)$, with $s_1(t,x) = \int_0^\infty \frac{\tau \mathrm{e}^{-\tau^2/4t}}{(4\pi t^3)^{1/2}} s_2(\tau, x)\,\mathrm{d}\tau$.

If we denote by P_t the transition semigroup with the kernel

$$P_t(x-y) = \mathrm{e}^{a^2 t} \frac{\mathrm{e}^{-d^2/4t}}{(4\pi t)^{n/2}},$$

where $d^2 = (x-y)^2$, then the solution to (7.12) is given by

$$u(t,x) = P_t g(x) + \int_0^t P_{t-t'} S_1(t', x)\,\mathrm{d}t'. \tag{7.18}$$

Note that it is only when $a^2 = 0$, does P_t describe a probability transition semigroup. Otherwise, it would be associated with a process where mass (probability) is created (resp., destroyed) when $a^2 > 0$ (resp., $a^2 < 0$).

To begin with, assume that $s_2(t, x)$ (and therefore $s_1(t, x)$) vanishes, and let us follow Bragg and Dettman to obtain the solution to (7.16) from solution (7.18) to problem (7.17) as indicated in Proposition 7.2.

For that, the first step is to recognize whether

$$\left(2\pi^{(n-1)/2}\right)^{-1} s^{(n-3)/2} e^{a^2/4s} e^{-d^2/s} \tag{7.19}$$

is a Laplace transform and then to invert it. The result we obtain will depend on the sign of a^2 and the parity of n.

If $n \geqslant 3$ and $n = 2k + 1$, $k \geqslant 1$, then $(n-3)/2 = k - 1$ and

$$s^{(n-3)/2} e^{a^2/4s} e^{-d^2/s} = s^{-1} e^{a^2/4s} s^k e^{-d^2 s},$$

with $s^k e^{-d^2 s}$ being an obvious Laplace transform, and $s^{-1} e^{a^2/4s}$ being the Laplace transform of $J_0(|a|\sqrt{t})$ or $I_0(a\sqrt{t})$ depending on whether $a^2 < 0$ or $a^2 > 0$ ($a^2 = (i|a|)^2$ or $a > 0$, if you prefer).

If $n = 2k + 2$ with $k \geqslant 1$, then $\frac{n-3}{2} = k - \frac{1}{2}$ and, since $s^{-1/2} e^{a^2/4s}$ is the Laplace transform of $(\pi t)^{-1/2} \cos(|a|\sqrt{t})$ or $(\pi t)^{-1/2} \cos h(a\sqrt{t})$ according to whether $a^2 < 0$ or $a^2 > 0$, we are in business: Proposition 2 of Section 1 can be applied.

The cases $n = 3, 2$, and 1 are better treated individually. Assume $a^2 < 0$ and $n = 3$. Then $e^{-|a|^2/4s} e^{-d^2 s}$ is the product of the Laplace transforms of $\delta(t - d^2)$ and that of $\delta(t) - \frac{|a| J_1(|a|\sqrt{t})}{2\sqrt{t}}$, therefore $e^{-|a|^2/4s} e^{-d^2 s}$ is the Laplace transform of their convolution, i.e.,

$$\delta(t - d^2) - \frac{|a| J_1(|a|\sqrt{t - d^2})}{t - d^2}$$

for $t > d^2$, i.e., according to Proposition 2 of Section 1,

$$v(t, x) = \frac{1}{2\pi} \int H(t^2 - d^2) \left\{ \delta(t^2 - d^2) - \frac{|a| J_1(|a|\sqrt{t^2 - d^2})}{2(t^2 - d^2)} \right\} g(y)\, dy, \tag{7.20}$$

where, recall $H(\xi)$ denotes the Heaviside function. Let us put

$$\int_{|\xi|=1} g(x + r\xi)\, d\Sigma(\xi) = \langle g \rangle (x, r) \quad \text{for any } r \geqslant 0.$$

We can shift the origin to x and introduce spherical coordinates to obtain

$$v(t, x) = \frac{1}{2\pi} \int_0^t \left\{ \delta(t^2 - r^2) - \frac{|a| J_1(|a|(t^2 - r^2)^{1/2})}{2(t^2 - r^2)^{1/2}} \right\} \langle g \rangle (x, r) r^2\, dr$$

and if we now use the identity $\delta(t^2 - r^2) = \delta(t - r)/2t$, the first term becomes

$$\frac{1}{4\pi t} \int_{\partial B(x,t)} g(x + y)\, \mathrm{d}\Sigma(y)$$

which, as $|a| = 0$, yields the standard solution to the three-dimensional wave equation. To obtain the solution to the first equation, just set $\bar{v}(t, x) = \mathrm{e}^{-\beta t} v(t, x)$, and to unscale t, just replace t by ct.

The case $a^2 = 0$ and $n = 2$ is easy. This time $s^{-1/2}\mathrm{e}^{-d^2 s}$ is the Laplace transform of $H(t - d^2)\frac{(t-d^2)^{1/2}}{\sqrt{\pi}}$ and, therefore, a trivial computation yields

$$v(t, x) = \frac{1}{2\pi} \int_{B(0,t)} \frac{g(x + y)\, \mathrm{d}y}{(t^2 - y^2)^{1/2}}.$$

When $n = 1$, another glance at a table of Laplace transforms yields

$$v(t, x) = \frac{1}{2} \int_{x-t}^{x+t} J_0\big(|a|(t^2 - (x - \xi)^2\big)\phi(\xi)\, \mathrm{d}\xi$$

for $a^2 < 0$.

As a side remark, note that the method developed in Chapter 4, when applied to (7.17), would lead to

$$\left(\frac{1}{2}(\alpha^2 - a^2) - \frac{1}{2}\Delta\right)\tilde{u}(\alpha, x) = \widetilde{S}(\alpha, x)/2.$$

If $a^2 < 0, \alpha^2 - a^2 > 0$ and what follows is valid for any $\alpha > 0$, whereas if $a^2 > 0$, we must consider $\alpha > |a|$. Anyway, we know that

$$\begin{aligned}
\tilde{u}(\alpha, x) &= \frac{1}{4\pi} \int \frac{\mathrm{e}^{-(\alpha^2 - a^2)^{1/2}|x-y|}}{|x - y|} \widetilde{S}(\alpha, y)\, \mathrm{d}y \\
&= \frac{1}{4\pi} \int_0^\infty r\mathrm{e}^{-(\alpha^2 - a^2)^{1/2} r} \langle \widetilde{S}(\alpha) \rangle (\alpha, r)\, \mathrm{d}r \\
&= \frac{1}{4\pi} \int_0^\infty \frac{\partial}{\partial r} \left(\frac{\mathrm{e}^{-r(\alpha^2 - a^2)^{1/2}}}{\alpha^2 - a^2} \right) r \langle \widetilde{S}(\alpha) \rangle (x, r)\, \mathrm{d}r \\
&= \frac{1}{4\pi} \int_0^\infty \frac{\mathrm{e}^{-r(\alpha^2 - a^2)^{1/2}}}{(\alpha^2 - a^2)^{1/2}} \frac{\partial}{\partial r} r \langle \widetilde{S}(\alpha) \rangle (x, r)\, \mathrm{d}r.
\end{aligned}$$

If we consider the case $a^2 < 0$ as above, and write $a^2 = -|a|^2$, and, since $(\alpha^2 + |a|^2)^{1/2} \exp\{-r(\alpha^2 + |a|^2)^{1/2}\}$ is the Laplace transform of $J_0(|a|(t^2 - r^2)^{1/2})H(t - r)$, we obtain

$$
\begin{aligned}
\tilde{u}(t,x) &= \frac{1}{4\pi}\int_0^t \mathrm{d}t' \int_0^\infty J_0\big(|a|(t'^2-r^2)^{1/2}\big)^{1/2}\frac{\partial}{\partial r} r\langle S(t-t')\rangle(x,r)H(t'-r)\,\mathrm{d}r \\
&= \frac{1}{4\pi}\int_0^t \mathrm{d}t' \int_0^{t'} J_0\big(|a|(t'^2-r^2)^{1/2}\big)\frac{\partial}{\partial r} r\langle S(t-t')\rangle(x,r)\,\mathrm{d}r \qquad (7.21)
\end{aligned}
$$

where, as the obvious notation suggests,

$$
\langle S(t-t')\rangle(x,r) = \int_{\partial B(0,1)} S(t-t', x+r\xi)\,\mathrm{d}\Sigma(\xi)
$$

which, as $|a| \to 0$, yields the result we obtained in Chapter 6, if instead of $\widetilde{S}(\alpha, x)$ we have just $g(x)$, then (7.21) becomes

$$
U(t,x) = \frac{1}{4\pi}\int_0^t J_0\big(|a|(t^2-r^2)^{1/2}\big)\frac{\partial}{\partial r}(r\langle g\rangle(x,r))\,\mathrm{d}r
$$

and we leave it up to the reader to verify that it matches (7.20).

7.4. More inversion techniques and simple examples

Let us now play a bit more with the transmutation operators connecting problems 1 and 2 in Section 1, but this time with Laplace transforms.

The Laplace transform of (P1) is

(P1)′ $\quad (\alpha - G)\tilde{u}(\alpha, x) = \tilde{s}_1(\alpha, x) + f(x)$

whereas that of problem (P2-2) is

(P2)′ $\quad (\alpha^2 - G)\tilde{v}(\alpha, x) = \tilde{s}_2(\alpha, x) + f(x).$

Notice now that, if $F(t)$ is any bounded function with the transform $\widetilde{F}(\alpha)$, then the Laplace transform of $(T_2F)(t)$ is $(T_2F)^{\sim}(\alpha) = F(\sqrt{\alpha})$. It is thus clear that T_2 transmutes solutions of (P2)′ (or (P2) into solutions of (P1)′ (or (P1)).

Therefore, the search for an operator inverting T_2, when viewed by its action at the level of Laplace transforms, consists in looking for a map T such that $(TF)^{\sim}(\alpha) = \widetilde{F}(\alpha^2)$.

For that consider the functions $B(t) = F_2(1, \frac{1}{2}, -\frac{t^2}{4}) = \sum_{n=0}^{\infty} \frac{(-1)^n t^{2n}}{2n!n!}$ and $J_0(t)$. Note that $B(t\sqrt{\tau})$ is bounded in t (for each fixed τ) and has the Laplace transform $(1/\alpha)\exp(-(\tau/\alpha^2))$, therefore the integral transform

$$
(T_BF)(t) = \int_0^\infty B\big(t\sqrt{\tau}\big)F(\tau)\,\mathrm{d}\tau \qquad (7.22)
$$

is such that

$$(T_B F)^{\sim}(\alpha) = \frac{1}{\alpha}\widetilde{F}\left(\frac{1}{\alpha^2}\right). \tag{7.23}$$

Since the transform

$$(T_J F)(t) = \int_0^\infty J_0\big(2\sqrt{\tau t}\big)F(\tau)\,\mathrm{d}\tau \tag{7.24}$$

is such that

$$(T_J f)^{\sim}(\alpha) = \frac{1}{\alpha}F\left(\frac{1}{\alpha}\right), \tag{7.25}$$

it is clear that the composition

$$(Tf)(t) = (T_J \circ T_B f)(t) \tag{7.26}$$

will have the Laplace transform

$$(Tf)^{\sim}(\alpha) = f(\alpha^2), \tag{7.27}$$

which makes it a good candidate to invert T_2.

The result is contained in

PROPOSITION 7.9. *Assume that $f(0+)$ exists, then $(Tf)(0+) = 0$ and $(Tf)'(0+) = f(0+)$, and whenever f and Tf have two bounded derivatives,*

$$\int_0^\infty \mathrm{e}^{-\alpha^2 t} f'(t)\,\mathrm{d}t = \int_0^\infty \mathrm{e}^{-\alpha t}(Tf)''(t)\,\mathrm{d}t.$$

PROPOSITION 7.10. *With the notation introduced above*

$$(T \circ T_2 f)^{\sim}(\alpha) = \tilde{f}(\alpha), \qquad (T_2 \circ T f_1)^{\sim}(\alpha) = \tilde{f}_1(\alpha).$$

The potential application of this inversion method for solving wave equations in heterogeneous media is the following: Consider (P1) with $G = \frac{1}{2}\sum a_{ij}\partial_{ij}^2 + \sum b_i \partial_i$. Then the solution to (P1) admits the path integral representation

$$u(t,x) = E^x\big[f(X(t))\mathrm{e}^{-\int_0^t q(X(s)\,\mathrm{d}s}\big]$$

and the result above asserts that the solution to the corresponding problem (P2) is given by

$$v(t,x) = (Tu(\cdot,x))(t).$$

Let us pass on to consider some simple examples.

Example 1: Begin with the degenerate equation

$$\frac{\partial^2 v}{\partial t^2} = c\frac{\partial v}{\partial x}, \quad t > 0, c > 0$$

with initial conditions $u(0,x) = 0$, $u_t(0,x) = \mathrm{e}^{-\lambda x}$. After transmutation that equation becomes

$$\frac{\partial u}{\partial t} = c\frac{\partial v}{\partial x}, \quad t > 0$$

with the initial condition $u(0,x) = \mathrm{e}^{-\lambda x}$. This equation has a solution $u(t,x) = \mathrm{e}^{-\lambda(x+ct)}$, which is the transmuted version of $v(t,x) = \mathrm{e}^{-\lambda x}\frac{\sin t\sqrt{\lambda c}}{\sqrt{\lambda c}}$.

Since for $e_\lambda(t) = \mathrm{e}^{-\lambda t}$ we obtain $(Te_\lambda)(t) = \frac{\sin\sqrt{\lambda}t}{\sqrt{\lambda}}$ if $\lambda > 0$, or $(Te_\lambda)(t) = \frac{\sinh\sqrt{|\lambda|}t}{\sqrt{|\lambda|}}$ if $\lambda < 0$, when the functions, eigenfunctions and eigenvalues of the operator G are known, the transmutation technology is easy to apply (but deep down it is unnecessary).

Example 2: Another example corresponds to the description of a linear chain of oscillators. Even though this is a situation to which the previous comment applies, the transmuted version of the problem has a nice probabilistic counterpart, hence its appeal to us.

Consider $N + 1$ oscillators, subject to mutual elastic forces. If $x_i(t)$ denotes the displacement of the i-th one from its equilibrium position, the combination of Newton's and Hooke's laws yields the following set describing the motion of the system:

$$\begin{aligned} &M_i x_i(t) = -k_i(x_i - x_{i-1}) - k_{i+1}(x_i - x_{i+1}), \quad 1 \leqslant i \leqslant N_1, \\ &M_0 x_0(t) = -k_1(x_0 - x_1), \\ &M_N x_N(t) = -k_N(x_0 - x_{N_1}), \end{aligned} \tag{7.28}$$

and assume that the initial conditions have been assigned so that $x_i(0) = 0$, $\dot{x}_i(0) = \eta_i$. Here M_i is the mass of the i-th particle and k_i the elastic constant of the "spring" between the $(i-1)$-th and the i-th particles.

Set

$$u_i(t) = (T_2 x_i)(t)$$

and transform (7.28) into

$$\dot{u}_i(t) = q_{i,i-1}u_{i-1}(t) + q_{i,i}u_i(t) + q_{i,i+1}u_{i+1}(t) \tag{7.29}$$

subject to $U_i(0) = \eta_i$. Here

$$q_{i,i-1} = \frac{ki}{M_i}, \qquad q_{i,i+1} = \frac{k(t)}{M_i}, \qquad q_{i,i} = -(q_{i,i-1} + q_{i,i+1})$$

and $q_{i,i} = 0$ for all the other values of j. Note that $q_{i,i} < 0$ and $\sum_j q_{i,j} = 0$, thus we can think $q_{i,j}$ as the elements of a transition rate matrix Q, describing a Markov chain $\xi(t)$ with a state space $\{0, \dots, N\}$.

The transition probabilities $P_t(i, j)$ evolve according to

$$\frac{\mathrm{d}P}{\mathrm{d}t} = QP, \qquad P(0) = I.$$

In terms of $P_t(i, j)$ we can define

$$u(i, t) = \sum_j P_t(i, j)\eta(j) = E^i\{\eta(\xi(t))\}. \tag{7.30}$$

This satisfies (7.29), and, to obtain and visualize the solution to (7.28), we have to apply the operator T, defined in (7.26), to (7.30).

To do a concrete example, assume that the springs get stiffer as we go down the line according to

$$q_{i,i-1} = \mu n, q_{i,i+1} = \lambda(n+1), \qquad q_{i,i} = -(\mu n + \mu(n+1)).$$

The transition matrix of this chain is

$$P_t(t, j) = \gamma^{-i} \sum_{k=0}^{N-1} g_j^{(k)} g_i^{(k)} \sigma^{\beta(k)} \left\{ \sum_{\ell=0}^{N-1} g_\ell \gamma^{-\ell} \right\}^{-1},$$

where $\gamma = \lambda/\mu$, $\sigma(t) = \mathrm{e}^{t(\lambda-\mu)}$, $\beta(k)$ and $g_j^{(k)}$'s are special constants described in the book by Goel and Richter-Dyn.

When we apply the operator T to $\sigma(t)^{\beta(k)}$, we obtain

$$\rho(k, t) = \frac{\sin\sqrt{\beta(k)(\mu-\lambda)}}{\sqrt{\beta(k)(\mu-\lambda)}} \quad \text{or} \quad \frac{\sinh\sqrt{\beta(\lambda)(\lambda-\mu)}}{\sqrt{\beta(k)(\lambda-\mu)}}$$

according to whether $\beta(k)(\lambda - \mu)$ is lesser or greater than zero. Therefore

$$x_i(t) = \gamma^{-i} \sum_{k=0} g_i^{(k)} g_j^{(k)} \rho(k, t) \left\{ \sum_{\ell=0}^{N-1} g_\ell^{(k)} \gamma^{\ell} \right\}^{-1} x_j(0)$$

is what we want. Simpler formulae crop up when the chain is unbounded.

Example 3: Let us now explore the passage from (P3) to (P1) when G is the n-dimensional Laplace operator. Now we start from the solution to

$$\frac{1}{2}\left(\frac{\partial^2}{\partial x_0^2} + \Delta\right) w(x_0, x) = 0, \qquad u(0+, x) = f(x)$$

which we saw in Chapter 2 to be

$$\begin{aligned} w(x_0, x) &= \frac{1}{\pi^{(n+1)/2}}\Gamma\left(\frac{n+3}{2}\right)\int \frac{x_0 f(y)\,\mathrm{d}y}{(x_0^2 + (x-y)^2)^{(n+1)/2}} \\ &= \frac{1}{\pi^{(n+1)/2}}\Gamma\left(\frac{n+3}{2}\right)\int_0^\infty \frac{x_0 r(r^{n-2}\langle f\rangle(x,r))}{x_0^2 + r^2}\,\mathrm{d}r, \end{aligned} \tag{7.31}$$

where $\langle f\rangle(x,r) = \int_{\partial b(x,y)} f(x + r\xi)\,\mathrm{d}\Sigma(\xi)$, as usual. Let us set $c(n) = \Gamma(\frac{n+3}{2})/\pi^{(n+1)/2}$.

The passage from (P3) to (P1) depends on whether n is even or odd and, after rearranging the integral a bit, we replace x_0 by $\tau + it$, and take the limit of the real part of $w(\tau + it, x)$ as $\tau \to 0$.

Case n even: Make use of the identity

$$\left(-\frac{\partial}{\partial x_0}\right)\left(-\frac{1}{r}\frac{\partial}{\partial r}\right)^{(n-2)/2}(x_0^2 + r^2)^{-1/2} = 3\cdot 5\cdots\left(\frac{n-3}{2}\right)x_0(x_0^2 + r^2)^{-(n+1)/2}$$

in (7.31) and rename the constant to obtain, after integration by parts,

$$w(x_0, x) = d(n)\left(-\frac{\partial}{\partial x_0}\right)\int_0^\infty\left(-\frac{\partial}{\partial r}\right)^{(n-2)/2}\left[r^{n-2}\langle f\rangle(x,r)\right]\frac{r\,\mathrm{d}r}{(x_0^2 + r^2)^{1/2}}. \tag{7.32}$$

There is no question about the analyticity of $(x_0^2 + r^2)^{-1/2}$ in the x_0-complex plane without the segment $(-ir, ir)$ along the imaginary axis. To get the real and imaginary parts of $((\tau + it) + r^2)^{-1/2}$, we write $X + iy = ((\tau + it)^2 + y^2)^{1/2} = ((r^2 + \tau^2 - t^2) + 2it\tau)^{1/2}$ from which we obtain

$$\begin{aligned} 2X^2 &= r^2 + \tau^2 - t^2 + \left((r^2 + \tau^2 - t^2)^2 + (2 + \tau)^2\right)^{1/2}, \\ 2Y^2 &= -(r^2 + \tau^2 - t^2) + \left((r^2 + \tau^2 - t^2)^2 + (2t\tau)^2\right)^{1/2}. \end{aligned}$$

Since $-\partial/\partial x_0 = i\partial/\partial t$, we have to consider the imaginary part of $((\tau + it)^2 + r^2)^{-1/2}$ as $\tau \to 0$. The result is

$$\begin{aligned} u(t, x) &= \lim\nolimits_{\tau\to 0}\operatorname{Re} w(\tau + it, x) \\ &= d(n)\frac{\partial}{\partial t}\int_0^t\left(\frac{1}{r}\frac{\partial}{\partial r}\right)^{(n-2)/2}\left[r^{n-2}\langle f\rangle(x,r)\right]\frac{r\,\mathrm{d}r}{(t^2 - r^2)^{1/2}}. \end{aligned} \tag{7.33}$$

Case n odd: As above, insert

$$\left(-\frac{\partial}{\partial x_0}\right)\left(-\frac{1}{r}\frac{\partial}{\partial r}\right)^{(n-3)/2}(x_0^2 + r^2)^{-1} = \left(\frac{n-3}{2}\right)!2^{(n-1)/2}x_0(x_0^2 + r^2)^{-(n+1)/2}$$

into (7.31), integrate by parts, rename constants and obtain

$$w(x_0, x) = d(n)\left(-\frac{\partial}{\partial x_0}\right)\int_0^\infty \left(\frac{1}{r}\frac{\partial}{\partial r}\right)^{(n-3)/2} \left[r^{(n-2)/2}\langle f\rangle(x,r)\right]\frac{r\,dr}{\xi^2+r^2}.$$

To invert the transform, set $x_0 = \tau + it$ and consider the analytic function $(x_0^2 + r^2)^{-1}$ in the x_0-complex plane punctured at $\pm ir$. As above, take $\partial/\partial x_0$ along the imaginary axis, and we have to worry about the imaginary part of

$$(\tau^2 + r^2 - t^2 + 2it\tau)^{-1} = (\tau^2 + r^2 - t^2 - 2it\tau)/A(t)^2,$$

where $A(t) = [(\tau^2 + r^2 - t^2) + (2t\tau)^2]^{1/2}$. Now split the domain of integration into 3 parts: $[0, t-\tau)$, $[t-\tau, t+\tau]$ and $[t+\tau, \infty)$. We leave for the reader to verify that the integrals over the two end parts tend to zero as $\tau \to 0$. To evaluate the integral over $[t-\tau, t+\tau]$, note that $A(\tau) \simeq 2\tau t$ and $\tau^2/A(t) \to 0$ and $-2it\tau/A(t) \to -i$, thus, as $\tau \to 0$

$$\begin{aligned} d(n)i\frac{\partial}{\partial t}\int_{t-\tau}^{t+\tau}\left(\frac{1}{r}\frac{\partial}{\partial r}\right)^{(n-3)/2} &\left[\gamma^{n-2}\langle f\rangle(x,r)\right]\frac{\gamma\,d\gamma}{(\tau+it)^2+r^2} \\ \to d(n)\frac{\partial}{\partial t}\left(\frac{1}{r}\frac{\partial}{\partial r}\right)^{(n-3)/2} &\left[r^{n-2}\langle f\rangle(x,r)\right]_{r=t} \\ = d(n)t\left(\frac{1}{t}\frac{\partial}{\partial t}\right)^{(n-3)/2} &\left[t^{n-2}\langle f\rangle(x,t)\right]. \end{aligned}$$

Therefore, for n even

$$u(t,x) = \lim_{\tau\to 0} \operatorname{Re} w(\tau + it, x) = d(n)t\left(\frac{1}{t}\frac{\partial}{\partial t}\right)^{n-1}\left[t^{n-2}\langle f\rangle(x,t)\right]. \tag{7.34}$$

These two results cover the passage from (P3) to (P1) as well as the pending computation announced in Chapter 2 under the heading of (de)-complexification. For some heuristics about these results, take a look at the last section of this chapter.

7.5. The ascent method

The transmutation game provides us with a clue as to how to build (some kinds of) waves in multi-dimensional spaces from waves in lower-dimensional spaces. It is a converse of the descent method, hence the name.

Due to some strange coincidence, "independence" of the degrees of freedom and statistical independence go hand in hand when building many-dimensional diffusions from one-dimensional diffusions: n-dimensional imposed motion $B(t)$ is made up of n (statistically) independent one-dimensional imposed motions $B(t) = (B_1(t), \dots, B_d(t))$. The statistical independence is reflected in the fact that the fundamental solution of

$$\frac{\partial\rho}{\partial t}(x,y) = \frac{1}{2}\Delta\rho_t(x,y)$$

is given by

$$\rho_t(x,y) = (2\pi t)^{-d/2}\exp -\left(\frac{(x-y)^2}{2t}\right) = \prod_{i=1}^{n}(2\pi t)^{-1/2}\exp\{-((x_i-y_i)^2/2t)\}.$$

But this does not happen to the solutions of $\frac{\partial^2 u}{\partial t^2} = c^2\Delta u$. Nevertheless, the operators we introduced in Section 1 still do allow us to get away with something. What comes below is taken from a nice paper by Bragg.

Let H_1 and H_2 be generators of continuous groups $Q_1(t)$ and $Q_2(t)$ on a Banach space B. We assume that there is a dense linear subspace $\mathcal{D}$ of B stable under arbitrarily large powers of $G_1 = H_1^2$ and $G_2 = H_2^2$, and let us assume that G_1 and G_2 generate holomorphic semigroups $P_1(t)$ and $P_2(t)$ on B. We have already seen that, for $\phi \in \mathcal{D}$,

$$P_i(t)\phi = \int_0^\infty \frac{e^{-t^2/4\sigma}}{(4\pi\sigma)^{1/2}} Q_i(\sigma)\phi\,d\sigma, \quad i = 1, 2. \tag{7.35}$$

Define $V_i(t)$ by

$$V_i(t) = \frac{1}{2}\{Q_i(t) + Q_i(-t)\}, \quad i = 1, 2, \tag{7.36}$$

and notice from the above that

$$s^{-1/2}U_i(1/4s)\phi = \frac{1}{\sqrt{\pi}}\int_0^\infty \frac{e^{-s\sigma}}{\sqrt{\sigma}} V_i(\sqrt{\sigma})\phi\,d\sigma. \tag{7.37}$$

Note now that in terms of $V_i(t)$ the (abstract) wave problems

$$\begin{aligned} &\frac{d^2 W_i}{dt^2} = G_i W_i, \quad t > 0, \\ &W_i(0) = \phi_i, \qquad \frac{dW_i}{dt}(0) = 0, \quad \phi_i \in \mathcal{D} \end{aligned} \tag{7.38}$$

have the obvious solution

$$W_i(t) = V_i(t)\phi_i, \qquad i = 1, 2,$$

the main result being contained in

PROPOSITION 7.11. *With the notation and assumptions introduced above, the solution to the problem*

$$\begin{aligned} &\frac{d^2 W}{dt^2} = (G_1 + G_2)W(t), \quad t > 0, \\ &W(0+) = \phi, \qquad \frac{dW}{dt}(0+) = 0 \end{aligned} \tag{7.39}$$

is given by

$$W(t)=\frac{\mathrm{d}}{\mathrm{d}t}\left\{\frac{1}{2\pi}\int_0^{t^2}(t^2-\xi)^{-1/2}\left[\int_0^{\xi}\frac{V_1(\sqrt{\xi-\sigma})V_2(\sqrt{\sigma})\phi}{\sqrt{\xi-\sigma}\sqrt{\sigma}}\,\mathrm{d}\sigma\right]\mathrm{d}\xi\right\}. \tag{7.40}$$

PROOF. First transmute (7.39) into

$$\frac{\mathrm{d}u(t)}{\mathrm{d}t}=(G_1+G_2)u(t), \qquad u(0+)=\phi$$

and note that the commutativity of G_i implies that

$$u(t)=P_1(t)P_2(t)\phi,$$

and according to Proposition 1 of Section 1, we can obtain $\widetilde{W}(t)$ from $u(t)$ by

$$\widetilde{W}(t)=\sqrt{\pi/4}\mathcal{L}^{-1}\left\{s^{-1/2}(s^{-1/2}U_1(1/4s))(s^{-1/2}U_2(1/4s))\phi\right\}(t^2), \tag{7.41}$$

where $\widetilde{W}(t)$ satisfies (7.38) with initial conditions $\widetilde{W}(0+)=0$, $\frac{\mathrm{d}}{\mathrm{d}t}\widetilde{W}(t)=\phi$. To obtain $W(t)$ from it, just differentiate. To get even closer to the answer, note that, setting $\Psi(s)=s^{-1}U_2(1/4S)\phi$ and using (7.37), we have

$$\Psi(s)=\pi^{-1}\int_0^{\infty}\mathrm{e}^{-\sigma s}\frac{1}{\sqrt{\sigma}}V_2(\sqrt{\sigma})\phi\,\mathrm{d}\sigma$$

and use this $\Psi(s)$ instead of ϕ for the corresponding identity for V_1. Thus,

$$\begin{aligned}
&s^{-1/2}U_1(1/4s)\big(s^{-1/2}U_2(1/4s)\phi\big)\\
&=\pi^{-1/2}\int_0^{\infty}\mathrm{e}^{-s\xi}\frac{V_1(\sqrt{\xi})}{\sqrt{\xi}}\Psi(\xi)\,\mathrm{d}\xi\\
&=\pi^{-1}\int_0^{\infty}\mathrm{e}^{-s\xi}\frac{V_1(\sqrt{\xi})}{\sqrt{\xi}}\left(\int_0^{\infty}\mathrm{e}^{-\sigma S}\frac{V_2(\sqrt{\sigma})}{\sqrt{\sigma}}\phi\,\mathrm{d}\sigma\right)\mathrm{d}\xi\\
&=\pi^{-1}\int_0^{\infty}\int_0^{\infty}\mathrm{e}^{-s(\sigma+\xi)}\frac{V_1(\sqrt{\xi})}{\sqrt{\xi}}\frac{V_2(\sqrt{\sigma}}{\sqrt{\sigma}}\,\mathrm{d}\sigma\,\mathrm{d}\xi
\end{aligned}$$

skipping a few technicalities, a simple change of variables yields

$$s^{-1/2}U_1(1/4s)s^{-1/2}U_2(1/4s)=\pi^{-1}\int_0^{\infty}\mathrm{e}^{-s\tau}\left(\int_0^{\tau}\frac{V_1(\sqrt{\tau-\sigma})V_2(\sqrt{\sigma})}{\sqrt{(\tau-\sigma)\sigma}}\phi\,\mathrm{d}\sigma\right)\mathrm{d}\tau,$$

and therefore

$$\mathcal{L}^{-1}\left\{s^{-3/2}U_1(1/4s)U_2(1/4s)\right\}(t)=\frac{1}{\pi}\int_0^{t}\frac{V_1(\sqrt{t-\sigma})V_2(\sqrt{\sigma})}{\sqrt{(t-\sigma)\sigma}}\phi\,\mathrm{d}\sigma.$$

Now, replace t by t^2 and differentiate to obtain the desired result in (7.39). □

Comments: To insist on the meaning of the symbols in (7.39): for each σ, first apply $V_2(\sqrt{\sigma})$, afterwards $V_1(\sqrt{t-\sigma})$, and then sum over σ.

To obtain a nicer-looking formula, set $\sigma = \xi \sin^2\theta$ and then $\xi = \lambda^{1/2}$, and (6) becomes

$$W(t) = \frac{\mathrm{d}}{\mathrm{d}t}\frac{2}{\pi}\int_0^t (t^2-\lambda^2)^{-1/2}\left(\int_0^{\pi/2} V_1(\lambda\cos\theta)V_2(\lambda\sin\theta)\,\mathrm{d}\theta\right)\mathrm{d}\lambda.$$

It is clear that Proposition 1 can be extended to cover the case

$$\frac{\mathrm{d}^2W}{\mathrm{d}t^2}(t) = \left(\sum_{i=1}^n G_i\right)W(t),$$

$$W(0+) = \phi, \qquad \frac{\mathrm{d}}{\mathrm{d}t}W(0+) = 0$$

with G_i commuting and having a common domain.

The solution will be

$$\begin{aligned} W(t) &= \frac{\mathrm{d}}{\mathrm{d}t}\left\{\sqrt{\pi/4}\mathcal{L}^{-1}\left(s^{-3/2}\prod_{i=1}^n U_i(1/4s)\right)(t^2)\right\} \\ &= \frac{\mathrm{d}}{\mathrm{d}t}\left\{\sqrt{\pi}/4\mathcal{L}^{-1}\left(s^{(n-3)/2}\prod_{i=1}^n (s^{-1/2}U_i(1/4s))\right)(t^2)\right\}. \end{aligned}$$

The special case $n = 3$ takes care of our world.

Assume $H_1 = \partial/\partial x_1$, $H_2 = \partial/\partial x_2$, $H_3 = \partial/\partial x_3$. These generate the groups of translation along the axes, and each $V_i(t)$ happens to be

$$V_i(t)f(x) = \frac{1}{2}\{f(x+te_i) + f(x-te_i)\},$$

where e_i denotes the unit vector along the i-th axis.

We leave it up to the reader to verify that

$$\begin{aligned} &\mathcal{L}^{-1}\left(\prod_{i=1}^3 (s^{-1/2}U_i(1/4s))\phi\right)(t^2) \\ &= 4t\int_0^{\pi/2}\int_0^{\pi/2} V_1(t\cos\theta_1)V_2(t\sin\theta_1\cos\theta_2)V_3(t\sin\theta_1\sin\theta_2)\phi\sin\theta_1\,\mathrm{d}\theta_1\,\mathrm{d}\theta_2. \end{aligned}$$

If we take ϕ as a continuously differentiable function $f(x)$, then the last identity becomes

$$\mathcal{L}^{-1}\left(\prod_{i=1}^3 (s^{-1/2}U_i(1/4s))\phi(x)\right)(t^2)$$

$$= \frac{t}{2} \int_0^{\pi/2} \int_0^{\pi/2} \Big\{ \sum_l f(x_1 + t\cos\theta_1, x_2 \pm t\sin\theta_1\cos\theta_2, \\ x_3 \pm t\sin\theta_1\sin\theta_2) \Big\} \sin\theta_1 \, d\theta_1 \, d\theta_2.$$

Here $\sum_\pm$ denotes the summation over all 8 possible sign combinations, and for each sign combination, the integral is over the corresponding orthant of the unit sphere. Thus (7.39) becomes

$$W(t) = \frac{\partial}{\partial t} \left\{ \frac{1}{4\pi t} \int_{B(0,t)} f(x+\xi)\, d\Sigma(\xi) \right\}$$

after appropriate scaling.

Final comments. The nice thing about the result contained in Proposition 1 is that $V_i(t)$ can be obtained from the groups $Q_i(t)$ generated by H_i, which are supposedly simpler to obtain, and then to use the fact that for independent degrees of freedom, the solution of the "parabolic" problem is a product of solutions. The challenging open problem is to attach a physical interpretation to this procedure.

7.6. The closing of the circle. Some heuristics

We saw in Section 2 of Chapter 3 that we could solve the wave equation by factorization or by complexifying the time parameter. Let us do a bit of symbol manipulation to examine how and why these cling together.

Consider a very simple case: $\mathcal{L} = d^2/dx^2 = D^2$. Certainly, $H = D$ is such that $H^2 = \mathcal{L}$, and $\widehat{G} = iG = -i\sqrt{-\mathcal{L}}$ also satisfies $\widehat{G}^2 = \mathcal{L}$ but $G = -\sqrt{-\mathcal{L}}$ satisfies $G^2 = -\mathcal{L}$. In terms of the Fourier decomposition $f(x) = \int e^{-ikx} \hat{f}(k)\, dk/2\pi$, with $\hat{f}(k)$ rapidly decaying, we have

$$P_t f(x) = e^{t\mathcal{L}} f(x) = \int e^{-k^2 t} e^{ikx} \hat{f}(k) \frac{dk}{2\pi},$$

$$K_t f(x) = e^{tG} f(x) = \int e^{-|k|t} e^{ikx} \hat{f}(k) \frac{dk}{2\pi},$$

$$U_t f(x) = e^{t\widehat{G}} f(x) = \int e^{i|k|t} e^{ikx} \hat{f}(k) \frac{dk}{2\pi},$$

$$Q_t f(x) = e^{tH} f(x) = \int e^{ikt} e^{ikx} \hat{f}(k) \frac{dk}{2\pi}.$$

From these it is also clear that

$$\frac{1}{2}(U(t) + U(-t))\, f(x) = \frac{1}{2}(Q(t) + Q(-t)) f(x) = v(t,x)$$

and that $v(t,x)$ satisfies $\frac{\partial^2 v}{\partial t^2} = \mathcal{L}v$ and $v(0+,x) = f(x)$, $\frac{\partial v}{\partial t}(0,x) = 0$.

To go from K_t to U_t, one either complexifies the time variables or considers the group generated by the self-adjoint operator $\sqrt{-\mathcal{L}}$.

The semigroup K_t has a transition kernel

$$k_t(x,y) = \frac{t}{\pi(t^2 + (x-y)^2)}$$

and we have seen repeatedly that $K_t f(x) = w(t,x)$ satisfies $\frac{\partial^2 w}{\partial t^2} + \mathcal{L}w = 0$, $w(0,x) = f(x)$, $w(\infty,x) = 0$.

References

There are several papers by Bragg and Dettman from which we extracted material for the chapter. The reader will profit from taking a look at

1. Bragg, L. R. and Dettman, J. W.: Related partial differential equations and their applications, *SIAM J. Appl. Math.* **16** (1968), 459–467.
2. Bragg, L. R. and Dettman, J. W.: Related problems in partial differential equations, *Bull. Amer. Math. Soc.* **74** (1968), 375–378.
3. Bragg, L. R. and Dettman, J. W.: A class of Dirichlet and initial value problems, *Proc. Amer. Math. Soc.* **31** (1969), 50–56.
4. Bragg, L. R. and Dettman, J. W.: An operator calculus for related partial differential equations *J. Math. Anal. Appl.* **22** (1968), 261–271.
5. Dettman, I. W.: Initial boundary value problems related through the Stieltjes transform, *J. Math. Anal. Appl.* **25** (1969), 341–349.
6. Bragg, L. R.: Singular non-homogeneous abstract Cauchy and Dirichlet type problems related by a generalized Stieltjes transform, *Indiana Univ. Math. J.* **24** (1974), 183–195.
7. Bragg, L. R.: The ascent method for abstract wave problems, *J. Differential Equations* **38** (1980), 413–421.

For full mathematical treatment of the transmutation operators introduced in the list above, consult with

8. Carroll, R.: *Transmutation and Operator Differential Equations*, Notas de Matemáticas 67, North-Holland, Amsterdam, 1979.

See as well the two nice consecutive exposes by Hersch

9. Hersch, R.: "The method of transmutation" and "Stochastic solutions of hyperbolic equations", in Lecture Notes in Math. 446, Springer-Verlag, Berlin, 1975, pp. 265–300.
10. Hersch, R.: Explicit solutions of higher order abstract Cauchy problems, *J. Differential Equations* **8** (1970), 570–579.

The key ideas for the stuff in Section 2 were taken from

11. Rosecranz, S. I.: Diffusion transforms, *J. Differential Equations* **13** (1973), 457–467.

Of independent interest is the book

12. Goel, N. S. and Richter-Dyn, N.: *Stochastic Models in Biology*, Academic Press, New York, 1974.

from which we took material for an example in Section 4.

CHAPTER

EIGHT

MORE CONNECTIONS

In this chapter, we explore a few other connections between "waves" and "diffusions". The quotation marks mean that we shall not be considering hyperbolic wave equations or diffusions, but the equations describing or associated with oscillatory phenomena and other kinds of stochastic processes.

8.1. Waves in discrete structures and Markov chains

We have already considered an instance of this situation in Chapter 7. Consider a system of N (finite or infinite) particles whose positions at rest are described by vectors x_i^0 in some $\mathbb{R}^d$. They are bound to those by forces derivable from potential functions $V_i(x_i)$ which have a minimum at x_i^0. Also, when displaced from their equilibria, they interact with each other via forces derivable from potentials $V_{ij}(x_i - x_j)$. Again, these have minima at $x_i^0 - x_j^0$, and are even and symmetric with respect to the positions of the particles.

Folklore has it, that when the deviations from equilibrium are small, the equation of motion of $\xi_i := x_i - x_i^0$ is

$$m_i \frac{\mathrm{d}^2\xi_i}{\mathrm{d}t^2} = -k_i\xi_i - \sum_{j\neq i} k_{ij}(\xi_i - \xi_j) \tag{8.1}$$

for $i \geqslant 1$. We shall consider only, without loss of generality, initial conditions like

$$\xi_i(0), \qquad \dot{\xi}_i(0) = V_0(i), \quad i \geqslant 1. \tag{8.2}$$

And we are not worrying about forcing terms either.

Note that $k_{ij} = k_{ji} > 0$ for $i \neq j$ (the term $i = j$ being automatically excluded from (8.1)), and write

$$q_i = k_i/m_i, \qquad \lambda_i = \sum_{j \neq i} k_{ij}/m_i, \qquad \pi_{ij} = k_{ij}/\sum_{\ell \neq i} k_{i\ell}.$$

We chose to use q_i instead of the traditional w_i^2 for obvious reasons. Thus, (8.1) becomes

$$\frac{\mathrm{d}^2\xi_i}{\mathrm{d}t^2} = -q_i\xi_i + \lambda_i \sum_{j \neq i} \pi_{ij}(\xi_i - \xi_j)$$

and setting

$$q_{ij} = \lambda_i \pi_{ij} \quad \text{if } i \neq j \quad \text{and} \quad q_{ii} = -\lambda_i \sum_{j \neq i} q_{ij}$$

we rewrite it further as

$$\frac{\mathrm{d}^2\xi_i}{\mathrm{d}t^2} = \sum_j q_{ij}\xi_j - q_i\xi_i. \tag{8.3}$$

We have introduced enough notation so that solving (8.3) probabilistically is natural enough. But let us recall a few basic facts about the construction of a Markov chain from $\{\lambda_i : i = 1, \ldots, N\}$ and $\{\pi_{ij} : 1 \leqslant i \neq j \leqslant N\}$.

For each i; the particle will stay there during the time exponentially distributed with the parameter λ_i. After that, it will jump to other states j with probabilities π_{ij}.

Thus, if we denote by $P_{ij}(t)$ the transition probabilities of this chain, i.e., $P_{ij}(t) = P^i(X(t) = j)$, then

$$\frac{\mathrm{d}}{\mathrm{d}t} P_{ij}(t) = \sum_k q_{ik} P_{kj}(t), \qquad P_{ij}(0) = \delta_{ij} \tag{8.4}$$

and if $F(i)$ denotes any bounded function and $q(i)$ any positive function, then the solution to

$$\frac{\partial u}{\partial t}(t, i) = \sum q_{ij} u(t, j) - q(i)u(t, i); \qquad u(0, i) = F(i) \tag{8.5}$$

is given by

$$u(i, t) = E^i \left\{ F(X(t)) \exp\left\{ -\int_0^t q(X(s))\,\mathrm{d}s \right\} \right\}. \tag{8.6}$$

$E^i(\cdot)$ denote averaging over a collection of all possible trajectories of the process, with respect to probabilities P^i constructed from $P_{ij}(t)$ as described in Chapter 1.

To finish this digression, if we denote by T the waiting time at a given state, i.e., $T = \inf\{t > 0 : X(t) \neq X(0)\}$, then $P^i(T > t) = \exp(-\lambda_i t)$, and the simple Markov property applied to the splitting of (8.5) as

$$\begin{aligned} u(t,i) &= E^i\{F(X(t))\mathrm{e}^{-\int_0^t q(X(s)\,\mathrm{d}s)}\} \\ &= E^i\{-; t < T\} + E^i\{-; T \leqslant t\} \end{aligned}$$

yields the following (renewal) equation for $u(t,i)$

$$u(t,i) = \mathrm{e}^{-(\lambda_i + q_i)t} F(i) + \lambda_i \sum_{j \neq i} \int_0^t \mathrm{e}^{-(q_i+\lambda_i)(t-s)} \pi_{ij} u(s,j)\,\mathrm{d}s. \tag{8.7}$$

That this satisfies (8.4) can be easily verified. To relate (8.4)–(8.6) to (8.3), just apply the transmutation operator (7.1) to $\xi(t,i)$ to obtain

$$v(t,i) = (T_2\xi)(t,i) = \int_0^t s \frac{\mathrm{e}^{-s^2/4t}}{(4\pi t^3)^{1/2}} \xi(s,i)\,\mathrm{d}s,$$

and $v(t,i)$ satisfy

$$\frac{\partial V}{\partial t}(t,i) = \sum_{j \neq 0} q_{ij} V(t,j) - q(i)V(t,i) \tag{8.8}$$

with the initial condition $V(0+,i) = V_0(i)$. That $\xi_i(t)$ and $V(t,i)$ take values in $\mathbb{R}^d$ should not bother us. According to (8.5), the solution to (8.8) can be represented as

$$v(t,i) = E^i\left\{V_0(X(t)) \exp\left\{-\int_0^t q(X(s))\,\mathrm{d}s\right\}\right\},$$

and, according to (8.6),

$$V(t,i) = V_0(i)\mathrm{e}^{-(q_i+\lambda_i)t} + \lambda_i \sum_{j \neq i} \int_0^t \mathrm{e}^{-(q_i+\lambda_i)(t-s)} \pi_{ij} V(s,j)\,\mathrm{d}s. \tag{8.9}$$

If we denote by $\widetilde{V}(\alpha,i)$ the Laplace transform of $V(t,i)$, then (8.9) becomes, with $f_i^2 = q_i + \lambda_i$, as follows:

$$\widetilde{V}(\alpha,i) = \frac{V_0(i)}{f_i^2 + \alpha} + \frac{\lambda_i}{f_i^2 + \alpha} \sum_{j \neq i} \pi_{ij} \widetilde{V}(\alpha,j) \tag{8.10}$$

and we saw in Section 4 of Chapter 7 that $\tilde{\xi}(\alpha, i) = \widetilde{V}(\alpha^2, i)$, so, since $\sin wt/w$ has the Laplace transform $1/(w^2+\alpha^2)$, it is clear that $\xi(t, i)$ satisfies the integral equations

$$\xi(t,i) = V_i(0)\frac{\sin f_i t}{f_i} + \lambda_i \int_0^t \sum_{j\neq i} \frac{\sin f_i(t-s)}{f_i}\pi_{ij}\xi(s,j)\,\mathrm{d}s, \tag{8.11}$$

the result which we challenge the reader to obtain directly from (8.3).

Naturally, if the eigenvalues and eigenvectors of k_{ij} or of q_{ij} are known, then we can integrate (8.3) or (8.8) easily.

Had we proceeded, as in Chapter 4, to integrate (8.3), we would have taken a Laplace transform on both sides of it and obtained

$$\alpha^2\tilde{\xi}(\alpha,i) - \sum_j (q_{ij} - q_i\delta_{ij})\xi_j = v_0(i),$$

the solution to which can be written as

$$\tilde{\xi}(\alpha,i) = \int_0^\infty \mathrm{e}^{-\alpha^2 t} E^i\left\{\mathrm{e}^{-\int_0^t q(X(s))\,\mathrm{d}s} V_0(X(t))\right\}\mathrm{d}t. \tag{8.12}$$

If we split the integral over $(0, T)$ and (T, ∞) and used the Markov property, we would obtain the same representation as (8.10) with α replaced by α^2.

Comment: Even though (8.12) can be explicitly computed in simple cases only, it can be used in computation by simulation. For that, you run a given number of trajectories starting from each point i. (Actually this makes the process shamelessly parallelizable.) You do this for a collection of values of α and then get your back broken by trying to invert the Laplace transform to obtain $\xi(t, i)$. (That's what happened to us.)

Apart from being (doubtfully) useful for direct modeling, i.e., given the medium, describe the waves, (8.12) can be used for inverse modeling as we shall see next chapter.

8.2. Approximate Laplacians, regular jump processes and random flights

8.2.1. Approximate Laplacians

We shall be considering wave fields in a variety of spaces. The wave character of the equations used to model the phenomenon being determined by the time dependence of the equation, in our case, the equations will be of second order in the time derivatives.

Consider, to begin with, a state space, i.e., our old friend $(E, \mathcal{E})$, a positive function $\lambda(x): E \to (0, \infty)$, and a probability kernel $\pi(x, \mathrm{d}y): E \times \mathcal{E} \to [0, 1]$. Think of E as $\mathbb{R}^d$, a lattice in $\mathbb{R}^d$, or a discrete structure as in the previous section. The functions $\lambda(x)$ and $\pi(x, \mathrm{d}y)$ will describe the jumps out of x.

A few examples may describe the title further. Consider $E = \mathbb{R}^d$ and $\sigma(\mathrm{d}\xi)$ to be a uniform measure on the unit sphere. Define, for each $h > 0$, the kernels

$$\pi_h f(x) = \int f(x + h\xi)\sigma(\mathrm{d}\xi)$$

and note that, if $f(x)$ is twice continuously differentiable, and $\lambda = 1/h^2$, independently of x, then

$$\Delta_h f(x) := \lambda(\pi_h f(x) - f(x)) \to \frac{1}{2}\Delta f(x)$$

as $h \to 0$. A similar situation is obtained when $\pi_h(x, \mathrm{d}y) = (1/2d)\sum_{i=1}^{d} \varepsilon_{x \pm he_i}(\mathrm{d}y)$, where e_i are the unit vectors along each direction and $\varepsilon_a(\mathrm{d}y)$ denotes point mass at a. Then the approximate Laplacian is given by

$$\Delta_{\mathrm{ap}} f(x) = \lambda(\pi_h f(x) - f(x)) = \frac{\lambda}{2d}\sum_{i=1}^{d}(f(x \pm he_i) - f(x))$$

and if $f(x)$ is twice continuously differentiable and $\lambda = d/h^2$, it is clear that

$$\lambda(\pi_f(x) - f(x)) \to \frac{1}{2}\Delta f(x)$$

as well.

To finish this list, consider now a probability measure $\pi(\mathrm{d}\xi)$, concentrated on the surface S of an ellipsoid, to be such that for all i, j:

$$\int \xi_i \pi(\mathrm{d}\xi) = 0 \quad \text{and} \quad \int \xi_i \xi_j \pi(\mathrm{d}\xi) = a_{ij},$$

and define $\pi_h f(x)$ by

$$\pi_n f(x) = \int_S f(x + h\xi)\pi(\mathrm{d}\xi);$$

clearly for $\lambda = 1/h^2$, this time

$$\lambda(\pi_h f(x) - f(x)) \to \sum a_{ij}\partial_{ij}^2 f(x).$$

We see below how to generalize so that a_{ij} depend on the point x. To further explain the connection with discretization, consider the one-dimensional example

$$\frac{1}{2}a(x)\frac{\partial}{\partial x}\left(b(x)\frac{\partial}{\partial x}\right).$$

If we were to replace $\partial/\partial x$ by the symmetric difference operator, i.e., $\nabla_h f(x) = \{f(x + h/2) - f(x - h/2)\}/h$, then, depending on whether we were to approximate the diffusion by random flights or by jump processes, we would have to replace $\mathbb{R}$ by a lattice of size h or not. So let us leave x undefined for the time being.

Define now $\pi_h(x, dy)$ to be the kernels

$$\pi_h(x, dy) = \frac{b(x+h/2)}{b(x+h/2)+b(x-h/2)}\varepsilon_{k+h/2}(dy) + \frac{b(x-h/2)}{b(x+h/2)+b(x-h/2)}\varepsilon_{x-h/2}(dy)$$

and $\lambda(x) = a(x)\{b(x+h/2)\}/h^2$. If we were to consider x to vary in a lattice of the step h, we would put $x = ih, i = 1, \pm 1, \pm 2$, etc.

8.3. Regular jump processes

In this section, we sketch the construction of the class of processes briefly mentioned in Section 1. We want processes that jump between states, and wait for an exponentially distributed time (a contingent condition) before jumping.

So let $(E, \mathcal{E})$ be our state space and form $E_a = E \times [0, \infty)$, $\mathcal{E}_a = \mathcal{E} \otimes \mathcal{B}([0, \infty))$. Set $(\Omega, \mathcal{J}) = (E_a, \mathcal{B}(E_a))$.

If $w = \{(x_n, t_n) : n \geqslant 0\}$, put $Y_n(w) = (Z_n(w), \tau_n(w))$ where $Z_n(w) = x_n$ and $\tau_n(w) = t_n$.

Z_n are to be interpreted as random positions in which the system is to be found at the random jump times τ_n.

The statistical nature of the process is to be put together from the data $\lambda(x) : E \to (0, \infty)$ and $\pi(x, dy) : E \times \mathcal{E} \to [0, 1]$. With these construct the transition kernel on E_a by means of

$$\pi_a(x, s; dy, dt) = \begin{cases} 0 & \text{if } s > t, \\ \pi(x, dy)\lambda(x)e^{-\lambda(x)(t-s)}\,dt & \text{if } s \leqslant t. \end{cases} \tag{8.13}$$

We also assume that $\pi(x, \{x\}) = 0$.

That is, at each jump, if the time spent at x is $t - s$, the probability of jumping within dt seconds and landing in dy is given by $\pi_a(x, s; dy, dt)$.

Now, for each initial measure μ on the auxiliary state space E_a, there exists a measure P^μ on $(\Omega, \mathcal{J})$ making $\{Y_n : n \geqslant 0\}$ a Markov chain with a transition probability

$$P^\mu(Y_{n+1} \in A | Y_0, Y_1, \ldots, Y_n) = \int_A \pi_a(Z_n, \tau_n; dy, ds), \tag{8.14}$$

$$P^\mu(Y_0 \in B) = \mu(B)$$

for any $B, A \in \mathcal{E}_a$. In particular, if $A = A_0 \times [0, \infty)$ and $\alpha > 0$

$$\begin{aligned} P^\mu[Y_{n+1} \in A | Y_0, \ldots, Y_n] &= \int_{A_0} \pi(Z_n, dy) = \pi(Z_n, A_0), \\ E^\mu\left[e^{-\alpha(\tau_{n+1}-\tau_n)}\right] &= E^\mu\left[\frac{\lambda(Z_n)}{\alpha + \lambda(Z_n)}\right]. \end{aligned} \tag{8.15}$$

Also, from the definition of π_a and the assumptions it is clear that

$$P^{\gamma}(Z_{n+1}=Z_n)=0, \qquad P^{\gamma}(\tau_0=0)=1, \qquad P^{\gamma}(\tau_{n+1}\leqslant \tau_n)=0.$$

When $\gamma(\mathrm{d}x,\mathrm{d}t)=\varepsilon_{x_0}(\mathrm{d}x)\varepsilon_0(\mathrm{d}t)$, we shall write $P^{x_0}(\cdot)$ for the corresponding measure. To go from the chain $\{Y_n : n\geqslant 0\}$ to a continuous time process $\{X(t); t\geqslant 0\}$, we proceed as follows: First define $\zeta(w)=\lim t_n(w)$. If $\zeta(w)<\infty$, adjoin a cemetery point Δ to E (do not mix up with the Laplace operator. Here the tradition is stronger than common sense) and define

$$X_t(w)=\begin{cases} Z_n(w) & \text{if } \tau_n(w)\leqslant t<\tau_{n+1}(w),\\ \Delta & \text{if } \zeta(w)\leqslant t.\end{cases} \tag{8.16}$$

$X_t(w)$, as defined in (8.16), is right continuous and has left limits on $[0,\zeta(w))$. As a matter of fact, $t\to X_t(w)$ is a step function (constant on (τ_n,τ_{n+1})). Defining the shift operators θ_t by $\theta_t(w)=\{(X_{n+k},(t_{n+k}-t)\vee 0); n\geqslant 0\}$ and skipping a few details, we have:

THEOREM 8.1. *$X=(\Omega,\mathcal{J},X_t,\theta_t,P^x)$ is a Markov process with a state space $(E,\mathcal{E})$.*

Let us introduce the notation

$$\pi_\alpha(x,\mathrm{d}y)=\frac{\lambda(x)}{\alpha+\lambda(x)}\pi(x,\mathrm{d}y) \tag{8.17}$$

and note that, if $k(x,\mathrm{d}y): E\times\mathcal{E}\to[0,1]$ is a kernel, its iterates can be defined by $k^0(x,\mathrm{d}y)=\varepsilon_x(\mathrm{d}y)$, $k^{n+1}(x,\mathrm{d}y)=\int k(x,\mathrm{d}z)k^n(z,\mathrm{d}y)$. With this notation and a simple induction argument, it is easy to see that

$$\int_0^\infty \mathrm{e}^{-\alpha t}\pi_a^n(x,0;A,\mathrm{d}t)=\pi_\alpha^n(x,A) \tag{8.18}$$

and we have

LEMMA 8.1. *Let $g\in\mathcal{E}_+$, $\alpha\geqslant 0$. Then for each x*

(i) $E^x\{g(Z_k)(\mathrm{e}^{-\alpha\tau_k}-\mathrm{e}^{-\alpha\tau_{k+1}})|Y_0,\ldots,Y_n\}$
$$=\alpha\mathrm{e}^{-\alpha\tau_n}\pi_\alpha^{n-k}\left(Z_n,\frac{g}{\alpha+\lambda}\right)$$

if $k\geqslant n$; as well as

(ii) $E^x\big[\mathrm{e}^{-\alpha\tau_{n+1}}g(Z_{n+1}); \tau_{n+1}>t|Y_0,Y_1,\ldots,Y_n\big]$
$$=\pi_\alpha(Z_n,g)\exp\{\lambda(x)\tau_n-(\alpha+\lambda(Z_n)(t\vee\tau_n)\}$$

for each $n\geqslant 0$ and $t\geqslant 0$.

This lemma plus the simple computation

$$\int_0^\infty e^{-\alpha t} g(X_t)\,dt = \int_0^\infty e^{-\alpha t} g(X_t)\,dt = \sum_{k\geqslant 0}\int_{\tau_k}^{\tau_{k+1}} e^{-\alpha t} g(X_t)\,dt$$
$$= \sum g(Z_k)\left(\frac{e^{-\alpha\tau_k} - e^{-\alpha\tau_{k+1}}}{\alpha}\right)$$

yield the important identity

$$\begin{aligned} U^\alpha g(x) &= \sum_{k=0}^{\infty} \pi_\alpha^k\left(x, \frac{g}{\alpha+\lambda}\right) \\ &= \frac{g(x)}{\alpha+\lambda(x)} + \pi_\alpha(x, U^\alpha g). \end{aligned} \tag{8.19}$$

This is important for it is related to the solutions of the wave equation

$$\frac{\partial^2}{\partial t^2} u(t,x) = \lambda(x)(\pi u(t,x) - u(t,x)) \tag{8.20}$$

with initial conditions $u(0,x) = 0$, $u_t(0,x) = g(x)$ by means of

$$\tilde{u}(\alpha, x) = U^{\alpha^2} g(x) \tag{8.21}$$

since, as must be clear from the examples above, the infinitesimal generator of $X(t)$ is just $\lambda(x)(\pi f(x) - f(x))$.

To tie all this with the extended Laplacians of the previous section, we note that when λ is constant, there is an easier description of the passage from the process Z_n to the process $X(t)$. Let $\{Z_n; n \geqslant 0\}$ denote the Markov chain with a transition probability given by the kernel $\pi(x, dy)$. Denote, as above, by $\pi^{(n)}(x, dy)$ the iterates of $\pi(x, dy)$, and let $N(t)$ denote a Poisson process of intensity λ, independent of $\{Z_n; n \geqslant 0\}$. Define, for $f \in \mathcal{E}_+$ or $f \in b\mathcal{E}$,

$$P_t f(x) = \sum \frac{(\lambda t)^n e^{-\lambda t}}{n!} \prod^{(n)} f(x) \tag{8.22}$$

and verify that P_t is a semigroup. For that just use the independence of the increments of $N(t)$. It is easy to see that

$$\left.\frac{dP_t}{dt} f(x)\right|_{t=0} = \lambda\left(\prod f(x) - f(x)\right)$$

and that the analogue of (8.17) holds, i.e., if $V^\alpha f(x)$ denotes the resolvent of (8.22), then

$$V^\alpha g(x) = \frac{g(x)}{\alpha+\lambda} + \prod_\alpha V^\alpha g(x)$$

and therefore $V^\alpha g(x) = U^\alpha g(x)$, or $P_t g(x)$ is indeed the transition semigroup of the process $X(t)$ constructed above. To tie with transmutation theory, if $V(t,x)$ is the transmuted version of $u(t,x)$ as in Section 1, then

$$\frac{\partial v}{\partial t}(t,x) = \lambda\Big(\prod v(t,x) - v(t,x)\Big), \qquad v(0,x) = f(x)$$

and according to (8.21), $v(t,x) = P_t f(x)$. Invoking the results of Chapter 7, we obtain $u(t,x)$.

8.4. Random flights

Let us begin by examining once more the example sketched at the end of the previous subsection. During any time interval $[0,t]$ we shall take $N(t)$ steps, where $N(t)$ is a given Poisson process of intensity λ. We begin at a random position ξ_0, distributed according to $\mu(\mathrm{d}x_0)$, and take independent steps of size ξ_k, distributed according to a common law $m(\mathrm{d}x)$. The position at time t is $X(t) = \xi_0 + \sum_{k=1}^{N(t)} \xi_n$, which has the characteristic function

$$\mathrm{e}^{t\psi(k)} = E^\mu\big[\mathrm{e}^{ik\cdot X(t)}\big] = \int \mathrm{e}^{ik\cdot x_0}\mu(\mathrm{d}x_0)\mathrm{e}^{\lambda t\int(\mathrm{e}^{iky}-1)m(\mathrm{d}y)},$$

where $E^\mu(\cdot) = \mu(\cdot)\otimes P_a(\cdot)$, and P_a is a given law on an auxiliary space $(\Omega_a, \mathcal{F}_a)$ on which $N(t)$ and $\{\xi_n : n \geqslant 1\}$ are defined. If $f(x)$ is defined on $\mathbb{R}^d$ and can be recovered from its Fourier transform $\hat{f}(k)$, then

$$P_t f(x) = \int \hat{f}(k)\mathrm{e}^{t\psi(k)}\,\mathrm{d}k/(2\pi)^d$$

is the transition semigroup of $X(t)$. By differentiation we obtain its infinitesimal generator to be

$$\begin{aligned} Gf(x) &= \lambda \iint \mathrm{e}^{ik\cdot x}(\mathrm{e}^{ik\cdot y} - 1)\hat{f}(k)\frac{\mathrm{d}k}{(2\pi)^d}m(\mathrm{d}y) \\ &= \lambda \int (f(x+y) - f(x))m(\mathrm{d}y). \end{aligned}$$

This is to be contrasted to the examples in Subsection 2.1, in which the size of the steps was fixed (equal to 1), but their direction was random.

A description that is more tied to the physical description of a free motion in any manifold is as follows. Consider a d-dimensional manifold, think of $\mathbb{R}^d$ with a metric $g_{ij}(x)$ if you will. Let M denote the manifold and E denote the tangent bundle to M, i.e., the manifold plus its tangent spaces at every point. Again, think of $\mathbb{R}^d$ plus its tangent planes at every point. If $z = (x, v)$ and $z' = (x, v')$ are two points on the tangent space to M at x, $g(z, z') := \sum g_{ij}(x)v_i v'_j$ is their scalar product. If $\gamma : [t_1, t_2] \to M$ is a differentiable curve on M such that $\gamma(t_1) = x_1$ and $\gamma(t_2) = x_2$, then $\ell(\gamma) = \int_{t_1}^{t_2} (g(\dot\gamma, \dot\gamma))^{1/2}\,\mathrm{d}t$ is the length of γ. For any x_1, x_2,

$$d(x_1, x_2) = \inf\{\ell(\gamma) : \gamma \text{ geodesic joining } x_1 \text{ to } x_2\},$$

where the geodesics are the curves satisfying

$$\ddot\gamma^i + \Gamma^i_{jk}\dot\gamma^j\dot\gamma^k, \qquad \gamma(0) = x_t; \qquad \dot\gamma(0) = v \tag{8.23}$$

as we saw in Chapter 5.

We assume that M is a complete manifold, i.e., any geodesic can be defined for $-\infty < t < \infty$. Again, let $N(t)$ be a given Poisson process of intensity λ. For a starting point $z = (x, v)$, define $Z(t) = (X(t), V(t))$ as follows: For $0 \leqslant t < \tau_1$, with τ_k denoting the successive jump times of $N(t)$, let $\gamma_t(x, v)$ be the geodesic satisfying (8.22). At τ_1 the position is $X(\tau_1) = \gamma_{\tau_1}(x, v)$ and the velocity is $V(\tau_1 -) = \dot\gamma_{\tau_1 -}(x, v) = \lim_{s\uparrow\tau_1} \gamma_s(x, v)$. At $t = \tau_1$ the velocity will receive a kick and jumps to $V(\tau_1)$ which is distributed uniformly on the geodesic unit sphere $S(X(\tau_1)) = \{v : \sum g_{ij}(X(\tau_1))v_i v_j = 1\}$ of the tangent space at $X(\tau_1)$. Now, for $\tau_1 \leqslant t < \tau_2$, we put $X(t) = \gamma_{t-\tau_1}(X(\tau_1), V(\tau_1))$ and $V(t) = \dot\gamma_{t-\tau_1}(X(\tau_1), V(\tau_1))$, and so on. The only issue we want to stress is that at each τ_k, $V(\tau_k -) = \lim_{t\uparrow\tau_k} V(t) \neq V(\tau_k)$, which is chosen afresh at a new sphere at the present position $X(\tau_k)$. This time the Markov process $Z(t)$ has two (d-dimensional components) $X(t)$ and $V(t)$, as expected from considering (8.23).

Actually, what we obtain in the construction sketched above, is (8.23) perturbed by random pulses. We leave it to the interested reader to play with this. We will only compute the infinitesimal generator of the $Z(t)$ process.

For that, let f be bounded and continuously differentiable. Then

$$P_t f(x, v) = E^{(x,v)} f(Z_t) = E^x[f(Z(t)); t < \tau_1] + E^{(x,v)}[f(Z(t)); \tau_1 \leqslant t]$$

and we have skipped the details about the construction of the measures $P^{(x,v)}$, but they are pretty much as we have described above in Section 3. Check with the book by Pinsky in the references, anyway. The first term on the right-hand side is easy to write down, it is just

$$E^{(x,v)}[f(Z(t)); t < \tau_1] = \mathrm{e}^{-\lambda t} f(\gamma_t(x, v), \dot\gamma_t(x, v)),$$

and, for the second one, use the strong Markov property at τ_1 to obtain

$$E_{x,v}[f(Z(t)); \tau_1 < \tau] = \lambda \int_0^t \mathrm{e}^{-\lambda s} \int_{S(\gamma_s(x,v))} P_{t-s} f(\gamma_s(x, v), \xi)\pi(\gamma_s(x, v), \mathrm{d}\xi).$$

If we denote $Df(x, v) = \frac{\mathrm{d}}{\mathrm{d}t} f(\gamma_t(x, v), \dot\gamma_t(x, v))|_{t=0}$, then

$$\frac{\mathrm{d}}{\mathrm{d}t} E^{(x,v)}[f(Z(t)); t < \tau_1] = Df(x, v) - \lambda f(x, v).$$

Clearly

$$\frac{\mathrm{d}}{\mathrm{d}t}E^{(x,v)}[f(Z(t));\tau_1 \leqslant t] = \lambda\pi f(x,v),$$

where, as usually said, by abuse of notation we have set

$$\pi f(x,v) = \int_{S(x)} f(x,\xi)\pi(x,\mathrm{d}\xi).$$

Therefore, putting all this into (8.23), we obtain

$$Gf(x,v) = Df(x,v) + \lambda(\pi - I)f(x,y). \tag{8.24}$$

How to obtain the Laplace–Beltrami operator from this is spelled out in Pinsky's book.

We leave it for the reader to explore what happens when $g_{ij} = \delta_{ij}$. The importance of (8.24) would lie in the possibility of considering wave equations like

$$\frac{\partial^2 u(t,x)}{\partial t^2} = Gu(t,x) + S(t,x), \quad t > 0,\ x \in \mathbb{R}^d$$

plus initial conditions, and then obtaining the path integral representation for the solutions, using the Markov process having G as a generator.

8.5. Random evolutions

Random evolutions are the name of an interesting method to piece together different systems evolving in time. As a motivation, consider the equation describing the propagation of waves in a stratified medium, which, after some changes of variables, was found to be

$$\frac{\partial^2 u}{\partial t^2} = \frac{\partial^2 u}{\partial x^2} + 2b(x)\frac{\partial u}{\partial x} + s(t,x) \tag{8.25}$$

which, when we introduce the new functions u_1 and u_2, by

$$u_1 = \frac{1}{2}\left(\frac{\partial u}{\partial x} + \frac{\partial u}{\partial t}\right), \qquad u_2 = \frac{1}{2}\left(\frac{\partial u}{\partial t} - \frac{\partial u}{\partial x}\right)$$

takes a trivial computation to verify that

$$\begin{aligned}\frac{\partial}{\partial t}\begin{pmatrix} u_1 \\ u_2 \end{pmatrix} &= \begin{pmatrix} \partial/\partial x & 0 \\ 0 & -\partial/\partial x \end{pmatrix}\begin{pmatrix} u_1 \\ u_2 \end{pmatrix} \\ &\quad + \begin{pmatrix} -(-b(x)) & (-b(x)) \\ (b(x)) & -(b(x)) \end{pmatrix}\begin{pmatrix} u_1 \\ u_2 \end{pmatrix} + \begin{pmatrix} S \\ S \end{pmatrix}.\end{aligned} \tag{8.26}$$

To generalize, note that the number of components may be arbitrary, but we keep them finite in number. Instead of the first term, consider

$$\widetilde{G}\tilde{u} = \begin{pmatrix} G_1 & 0 & \dots & 0 \\ 0 & G_2 & \dots & 0 \\ \vdots & \vdots & \dots & \vdots \\ 0 & 0 & \dots & G_n \end{pmatrix} \begin{pmatrix} u_1 \\ u_2 \\ \vdots \\ u_n \end{pmatrix}.$$

That is, $\widetilde{G}$ is a diagonal matrix-operator, and each u_i belongs to the domain $\mathcal{D}(G_i)$ of the corresponding G_i. We assume that $\mathcal{D}_c = \bigcap \mathcal{D}(G_i)$ is a dense linear subspace of a Banach space B.

Also, we assume that G_i generate contraction semigroups $S_i(t)$ on B. The second term can be generalized to

$$\sum_{j=1}^{n} q_{ij} T_{ij} u_j \quad \text{for each } i = 1, \dots, n,$$

where q_{ij} is the infinitesimal generator (or rate matrix) of a Markov chain $\xi(t)$ that has the state space $[1, n] = \{1, 2, \dots, n\}$, and T_{ij} are bounded operators on B.

To finish, instead of the last term, we could have an arbitrary vector $\tilde{s}$ with components in B. Spelling things out in terms of components, the generalization of (8. 26) looks like

$$\frac{\partial u_i}{\partial t} = (G_i - a_i)u_i + \sum_j q_{ij} T_{ij} u_j + s_i(t). \tag{8.27}$$

Recall that the infinitesimal generator of a (conservative, i.e., $\sum_j q_{ij} = 0$) Markov chain has some inner structure. It can be written as $q_{ij} = \lambda_i(\pi_{ij} - \delta_{ij})$, where λ_i is the parameter of the (exponential) distribution of the holding time at i, i.e., $P^i(T > t) = \exp(-\lambda_i t)$ for $T = \inf\{t > 0; \xi(t) \neq \xi(0)\}$, and $P^i(X(T) = j) = \pi_{ij}$ when $j \neq i$.

Therefore, (8.27) can be rewritten as

$$\frac{\partial u}{\partial t} = (G_i - a_i - \lambda_i T_{ii})u_i + \sum_{j \neq i} \lambda_i \pi_{ij} T_{ij} u_j + s_i \tag{8.28}$$

and the reason for the appearance of a_i in (8. 27) becomes clear: we can gain some generality or compensate the other term away by an appropriate choice of a_i.

There are different ways of representing the solutions to (8.28), depending on the nature of the operators G_i.

Let us collect some assumptions out of the context of the statement of the representation theorem.

ASSUMPTION 1. Let $S_i(t); i = 1, \dots, n$ be C_0 contraction semigroups on a common Banach space $\mathbf{B}$ such that $\mathcal{D}_c = \bigcap \mathcal{D}(G_i)$ is linear and dense in $\mathbf{B}$.

ASSUMPTION 2. For each $j \neq i$, $T_{ij} : \mathbf{B} \to \mathbf{B}$ is bounded.

ASSUMPTION 3. Assume that $T_i \equiv T_{ii} : \mathbf{B} \to \mathbf{B}$ are all bounded, commute among themselves, and that the solution to

$$\frac{\mathrm{d}R}{\mathrm{d}t} = -\lambda_{\xi(t)} T_{\xi(t)} R, \quad t > 0$$

and $R(0) = I$, exists and is a bounded operator for all $t > 0$.

ASSUMPTION 4. Assume that $R(t)$ commutes with T_{ij}'s and $S_i(t)$'s.

THEOREM 8.2. *With the notation and assumptions introduced above, let*

$$m(t) = \exp - \left\{ \int_0^t k(\xi(s))\,\mathrm{d}s \right\}$$

for a bounded function $k(i)$. *Now, set*

$$M(t) = m(t) R(t) S_{\xi_0}(\tau_1) T_{\xi_0 \xi_1} S_{\xi_1}(\tau_2 - \tau_1) T_{\xi_1 \xi_2} \ldots T_{\xi_{n-1} \xi_n} S(t - \tau_n)$$

if $\tau_n \leqslant t < \tau_{n+1}$, *where* $\tau_0 = 0, \tau_1 = \inf\{t > 0 : \xi(t) \neq \xi(0)\}$ *and by recurrence* $\tau_{n+1} = \tau_n + \tau_1 \circ \theta(\tau_n)$. *Then*

$$u(i, t) = E^i[M(t) f(\xi(t))] \tag{8.29}$$

solves (8.28) *with* $s_1 \equiv 0$ *and* $u(i, 0) = f(i)$.

Comment: We put $\xi_n = \xi(\tau_n)$. Clearly $\xi_{n-1} = \xi(\tau_{n^-}) = \xi(\tau_{n-1})$.

PROOF OF (8.29). Let us outline the basic ideas. Decompose the right-hand side of (8.29) into $E^i[M(t) f(\xi(t)); t < T_1]$ and $E^i[M(t) f(\xi(t)); T_1 \leqslant t]$. Clearly

$$E^i[M(t) f(\xi(t)), t < T_1] = \exp\bigl(-(\lambda_1 + k_i + \lambda_1 T) S_i(t) f(i)\bigr)$$

and an application of the strong Markov property at time T_1 yields

$$E^i[M(t) f(\xi(t)); T \leqslant t] = \lambda_i \int_0^t \mathrm{e}^{-(\lambda_i + k_i + \lambda_i T_1)s} \sum_j \pi_{ij} T_{ij} U(j, t - s)\,\mathrm{d}s.$$

□

We leave for the reader to verify that the right-hand side of (8.29) defines a semigroup, or check with references [2] or [3] of this chapter. Therefore, to verify that $u(i, t)$ satisfies (8.28), it suffices to do so at $t = 0$. This is easy from the results we obtained and setting $k_i + \lambda_i = a_i$ the result follows.

Comment: If you do not like to fudge with a_i's, you can solve $\mathrm{d}R/\mathrm{d}t = -\lambda_{\xi(t)}(T_{\xi(t)} - I)R$ instead and, again, verify that the corresponding $E^i[M(t)f(\xi(t))]$ satisfies (8.28).

A particular case, but directly related to (8.26) occurs in the following case. Let $V_i(x)$ be vector fields on $\mathbb{R}^d$ and assume that the corresponding flows $\phi_i(t,x)$ on $\mathbb{R}^d$ that solve $\frac{\mathrm{d}\phi_i}{\mathrm{d}t}(t,x) = V_i(\phi_i(t,x))$ with $\phi_i(0,x) = x$ can be defined on an interval containing $[0,\infty)$. Let $B = C_0(\mathbb{R}^n)$ and $\mathcal{D}(G_i) = C_0^1(\mathbb{R}^n)$ be the functions, with at least one continuous derivative, vanishing at infinity.

Put $S_i(t)f(x) = f(\phi_i(t,x))$ and consider the solution of the stochastic equation

$$\mathrm{d}X(t) = V_{\xi(t)}(X(t)), \qquad X(0) = x. \tag{8.30}$$

Then the pair $(\xi(t), X(t))$ is a Markov process having the state space $E = [1,n] \times \mathbb{R}^d$ with the semigroup

$$\begin{aligned} P_t f(i,x) &= E^{(i,x)}[f(\xi(t), X(t))] \\ &= \sum_{n=0}^{\infty} E^i[S_{\xi_n}(t - T_n)\ldots S_{\xi_0}(T_1);\, N(t) = n]. \end{aligned} \tag{8.31}$$

$\{N(t) = n\} = \{\tau_n \leqslant t < \tau_{n+1}\}$ and $P^{(i,x)} = \varepsilon_x \otimes P^i$, as must be obvious. Also, if must be clear that the notation $S_i(t)f(j,x)$ suggests that $S_i(t)$ acts on the second variable of $f(j,x)$.

Definition (8.31) is basically the unperturbed version of (8.29). It is easy to see that, if $f(i,x) \in C_0^1(\mathbb{R}^d)$, then $w_t(i,x) = P_t f(i,x)$ satisfies

$$\frac{\partial}{\partial t} w_t(i,x) = V_i(x) \cdot \nabla w_t(i,x) + \sum_j q_{ij} w_t(j,x)$$

with $w_0(i,x) = f(i,x)$.

Instead of writing the analogue of (8.29), let us just consider a particular case that leads to the solution of (8.26). Let $v_1 = 1$ and $v_2 = -1$. The solution to $\mathrm{d}X(t) = v_{\xi(t)}\,\mathrm{d}t$ is now

$$X(t) = x + \int_0^t V_{\xi(s)}\,\mathrm{d}s = x + \int_0^t I_{\{1\}}(\xi(s))\,\mathrm{d}s - \int_0^t I_{\{-1\}}(\xi(s))\,\mathrm{d}s,$$

and therefore

$$P_t f(i,x) = E^i\left[f\left(x + \int_0^t I_{\{1\}}(\xi(s))\,\mathrm{d}s - \int_0^t I_{\{-1\}}(\xi(s))\,\mathrm{d}s\right)\right].$$

In order to obtain $b(x)$'s, we must perturb (8.31). For that, define

$$m(t) = \exp\left\{-\int_0^t a_{\xi(s)}(X(s))\,\mathrm{d}s\right\}$$

and, when $\tau_n \leqslant t < t$, put

$$M(t) = m(t)b_n(X_n)S_{\xi_n}(t - \tau_n)\dots b_1(X_1)S_{\xi_0}(\tau_1)$$

which is to act on functions in $C_0(\mathbb{R}^n)$. We set $b_n(X_n) = b_{\xi_n}(X(T_n))$. We put $b_1(x) = -b(x)$ and $b_2(x) = b(x)$. As above, for bounded $f(i, x)$, let

$$u_i^0(t, x) = E^i[M(t)f(\xi(t), x)], \tag{8.32}$$

where $M(t)$ acts on the second variable in $f(i, x)$.

The same argument as in the proof of Theorem 8.2 leads to

$$\begin{aligned} u_i^0(t, x) &= f(1, x + t)\mathrm{e}^{-\lambda_1 t}\mathrm{e}^{-\int_0^t a_1(x+s)\,\mathrm{d}s} \\ &\quad + \lambda_1 \int_0^t \mathrm{e}^{-\lambda_1 s}\mathrm{e}^{-\int_0^s a_1(x+s')\,\mathrm{d}s'} b_1(x + s)u_2(t - s, x + s)\,\mathrm{d}s \end{aligned}$$

and a similar result for $u_2^0(t, x)$. Differentiating with respect to t, we obtain

$$\frac{\partial u_1^0}{\partial t}(t, x) = \frac{\partial u_1^0}{\partial x} - (\lambda_1 + a_1) - \lambda_1 b(x)u_2^0(t, x)$$

when $f(i, x)$ is in $C_0^1(\mathbb{R}^n)$ for each i. If we want $-(\lambda_1 + a_1) = b$, then $a_1 = -(b - \lambda_1)$. Similarly $\lambda_2 + a_2 = b$ or $a_2 = b - \lambda_1$. In (8.26) $\lambda_1 = 1$. Thus, (8.32) satisfies (8.26) with $s(t, x) = 0$.

To add the effect of the last term is easy: note that

$$w_i(t, x) = \int_0^t E^i[M(t - t')s(t', x)]\,\mathrm{d}t' \tag{8.33}$$

satisfies (8.26) with a vanishing initial condition.

Thus, $u_i(t, x) = u_i^0(t, x) + w_i(t, x)$ satisfies (8.26) with the initial condition $u_i(0, x) = f(i, x)$. To obtain the solution to the wave equation (8.25), note that

$$\frac{\partial u}{\partial t} = u_1(t, x) + u_2(t, x)$$

has the initial condition $f(1, x) + f(2, x)$. Integrating with respect to t, we obtain

$$u(t, x) = \int_0^t (u_1(s, x) + u_2(s, x))\,\mathrm{d}s + u(0, x).$$

If we wanted to find $f(1, x)$ and $f(2, x)$ from the data $u(0, x)$ and $\frac{\partial u}{\partial t}(0, x)$ for u, note that

$$\frac{\partial u}{\partial x}(0, x) = f(1, x) - f(2, x), \qquad \frac{\partial u}{\partial t}(0, x) = f(1, x) + f_2(x).$$

To close this section, we must refer the reader to Pinsky's book and to the papers by Dunbar and Othmer, Dunbar and Alt, where interesting applications are discussed. Also another point of connection between stochastics and transmutation is further explored in these references.

8.6. First-order hyperbolic systems

We have come across two ways of transforming the wave equation into a system of two first-order equations.

First, in Section 3.18, motivated by D'Alambert factorization of $(\partial_t^2 - c^2\partial_x^2)u = s$ into $(\partial_t + c\partial_x)(\partial_t - c\partial_x)u = s$, we examined the factorization of the wave equation into a first-order system. There the resulting equations involved extracting square roots that turned the (spatial) local, differential operator into an integral operator.

Worse than that, when the medium is isotropic and homogeneous, the problem has symmetries that were destroyed by the factorization. In compensation we came up with a procedure that had ties to probability and to a complexification technique.

A different way of turning the (one-dimensional) wave propagation problem into a system of equations was outlined in the previous section. Here we recall Dirac's procedure to convert the wave equation into a system of first-order equations, and we also recall the usual procedure to convert the wave equation

$$\frac{\partial^2 u}{\partial t^2} = \sum a_{ij}(x)\frac{\partial^2 u}{\partial x_i \partial x_j} + \sum b_j(x)\frac{\partial u}{\partial x} + c(x)\frac{\partial u}{\partial t} + q(x)u + s \tag{8.34}$$

into a system of first-order equations.

Several path integral approaches were proposed during the 1980's to represent the solutions of the resulting systems, but here we apply one of the procedures proposed for Dirac's equation to examine a variant of the problem dealt with in the previous section. Later we present a nice way of dealing with (one-dimensional) Dirac's equation due to E. Thomas.

The standard routine, outlined in F. John's book [3-1], goes as follows: Put $\phi_i = \partial u/\partial x_i, i = 1, 2, 3; \phi_4 = \partial u/\partial t, \phi_5 = u$ and put together symmetric matrices A, A^1, A^2, A^3, and an appropriate matrix B such that

$$A\frac{\partial}{\partial t}\phi = \sum_{i=1}^{3} A^i \frac{\partial}{\partial x_i}\phi + B\phi + f \tag{8.35}$$

is equivalent to (8.34). The matrix B is not required to be symmetric.

The consistency conditions to keep the track are

$$\sum_{k=1}^{3} a_{ik}\frac{\partial}{\partial t}\phi_k = \sum_{k=1}^{3} a_{ik}\frac{\partial}{\partial x_k}\phi_4 = 0, \quad i = 1, 2, 3,$$

$$\frac{\partial}{\partial t}\psi_4 = \sum_{k=1}^{3} a_{ik}\frac{\partial}{\partial x_k}\psi_i + \sum_{i=1}^{3} b_i\psi_i + c\psi_4 + q\phi_s, \tag{8.36}$$

$$\frac{\partial \phi_5}{\partial t} = \phi_4.$$

With this, it is a routine exercise to find A's and B such that (8.35) is equivalent to the set of consistency conditions (8.36). We considered only 3-spaces to compare with Dirac's trick.

When the second, third and fourth terms on the right-hand side of (8.34) are absent, we can suppress the 5-th element ϕ_5 in the vector ϕ and we are left with an equation like

$$A\frac{\partial \phi}{\partial t} = \sum_{i=1}^{3} A^i \frac{\partial}{\partial x_i}\phi + f,$$

where only one entry of f is non-null and equals $s(t,x)$.

During the early 1930's, Dirac was trying to give a probabilistic interpretation to solutions to the Klein–Gordon equation $\partial^2 u/\partial t^2 = c^2\Delta u + m^2 u$ and simultaneously preserve the invariance of the equation with respect to the Lorentz transformations. For the first task, and proceeding by analogy with non-relativistic quantum mechanics, he searched for a first-order factorization of the Klein–Gordon equation and he came up with something like

$$\left(\gamma_0\frac{\partial}{\partial t} - \sum \gamma_i \frac{\partial}{\partial x_i} - im\right)\psi = 0 \tag{8.37}$$

for a vector function $\psi(t,x)$ which turned out to be four-dimensional and the matrices $\gamma_0, \gamma_1, \gamma_2, \gamma_3$ had to be (4×4)-Hermitian matrices that satisfied

$$\gamma_\mu^2 = I, \qquad \gamma_\mu\gamma_\nu + \gamma_\nu\gamma_\mu = 0 \quad \text{if } \mu \neq \nu.$$

At this point, physical interpretation forced a new kind of particles out of mathematics and relativistic quantum mechanics was born. The fact that each component $\psi_i(t,x)$ had to satisfy the Klein–Gordon equation became irrelevant. Equation (8.37) can be modified to include electromagnetic fields via their vector potential A_0, A_1, A_2, A_3 by replacing $\partial/\partial x_\mu$ by $\partial/\partial x_\mu - ieA_\mu$, $\mu = 0,1,2,3$ and $x^0 = t$ (in units where $c = 1$). For more on Dirac's equation check in [5-2].

During the eighties several authors (see references at the end of the chapter) proposed the ways of representing the solutions to the Dirac equation with one spatial dimension, in terms of some associated Poisson process. Different path integrals were proposed by Ichinose for general symmetric hyperbolic systems, and in 1988 T. Zastawniak proved that for the full Dirac equation in the four-dimensional space time a path-space measure was non-existent. It still remains to be seen whether this impossibility is related to the impossibility of Lorentz invariant diffusions found earlier by R. Dudley.

8.6.1. One-dimensional stratified media a la Gaveau

Consider equation (8.25) with $s=0$, and put $\phi_1 = u_t, \phi_2 = u_x$. Applying the standard technique described above, one ends up with

$$\frac{\partial}{\partial t}\phi = \begin{pmatrix} 0 & 1 \\ 1 & 0 \end{pmatrix} \frac{\partial}{\partial x}\phi + \begin{pmatrix} 0 & 2b \\ 0 & 0 \end{pmatrix}\phi. \tag{8.38}$$

Assume for simplicity, that $2b(x) = \int e^{-ipx} m(dp)$ for some positive, finite measure on $\mathbb{R}$. Taking the Fourier transform one can write (8.38) as

$$\frac{\partial \hat{\phi}}{\partial t}(p) = i \begin{pmatrix} 0 & -p \\ p & 0 \end{pmatrix} \hat{\phi}(p) + \begin{pmatrix} 0 & 1 \\ 0 & 0 \end{pmatrix} m * \hat{\phi}(p), \tag{8.39}$$

where $m * \hat{\phi}(p)$ is a vector with components $m * \hat{\phi}_i(p)$, $i = 1, 2$; and, certainly, $\hat{\phi}_i(p) = \int e^{ipx} \phi_i(x)\, dx$. Our notation means

$$m * f(y) = \int f(x-y) m(dy),$$

as usual. Let $A(p)$ and B be the matrices $\left(\begin{smallmatrix} 0 & -p \\ -p & 0 \end{smallmatrix}\right)$ and $\left(\begin{smallmatrix} 0 & 1 \\ 0 & 0 \end{smallmatrix}\right)$, respectively.

Put $E = [0, \infty) \times \mathbb{R}$ and provide it with the usual Borel σ-algebra $\mathcal{E}$. Define the measure $d\lambda = dt \times m(dp)$ on $\mathcal{E}$.

On an auxiliary probability space $(\Omega, \mathcal{F}, P)$, we have defined a random measure (kernel) $\nu : \Omega \times \mathcal{E} \to [0, \infty)$ such that

(i) For any C in $\mathcal{E}$, $P(\nu(C) = k) = e^{-\lambda(C)} \lambda(C)^k / k!$

(ii) If $C_1, C_2, \ldots, C_n$ are in $\mathcal{E}$, the random variables $\nu(C_1), \ldots, \nu(C_n)$ are independent.

Now define the R-valued process $p(t)$ by

$$p(t) = p(0) + \int_0^t \int_{\mathbb{R}} p\nu(ds, dp).$$

This is a process with independent increments, with jumps of size 1 occurring at times $T_i, i \geqslant 1$, such that $0 \equiv T_0 < T_1 < \cdots < T_n < \cdots$.

If $\tau_i = T_i - T_{i-1}$, for $i \geqslant 1$, τ_i are independent and exponentially distributed, $P(\tau_i > t) = \exp(-Mt)$, with $M = m(\mathbb{R})$.

Now, for $s \leqslant t$, denote by $M_{[s,t]}$ the solution to the matrix equation

$$d_t M_{[s,t]} = M_{[s,t]}(iA(P(t)))\, dt + (B - I) \int_{\mathbb{R}} \nu(dt, dp)$$

with the initial condition $M_{[s,s]} = I$. If we denote by $M_t^{(k)}$ the solution to $d_t M_{[T_k,t]} = iM_{[T_k,t]} A(p(t))\, dt$ on (T_k, T_{k+1}) which satisfies $\lim_{t \downarrow T_k} M_{[T_k,t]} = I$, then, denoting by R the matrix $(B - I)$, we can write, for $T_n \leqslant t < T_{n+1}$, that

$$M_{[0,t]} = M_{T_1}^{(0)} R M_{T_2}^{(1)} R \ldots R M_{[T_n,t]}.$$

For any initial $\phi_0(p_0)$, we put

$$P_t\phi_0(p_0) = E[M_{[0,t]}\phi_0(p(t)) \mid p(0) = p_0]. \tag{8.40}$$

Once we have verified that $M_{[0,t]}$ defines a multiplicative functional (i.e., $M_{[0,t]} = M_{[0,s]}M_{[s,t]}$) of $p(t)$, we can verify that (8.40) defines a semigroup and $P_t\phi_0(p)$ satisfies (8.39).

Even though the technique to obtain (8.40) is reminiscent of what we did in Section 3, the processes involved are certainly very different.

8.6.2. *Thomas' completion of Feynmann's suggestion*

Let $a = (0,0), b = (t,x)$ be two points in the two-dimensional space time. In their book "Quantum Mechanics and Path Integrals", Feynmann and Hibbs propose the following way to compute the propagator (the quantum mechanical analogue of the probabilistic transition semigroup) of a relativistic particle.

Split $[0,t]$ into n intervals $[(k-1)\varepsilon, k\varepsilon], \varepsilon = t/n$ and $k = 1, 2, \ldots, n$. Put $\sigma_n = \{-n\varepsilon, \ldots, n\varepsilon\}$. Place a mesh of size ε on the (t,x)-plane, start at $(0,0)$ and take n-steps.

For $x = m\varepsilon = mt/n \in \sigma_n$, and R any integer, let

$$N_{\alpha,\beta}(\xi, R) = N_{\alpha,\beta}(n, m, R)$$

with $\alpha = +$ or $-$, be the numbers of paths with R-reversals, initial slope α and final slope β which join $(0,0)$ to $b = (t,\xi)$.

Then the following limit exists

$$K_{\alpha,\beta}(b,a) = \lim_{\varepsilon\to 0} \sum_{R\geqslant 0} N_{\alpha\beta}(x,R)(i\varepsilon)^R. \tag{8.41}$$

In his nice note [11], Thomas verifies this claim. Actually he proves:

THEOREM 8.3. *There exist kernels $K_t^{\alpha,\beta}(\varphi)(\xi)$, represented by functions $k_{\alpha,\beta}(t,\xi)$ such that for every continuous function φ on $\mathbb{R}$*

$$K_t^{\alpha,\beta}(\varphi)(\xi) = \lim_{n\to\infty} \sum_{\xi\in\sigma_n} \varphi(\xi) \sum_{R\geqslant 0} N_{\alpha,\beta}(\xi,R)\left(i\frac{t}{n}\right)^R. \tag{8.42}$$

Also, $k_{\alpha,\beta}(t,x)$ are given by formulae like:

$$\begin{aligned}
k_{++}(t,x) &= \frac{1}{2\pi}\int_{-\pi}^{\pi} \mathrm{e}^{i(t\cos\theta + ix\sin\theta)}\mathrm{e}^{-i\theta}\,\mathrm{d}\theta,\\
k_{+-}(t,x) &= \frac{1}{2\pi}\int_{-\pi}^{\pi} \mathrm{e}^{i(t\cos\theta + ix\sin\theta)}\,\mathrm{d}\theta = J_0\big((t^2-x^2)^{1/2}\big).
\end{aligned}$$

Actually, $K_t^{++}(\varphi)(x)$ and $K_t^{+-}(\varphi)(x)$ are

$$\begin{aligned} K_t^{++}(\varphi)(x) &= \frac{i}{2}\int_{-t}^{t} k_{++}(t,x)\varphi(x)\,\mathrm{d}x + \varphi(t), \\ K_t^{+-}(\varphi)(x) &= \frac{i}{2}\int_{-t}^{t} k_{+-}(t,x)\varphi(x)\,\mathrm{d}x. \end{aligned} \tag{8.43}$$

PROOF. (Sketch) To compute (8.41), one first notices that for $R = 2r > 0$

$$N_{++}(n,m,R) = \binom{u-1}{r}\binom{v-1}{r-1}, \qquad N_{--}(n,m,R) = \binom{u-1}{r-1}\binom{v-1}{r},$$

where $u = (n+m)/2$, $v = (n-m)/2$. Also, $N_{+-}(n,m,R) = 0$. When $R = 2r+1$, the counting of paths yields $N_{++}(n,m,R) = N_{--}(n,m,R) = 0$ and

$$N_{+-}(n,m,R) = N_{-+}(n,m,R) = \binom{u-1}{r}\binom{v-1}{r}.$$

For the next step, use identities like

$$\sum_{r>0}\binom{n}{r}\binom{m}{r+b}s^{2r+b} = \frac{1}{2\pi}\int_{-\pi}^{\pi}(1+s\mathrm{e}^{i\theta})^n(1+\mathrm{e}^{-i\theta})^m\mathrm{e}^{ib\theta}\mathrm{d}\theta$$

to obtain identities (8.42). Also note that the matrix kernel

$$K_t = \begin{pmatrix} K_+^{++} & K^{+-} \\ K_t^{-+} & K_t^{--} \end{pmatrix}$$

satisfies

$$\frac{\partial}{\partial t}K_t - \left(A\frac{\partial}{\partial x} + iB\right)H(t)K_t = \delta(t)\delta(x),$$

where $A = \left(\begin{smallmatrix}-1 & 0\\ 0 & 1\end{smallmatrix}\right)$ and $B = \left(\begin{smallmatrix}0 & 1\\ 1 & 0\end{smallmatrix}\right)$. Also, from (8.42) t is clear that K_t is concentrated on $\{(t,x) : |x| \leqslant t\}$. □

About the matrix kernels K_t, acting on

$$\mathcal{H} = \{(f^+, f^-) : f^\alpha \in L^2(\mathbb{R}, \mathbb{C}), \alpha = +, -\},$$

the following can be established:

THEOREM 8.4. *The kernels* $\{K_t : t \in \mathbb{R}\}$ *satisfy*

$$K_{t+s} = K_t * K_s, \quad t, s \in \mathbb{R}.$$

The operator $f \to U_t f = K_t * f$ *is unitary on* $\mathcal{H}$, *and* $\{U_t : t \in \mathbb{R}\}$ *is a strongly unitary group of operators on* $\mathcal{H}$, *having the generator*

$$H_0 = \begin{bmatrix} -(1/i)\partial/\partial x & 1 \\ 1 & (1/i)\partial/\partial x \end{bmatrix}$$

with the domain $\mathcal{D}(H_0) = \{(f^+, f^-) : f^\alpha$ *has a derivative (in the sense of distributions) in* $L^2(\mathbb{R}, \mathbb{C})\}$.

To go from the kernels K_t to path integrals, we proceed in two steps. First define the transition operators $K_{t,s}(\mathrm{d}x', x)$ by

$$K_{t,s}\varphi(x) = \int K_{t,s}(\mathrm{d}x', x)\varphi(x) = \int K_{t-s}(\mathrm{d}x')\varphi(x' + x)$$

for $\varphi = (\varphi^+, \varphi^-)$, with φ^α in $C_0(\mathbb{R})$. Extend to φ^α's in $C_0(\mathbb{R}^n)$ by

$$K_{t_n, t_{n-1}, \ldots, t_1, t_0}\varphi(x) = \int K_{t_n - t_{n-1}}(\mathrm{d}x_n, x_{n-1}) \ldots K_{t_1 - t_0}(\mathrm{d}x_1, x)\varphi(x_1, x)\varphi(x_1, \ldots, x_n).$$

To complete the circle of ideas, Thomas relates these kernels to measures on the class of continuous paths as follows. First define the following measures on $\Omega_p[0, t] = \{w : (0, t] \to \mathbb{R},\ w$ piecewise linear, affine, with slope ± 1, $w(0) = 0\}$. For $\phi : \Omega_p[0, t] \to \mathbb{R}$ continuous and bounded, put

$$F_{t,n}^{\alpha,\beta}\phi = \sum_{w \in \Omega_n^{\alpha,\beta}(0,t]} \phi(w)\left(i\frac{t}{n}\right)^{R(w)},$$

where $R(w)$ is the number of velocity reversals along w, $\Omega_n^{\alpha,\beta}[0, t]$ is the class of paths having the initial slope α, the final slope β, and n possible points at which changes of slope may occur. As usual: $\alpha, \beta = \pm 1$.

Combine these measures in a matrix measure

$$F_{t,n} = \begin{pmatrix} F_{t,n}^{++} & F_{t,n}^{+-} \\ F_{t,n}^{-+} & F_{t,n}^{--} \end{pmatrix}$$

and verify

THEOREM 8.5. *The measures* $F_{t,n}^{\alpha,\beta}$ *converge narrowly (or weakly) to measures* $F_t^{\alpha,\beta}$, *concentrated on* $\Omega_p[0, t]$. *If there exists* $\varphi(x_1, \ldots, x_n)$ *such that* $\phi(w) = \varphi(w(t_1), \ldots, w(t_n))$ *for* $0 < t_1 < \cdots < t_n \leqslant t$, *then*

$$F_t\phi = K_{t, t_n, t_{n-1}, \ldots, t_1}\varphi(0).$$

8.7. Pseudo processes and Euler's equation

We relate here still another equation describing oscillatory phenomena, namely, Euler's equation describing the oscillations of an infinite bar, i.e.,

$$\mu \frac{\partial^2 u}{\partial t^2} + 2\mu w \frac{\partial v}{\partial t} + EI \frac{\partial^4 v}{\partial x^2} + cv = s(t, x) \tag{8.44}$$

(plus appropriate initial conditions), to the equations associated with the generalized diffusion described by

$$\frac{\partial u}{\partial t} + \sigma \frac{\partial^4 u}{\partial x^2} = 0. \tag{8.45}$$

For a recent study involving (8.44) see reference [14]. That equation is used to model oscillations of train rails. The source is supposed to be random, and it includes damping and absorption modeled by the second and fourth terms.

Less standard is the pseudo-Markov process associated to (8.45). Let us describe it and send the reader to [15] and [16] for much more about these matters.

Consider, to begin with, the solution to (8.45) supplemented with the initial condition $u(0, x) = f(x)$. It is easy to see that

$$u(t, x) = \int p(t, x - y) f(y) \, \mathrm{d}y, \tag{8.46}$$

where the kernel $p(t, x)$ is given by

$$p(t, x) = \int \mathrm{e}^{-ikx} \mathrm{e}^{-\sigma t k^4} \mathrm{d}k / 2\pi. \tag{8.47}$$

It is easy to verify that $p(t, x)$ has the following properties: it is infinitely differentiable on $(0, \infty) \times (-\infty, \infty)$, scales as $p(t, x) = p(1, t^{-1/4} x) t^{-1/4}$, satisfies $p(t, x) \to \delta(x)$ as $t \downarrow 0$, and for every $t > 0$ integrates to 1 over $(-\infty, \infty)$. Even more, it satisfies

$$P(t + s, x) = \int p(t, x - z) p(s, z) \, \mathrm{d}z$$

for every t, s. That is, it could be used to define a semigroup.

Representation (8.47) allows us to conclude that $x^k p(t, x)$ decays to zero as $|x| \to \infty$, thus the semigroup property allows us to define a consistent family of finitely additive (non-positive) measures P^x on $(\Omega, \mathcal{F})$, where, again, $\Omega = \{w : [0, \infty) \to \mathbb{R}, w \text{ continuous}\}$ and $\mathcal{F}$ denotes the σ-algebra generated by the coordinate mappings $X(t) : \Omega \to \mathbb{R}$, $X(t)(w) = X_t(w) = w(t)$. This is done in the usual way, i.e., define for $F \in bB(\mathbb{R}^n)$, $0 < b_1 < \cdots < t_n$

$$E^x\{F(X(t_1), \ldots, X(t_n))\}$$
$$= \int \cdots \int F(x_1, \ldots, x_n) p_{t_1}(x_1 - x) \cdots p_{t_n - t_{n-1}}(x_n - x_{n-1}) \, \mathrm{d}x_1 \cdots \mathrm{d}x_n.$$

Since $p(t, x)$ are non-positive, one cannot obtain a σ-additive measure on the path space, thus some interesting functionals (functions on $(\Omega, \mathcal{F})$) could be hard to integrate.

Also, even though the finite-dimensional Markov property can be obtained, i.e.,

$$E^y[F(X(t))|X_{t_1}, \ldots, X_{t_n}] = \int p_{t-t_n}(x - X(t_n))F(x)\,\mathrm{d}x$$

for bounded F and $t > t_n > \cdots > t_1 \geqslant 0$, but the general version

$$E^y[H \circ \theta_t|\mathcal{F}_t] = E^y[H \circ \theta_t|X(t)]$$

may not be always true.

Thus, since $p(t, x)$ may take negative values, and P^x are not σ-additive, the name pseudo-Markov is appropriate for $X(t)$.

But not all is lost, for much of the stochastic calculus can be extended to $X(t)$.

The Feynmann–Kac formula can also be obtained for $X(t)$. That is, it can be proved that the solution to

$$\frac{\partial u}{\partial t} + \sigma \frac{\partial^4 u}{\partial x^4} + q(x)u = 0, \qquad u(0, x) = f(x) \tag{8.48}$$

is given by the path integral

$$u(t, x) = E^x\left[f(X(t)) \exp\left\{-\int_0^t q(X(s))\,\mathrm{d}s\right\}\right] \tag{8.49}$$

when $q(x)$ is continuous and bounded. From this we, finally, reach the path integral representation of the solutions to

$$\left(\alpha + \sigma \frac{\partial^4}{\partial x^4} + q(x)\right)U(x) = f(x), \tag{8.50}$$

$U(x)$ being represented by means of the path integral

$$U(x) = \int_0^\infty \mathrm{e}^{-\alpha t} E^x\left[f(X(t) \exp\left\{-\int_0^t q(X(s))\,\mathrm{d}s\right\}\right] \mathrm{d}t. \tag{8.51}$$

We now present two ways of obtaining path integral representations of solutions to (8.43).

First, note that $w(t, x)$ is a solution to

$$\frac{\partial^2 w}{\partial t^2} + \left(\frac{EI}{\mu}\right)\frac{\partial^4 w}{\partial x^4} + \left(\frac{c}{\mu}\right)w = s(t, x)/\mu \tag{8.52}$$

with initial conditions $w(0, x) = f(x)$, and $\frac{\partial W}{\partial t}(0, x) = g(x)$. Now, let $N(t)$ be a given Poisson process of intensity $w(> 0)$, and, if we put $T(t) = \int_0^t (-1)^{N(s)}\,\mathrm{d}s$, then

$$v(t, x) = \langle w(T(t), x)\rangle \tag{8.53}$$

satisfies (8.44). The brackets denote the average overall realizations of $N(t)$. Thus, we only have to solve (8.52).

For that we use the transmutation theory developed in Chapter 7. We only have to be careful about the choice of initial conditions for (8.52). For that assume $g(x) \equiv 0$, and let $u(t,x) = (T_1 W)(t,x)$ as in (7.1). Therefore $u(t,x)$ satisfies (8.48) with a different source term, i.e.,

$$\frac{\partial u}{\partial t} + \sigma \frac{\partial^4 u}{\partial x^4} + qu = s_1(t,x)/\mu, \qquad u(0,x) = f(x), \tag{8.54}$$

where $\sigma = EI/\mu$ and $q = c/\mu$ with $s_1(t,x) = (T_1 s(\cdot, x))(t)$.

If we denote the right-hand side of (8.48) by $Q_t f(x)$, it is easy to see that

$$u(t,x) = Q_t f(x) + \frac{1}{\mu} \int_0^t Q_{t-t'} s_1(t', x)\, \mathrm{d}x.$$

To go from this to $w(t,x)$, apply the techniques expounded in Chapter 7. To go from $w(t,x)$ to $v(t,x)$, apply (8.53).

Another way of obtaining a path integral representation of the solution to (8.44) consists in taking Laplace transforms to both sides of it

$$(\alpha^2 + 2w\alpha)\tilde{v} + \sigma \frac{\partial^4 \tilde{v}}{\partial x^4} + q\tilde{v} = \widetilde{S}(\alpha, x). \tag{8.55}$$

Note that we are keeping the last term separated from the first one to allow for non-constant q. Also we put $\underset{\sim}{S}(\alpha, x) = (\frac{1}{\mu})\widetilde{S}(\alpha, x) + (\alpha + 2w) f(x) + g(x)$. According to (8.51)

$$\tilde{u}(\alpha, x) = \int_0^\infty \mathrm{e}^{-(\alpha^2 + 2w\alpha)t} E^x \left[\widetilde{S}(\alpha, X(t)) \exp\left\{ - \int_0^t q(X(s))\, \mathrm{d}s \right\} \right] \mathrm{d}t. \tag{8.56}$$

If $q(x)$ were independent of x, the exponential factor inside $E^x[\cdot]$ goes out, and if w were to depend on x, we would have to add $-2\alpha \int_0^t w(X(s))\, \mathrm{d}s$ in the exponent inside $E^x[\cdot]$. Certainly, if q and w are constants and $\widetilde{S}(\alpha, x)$ can be recovered from its Fourier transform $\widehat{S}(\alpha, k)$, then

$$\begin{aligned} \tilde{u}(\alpha, x) &= \int_{-\infty}^\infty \mathrm{e}^{-ikx} \left(\int_0^\infty \mathrm{e}^{-(\alpha^2 + 2\alpha w + q)t} \mathrm{e}^{-\sigma k^4 t}\, \mathrm{d}t \right) \widehat{S}(\alpha, k) \frac{\mathrm{d}k}{2\pi} \\ &= \int_{-\infty}^\infty \frac{\mathrm{e}^{-ikx} \widehat{S}(\alpha, k)}{(\alpha^2 + 2\alpha w + q + \sigma k^4)} \frac{\mathrm{d}k}{2\pi}. \end{aligned}$$

Let us now take a look at the case considered in [14] where $w = w_0 + w_1(x)$ with $\sup |w_1(x)| \ll w_0$ and $c = c_0 + c_1(x)$, again, with $\sup |c_1(x)| \ll c_0$. In this case, (8.52) becomes

$$\tilde{u}(\alpha, x) = \int_0^\infty \mathrm{e}^{-at} E^x \left[\widetilde{S}(\alpha, X(t)) \exp\left\{ - \int_0^t r(X(s))\, \mathrm{d}s \right\} \right].$$

Assume that $f=g=0$, thus $\widetilde{S}(\alpha,x)=\widetilde{S}(\alpha,x)/\mu$. We have put $a=\alpha^2+q_0+2\alpha w_0$ and $r(x)=2\alpha w_1(x)+q_1(x)$. Assume that $w_1(x)$ and $q_1(x)$ are random. Moreover, assume that there exist two Gaussian spectral measures $R_1(\mathrm{d}\lambda)$ and $R_2(\mathrm{d}\lambda)$ such that

$$w_1(x)=\int \mathrm{e}^{-i\lambda x} R_1(\mathrm{d}\lambda), \qquad q_1(x)=\int \mathrm{e}^{-i\lambda x} R_2(\mathrm{d}\lambda), \tag{8.57}$$

where $R_i(\mathrm{d}\lambda)$ have the covariance

$$\mathcal{E}(R_i(I)R_j(I'))=\delta_{ij}m_i(I\cap I')$$

which amounts to say that for different indices i, j and/or non-overlapping intervals I and I', $R_i(I)$ are independent. From this it follows that $R(\mathrm{d}\lambda)=(2\alpha)^2R_1(\mathrm{d}\lambda)+R_2(\mathrm{d}\lambda)$ has the covariance $m(\mathrm{d}\lambda)=2\alpha m_1(\mathrm{d}\lambda)+m_2(\mathrm{d}\lambda)$. Assume, furthermore, that $m_i(\mathrm{d}\lambda)=f_i(\lambda)\,\mathrm{d}\lambda$ for appropriate, continuous $f_i(\lambda)$. Thus $m(\mathrm{d}\lambda)=f(\lambda)\,\mathrm{d}\lambda$ and $f(\lambda)=(2\alpha)^2f_1(\lambda)+f_2(\lambda)$.

Let us denote by $\mathcal{E}(\cdot)$ the average over the realizations of $r(x)$. If we want to compute the average of $\tilde{u}(\alpha,x)$, we have to evaluate

$$\mathcal{E}(\tilde{u}(\alpha,x))=\frac{1}{\mu}\int_0^\infty \mathrm{e}^{-at}E^x\big[\widetilde{S}(\alpha,X(t))\mathcal{E}\big[\mathrm{e}^{-\int_0^t r(X(s))\,\mathrm{d}s}\big]\big]\,\mathrm{d}t, \tag{8.58}$$

which we do as follows: Note that

$$\mathrm{e}^{-\int_0^t r(X(s))\,\mathrm{d}s}=\lim_n \prod_{m=-\infty}^{\infty} \mathrm{e}^{R[m,n]\int_0^t \mathrm{e}^{-i\frac{m}{n}X(s)}\mathrm{d}s},$$

where we used the spectral representation (8.53) and put $R[m,n]=R([\frac{m-1}{n},\frac{m}{n}])$. From the independence and Gaussian character of $R[m,n]$ mentioned above, it is clear that

$$\begin{aligned}\mathcal{E}\big[\mathrm{e}^{-\int_0^t r(X(s))\,\mathrm{d}s}\big] &= \exp\int_{-\infty}^{\infty} f(\lambda)\Big(\int_0^t \mathrm{e}^{-i\lambda X(s)}\,\mathrm{d}s\Big)^2\,\mathrm{d}\lambda\\ &= \exp\int_0^t\int_0^t \hat{f}(X(s)+X(s'))\,\mathrm{d}s\,\mathrm{d}s',\end{aligned}$$

where $\hat{f}(\xi)=\int_{-\infty}^{\infty} f(\lambda)\mathrm{e}^{-i\lambda\xi}\,\mathrm{d}\lambda$, which we are forced to assume to exist. Substituting in (8.54), we would be through, once we evaluate the integrals.

8.8. Damped waves: Playing with a simple model

In this section, we explore some other ways of solving

$$\frac{\partial^2 u}{\partial t^2}(t,x)+2b\frac{\partial u}{\partial t}(t,x)=c^2\Delta u(t,x)+s(t,x), \quad t>0,\ x\in D. \tag{8.59}$$

We saw in Chapter 7 several ways of dealing with this problem, and here we explore two more: the first of which is in line with our basic Laplace transform philosophy where as the second one relates to the results in the previous sections.

As usual, we assume that $u(0,x) = f(x)$ and $u_t(0,x) = g(x)$ are appropriate boundary conditions, which we may assume to be of Dirichlet type to have a concrete example, i.e., $u(t,x) = h(t,x)$ for $t > 0, x \in \partial D$.

Example 1: Assume that $D = \mathbb{R}$ or $\mathbb{R}^3$ and begin by taking the Laplace transform of (8.59) to obtain

$$\left(\frac{1}{2}\left(\frac{\sqrt{\alpha^2+2\alpha b}}{c}\right)^2 - \frac{1}{2}\Delta\right)\tilde{u}(\alpha,x) = \widetilde{S}(\alpha,x) \tag{8.60}$$

where, as usual, $\widetilde{S}(\alpha,x) = \frac{1}{2}(\tilde{s}(\alpha,x)) + (\alpha+2b)f(x) + g(x)$. We have already seen that in one spatial dimension (8.60) has the solution

$$\tilde{u}(\alpha,x) = c\int \frac{\mathrm{e}^{-\sqrt{\alpha^2+2\alpha b}\frac{|x-y|}{c}}}{\sqrt{\alpha^2+b^2}}\widetilde{S}(\alpha,y)\,\mathrm{d}y \tag{8.61}$$

and all boils down to invert the Laplace transforms. To compare with the computations in [17], assume that $\tilde{s}(\alpha,x) = g(x) = 0$. Since

$$\int_0^\infty \mathrm{e}^{-\alpha t} I_0\big(b\sqrt{t^2-y^2}\big)\theta(t-|y|)\,\mathrm{d}t = \frac{\mathrm{e}^{-|y|\sqrt{\alpha^2-b^2}}}{\sqrt{\alpha^2-b^2}}, \tag{8.62}$$

therefore $\mathrm{e}^{-bt} I_0(b\sqrt{t^2-y^2})\theta(t - |y|)$ has the Laplace transform $\mathrm{e}^{-|y|\sqrt{\alpha^2+2\alpha b}}/\sqrt{\alpha^2+2\alpha b}$, and we can now rewrite (8.61) as

$$\tilde{u}(\alpha,x) = \frac{1}{2}\int_{-\infty}^\infty \frac{\mathrm{e}^{-|y|\sqrt{\alpha^2+2\alpha b}}}{\sqrt{\alpha^2+2\alpha b}}(\alpha+2b)f(x+cy)\,\mathrm{d}y.$$

And, bringing in our result about Laplace transforms, we rewrite it as

$$\tilde{u}(\alpha,x) = \frac{1}{2}\int_0^\infty \mathrm{e}^{-\alpha t}\left(\mathrm{e}^{-bt}\int_\infty^\infty I_0\big(b\sqrt{t^2-y_2}\big)\right)\theta(t-|y|)(\alpha+2b)f(x+(y)\,\mathrm{d}y)\,\mathrm{d}t$$

from which, after the elementary calculation, we obtain

$$\begin{aligned} w(t,x) &= \frac{\mathrm{e}^{-bt}}{2}\{f(x+ct)+f(x-ct)\} + b\mathrm{e}^{-bt}\int_{-t}^t I_0\big(b\sqrt{t^2,y^2}\big)f(x+cy)\,\mathrm{d}y \\ &\quad + \frac{\mathrm{e}^{-bt}}{2}\int_{-t}^t \frac{tI_1(b\sqrt{t^2-y_2})}{\sqrt{t^2-y^2}}f(x+cy)\,\mathrm{d}y \end{aligned}$$

which is the result in [17].

Let us now consider the case $D \subset \mathbb{R}^3$, and assume, this time, that $\tilde{s}(\alpha, x) = f(x) = 0$. Go back to (8.61), which we rewrite as

$$\tilde{u}(\alpha, x) = \int_0^\infty e^{-\rho\sqrt{\alpha^2+2\alpha b}} \rho\langle g\rangle(x, cp)\,dp, \tag{8.63}$$

where, recall, $\langle g\rangle(x, \tau) = \frac{1}{4\pi}\int_{|\xi|=1} g(x + \tau\xi)\,d\sigma(\xi)$.

To begin with, note that differentiating the identity written right after (8.62), we obtain

$$-\frac{\partial}{\partial\rho}\int_0^\infty e^{-\alpha t}e^{-bt} I_0\big(b\sqrt{t^2-\rho^2}\big)\theta(t-\rho)\,dt = -\frac{\partial}{\partial\rho}\frac{e^{-\rho\sqrt{\alpha^2+2\alpha b}}}{\sqrt{\alpha^2+2\alpha b}} = e^{-\sqrt{\alpha^2+2\alpha b}}$$

and therefore the inverse Laplace transform of (8.63) is

$$u(t, x) = \int_0^\infty -\frac{\partial}{\partial\rho}\{e^{-bt}\}. \tag{8.64}$$

Example 2: We now consider problem (8.59), but this time no special assumptions are made on the domain. We are going to simplify it using the technique proposed by Kac in [18] and used in [17] as well as in [2] and [5,6].

Consider the related problem

$$\frac{\partial^2 v}{\partial t^2}(t, x) = c^2\Delta v(t, x) + s_0(t, x), \quad t > 0,\ x \in D \tag{8.65}$$

subject to the same initial conditions as (8.59), and below we shall see how to relate the sources and boundary conditions of both problems.

As in Section 7 above, consider a Poisson process $\{N(t): t \geqslant 0\}$ of intensity λ (i.e., $E[N(t)] = \lambda t$, now E denoting the average over the realizations of $N(t)$), and define $T(t) = \int_0^t (-1)^{N(s)}\,ds$ and put

$$w(t, x) = E[v(t(t), x)] \tag{8.66}$$

for $t \geqslant 0$ and $x \in D$.

Also assume that $s_0(t, x)$ and the Dirichlet boundary condition $h_0(t, x)$ for (8.64) are such that

$$s(t, x) = E[s_0(T(t), x)], \qquad h(t, x) = E[h_0(T(t), x)]. \tag{8.67}$$

To verify that $w(t, x)$ satisfies (8.59), it suffices to assume that $v(t, x)$ can be recovered from its Fourier transform according to $v(t, x) = \int e^{iwt}\hat{v}(w, x)\frac{dw}{2\pi}$, and then compute the differential equation satisfied by $E[e^{iwT(t)}]$.

PROPOSITION 8.1. *With the notation just introduced, set* $h_+(t) = E[e^{iwT(t)}]$ *and* $h_-(t) = E[e^{-iwT(t)}]$. *Then*

$$\frac{d^2h_+}{dt^2} + 2\lambda\frac{dh}{dt} = -\omega^2 h_t \tag{8.68}$$

and similarly for $h_{(t)}$.

PROOF. Let $T = \inf\{t > 0 : N(t) > 0\}$ be the time of occurrence of the first jump of $N(t)$. Then

$$E\big[e^{iwT(t)}\big] = E\big[e^{iwT(t)}; T > t\big] + E\big[e^{iwT(t)}; T \leqslant t\big].$$

Now, since for $T > t$, which occurs with probability $P(T > t) = e^{-(\lambda - iw)t}$.

Writing $N(s) = 1 + N(s) - N(T)$, since $T(t) = T - \int_T^t (-1)^{N(s)-N(T)}\,ds$, and $N(t) - N(T)$ is independent of T and since the density of $P(T \leqslant t)$ is $\lambda e^{-\lambda t}$, we have

$$\begin{aligned} E\big[e^{iwT(t)}, T \leqslant t\big] &= \lambda \int_0^t e^{(\lambda - iw)s} E\big[e^{-iwT(t-s)}\big]\,ds \\ &= \lambda \int_0^t e^{-(\lambda - iw)(t-s)} E\big[e^{-iws}\big]. \end{aligned}$$

Putting all of this together, we obtain

$$h_+(t) = e^{-(\lambda - iw)t} + \lambda \int_0^t e^{-(\lambda + iw)(t-s)} h_+(s)\,ds$$

and replacing ω by $-\omega$, we obtain

$$h_-(t) = e^{-(\lambda + iw)(t)} + \lambda \int_0^t e^{-(\lambda + iw)(ts)} h_+(s)\,ds.$$

By differentiation we obtain the system

$$\frac{d}{dt}\begin{pmatrix} h_+(t) \\ h_-(t) \end{pmatrix} = iw\begin{pmatrix} h_+(t) \\ h_-(t) \end{pmatrix} + \begin{pmatrix} -\lambda & \lambda \\ \lambda & -\lambda \end{pmatrix}\begin{pmatrix} h_+(t) \\ h_-(t) \end{pmatrix} \tag{8.69}$$

which ties all of this with the subject matter of [2]. Now, differentiating (8.68) once more, we obtain the conclusion of the proposition.

Observe now that, if

$$w(t, x) = \int E\big[e^{iwT(t)}\big]\widehat{V}(w, x)\frac{dw}{2\pi}$$

and any assumption on the integrability of $\omega^2\widehat{V}(\omega,x)$ allows us to exchange derivatives with integrals to conclude that

$$\frac{\partial^2 w}{\partial^2 t^2}(t,x)+2\lambda\frac{\partial w}{\partial t}(t,x)=\int E\left[e^{-iwT(t)}\right]\left(-w^2\widehat{V}(w,x)\right)\frac{dw}{2\pi},$$

and, since $-w^2\widehat{V}(w,x)=(\frac{\partial^2}{\partial t^2}V(w,x))\widehat{V}(w)=c^2\Delta\widehat{V}(w,x)+\widehat{S}_0(w,x)$, by substituting in the identity above, we obtain

$$\frac{\partial^2 w}{\partial t^2}+b\frac{\partial w}{\partial t^2}=c^2\Delta w(t,x)+s(t,x),\quad t>0,\ x\in D.$$

Clearly, $w(t,x)=h(t,x)=E[h(T(t),x)]$ for $t>0$ and $x\in\partial D$, and we have to verify that the initial conditions are satisfied. These are two issues: first, that the values of $w(t,x)$ and $\frac{\partial w}{\partial t}(t,x)$ at $t=0, x\in D$ are what they are supposed to be, and second, the identities hold in the limit as $t\downarrow 0$ as well. To take care of the first issue, note that

$$w(0,x)=E[V(T(0),x)]=f(x)$$

since $T(0)=0$. Making use of (8.68) and assuming that exchanging derivatives and integrals is taken care of, we see that

$$\frac{\partial w}{\partial t}(t,x)=\int\left\{i\omega E\left[e^{i\omega T(t)}\right]+\lambda(h_-(t)-h_+(t))\right\}\widehat{V}(\omega,x)\frac{d\omega}{2\pi}$$

and, since $i\omega\widehat{V}(w,x)=(\frac{\partial V}{\partial t}(\cdot,x))^{\wedge}(\omega)$, we have for $t=0$,

$$\frac{\partial w}{\partial t}(0,x)=\int_{-\infty}^{\infty}i\omega\widehat{V}(\omega,x)\frac{d\omega}{2\pi}=\frac{\partial V}{\partial t}(0,x).$$ □

PROPOSITION 8.2. *Assume that $v(t,x)$ satisfies* (8.64) *with initial conditions $u(0,x)=f(x)$, $u_t(0,x)=g(x)$, and that* (8.66) *holds. Then $w(t,x)$, given by* (8.65)*, satisfies* (8.59) *with the same initial conditions and boundary conditions related according to* (8.66).

PROOF. We have already seen that (8.59) holds, and that the values of $u(t,x)$ and $u_t(t,x)$ at $t=0$ are as they should be.

Assume now that $\sup_{x\in D}\|v(t,x)\|=B$; $\sup_{x\in D}\|f(x)\|<B$ and that $\|v(t,x)-f(\cdot)\|_\infty\to 0$ as $t\downarrow 0$.

Then

$$\begin{aligned}|w(t,x)-f(x)| &\leqslant E[|v(T(t),\cdot)-f(x)|]\\ &\leqslant E[|V(T(t),x)-f(x)|;T>t]+E[|V(T(t),x)-f(x)|,T\leqslant t]\\ &\leqslant \|V(t,\cdot)-f\|_\infty e^{-\lambda t}+2B(1-e^{-\lambda t})\end{aligned}$$

from which it follows that $\|w(t,\cdot) - f(\cdot)\|_\infty \to 0$ as $t \downarrow 0$. We leave for the reader to do a similar analysis for the convergence of the first time derivative.

Assume now that $\sup_{t\in[-A,A]} |V(t,y) - h_0(t,x)| \to 0$ as $y \in D \to x \in \partial D$. Since $|T(t)| \leqslant t$, clearly

$$|w(t,y) - h(t,x)| \leqslant E[|V(T(t),y) - h_0(T(t),x)|]$$

from which the desired conclusion holds. □

Comments: The same reasoning yields similar conclusions for any type of convergence that would involve a convex norm. That is why we did not blink at throwing assumptions in the middle of the proof instead of making them explicit in the statement of the theorem.

Anyway, the gist of the example was to explore one more way of looking at (8.59) and to connect to (8.68).

References

A seemingly unexplored connection between waves, circuits and Markov chains seems to exist. For connections between random walks and resistor networks see

1. Doyle, P. G. and Snell, J. L.: *Random Walks and Electric Networks*, Carus Math. Monograph 22, M.A.A., 1984.

And connections between systems of oscillator and L-R-C circuits appear in almost any elementary physics text book, but nobody seems to have played with the connection waves-circuits-Markov chains. For the material in Sections 2 and 3 see the nice:

2. Pinsky, M.: *Lectures on Random Evolution*, World Sci. Publ., Singapore, 1991.

and for full fledged mathematical analysis and some other connections

3. Jefferies, B.: *Evolution Processes and the Feynmann–Kac Formula*, Kluwer Acad. Publ., Dordrecht, 1996.

For material related to Section 3, try to locate the doctoral dissertation by M. Keepler and D. Heath mentioned in [7–9], as well as

4. Hersch, R.: Random evolutions: A survey of results and problems, *Rocky Mountain J. Math.* **4** (1974), 443–476.
5. Dunbar, S. R.: A branching random evolution and a nonlinear hyperbolic equation, *SIAM J. Appl. Math.* **48** (1988), 1510–1526.
6. Othmer, H. G. et al.: Models of dispersal in biological systems, *J. Math. Biol.* **26** (1988).

For Section 4, a few of the key references are:

7. Gaveau, B.: Representation formulas of the Cauchy problem for hyperbolic systems generalizing Dirac system, *J. Funct. Anal.* **58** (1984), 310–319.

8. Ichinose, T.: Path integral for a hyperbolic system of first order, *Duke Math. J.* **51** (1984), 1–36.
9. Zastawniak, T.: The non-existence of the path-space measure for the Dirac equation, *J. Math. Phys.* **30** (1989), 1354–1358.
10. Pereira, P. and Quezada, R.: Probabilistic representation formulas for the time evolution of quantum systems, *J. Math. Phys.* **34** (1993), 59–68.
11. Thomas, E. G. T.: Finite path integrals, *Proc. Sympos. Appl. Math.* **52** (1997), 225–232.

To see a proof of impossibility of Lorentz invariant diffusions consult (see Theorem 11.3)

12. Dudley, R. M.: Lorentz-invariant Markov processes in relativistic phase space, *Ark. Math.* **6** (1965), 241–268.
13. Dudley, R. M.: Recession of some relativistic Markov processes, *Rocky Mountain J. Math.* **4** (1974), 401–406.

A few references devoted to higher order processes are:

14. Fryba, L., Nakagiri, S. and Yoshikawa, L.: Stochastic analysis of a beam on random foundation with uncertain foundation subjected to moving load, in N. Bellomo and F. Casciati (eds), *Proc. IUTAM Symp.*, Springer-Verlag, Berlin, 1992.
15. Hochberg, K. J.: A signed measure on path space related to Wiener measure, *Ann. Probab.* **6** (1975), 433–458.
16. Nishioka, K.: Stochastic calculus for a class of evolution equations, *Japan. J. Math.* **11** (1985), 9–102.
17. Dewitt-Morette, C. and Foong, S. K.: Path integral solutions of wave equations with dissipation, *Phys. Rev. Lett.* **62** (1989), 2201–2204.
18. Kac, M.: *Some Stochastic Problems in Physics and Mathematics*, Colloquium Lect. in Pure and Applied Sciences, Magnolia Petroleum Co. and Socono-Mobil Oil Cs, Oct. 1956.

After the manuscript was processed I discovered

19. Ratanov, N. E.: Telegraph evolutions in inhomogeneous media, *Markov Processes and Rel. Fields* **62** (1999), 53–68.

CHAPTER

NINE

APPLICATIONS

This chapter deals with some applications, by which we mean problems in wave propagations in which the probabilistic connection plays a role. Most of the issues we consider come from the field of inverse problems. To be consistent with the spirit of this book, and to use "modern" jargon, this chapter should be regarded as a site under permanent construction.

9.1. An inverse source problem

9.1.1. A basic result

Consider a medium, consisting of the half-space $D = \{x \in \mathbb{R}^3 : x_3 > 0\}$, in which there is a wave propagating according to

$$\frac{\partial^2 u}{\partial t^2} = \sum a_{ij}(x) \frac{\partial^2 u}{\partial x_i \partial x_j} + s_0(t, x) \tag{9.1}$$

for $t > 0$, $x \in D$. We assume that the medium was unperturbed and at rest at $t = 0$, and that the boundary is free; i.e.,

$$\frac{\partial u}{\partial x_3}(t, x) = 0 \tag{9.2}$$

for $t > 0$, $x = (x_1, x_2, x_3 = 0) \in \partial D$.

A standard inverse problem in geophysics consists of finding $a_{ij}(x)$ for $1 \leqslant i, j \leqslant 3$ and $x \in D$, from the measurement of $u(t, x)$ at $x \in \partial D$, for $t > 0$.

To see what is involved in this problem, consider its Laplace transformed version of equation (9.1):

$$\left(\frac{\alpha^2}{2} - \frac{1}{2}\sum a_{ij}\frac{\partial^2}{\partial x_i \partial x_j}\right)\tilde{u}(\alpha, x) = \frac{1}{2}\tilde{s}_0(\alpha, x) \tag{9.3}$$

for $\alpha > 0$, $x \in D$, subject to the boundary condition

$$\frac{\partial \tilde{u}}{\partial x_3}(\alpha, x) = 0 \tag{9.4}$$

for $\alpha > 0$, $x = (x_1, x_2, 0) \in \partial D$.

We already know that if $X(t)$ is the diffusion in $\overline{D}$, reflected upon reaching the boundary ∂D, the solution to (1.3) and (1.4) can be represented by the path integral

$$\tilde{u}(\alpha, x) = E^x \int_0^\infty e^{-\alpha^2 t/2}\tilde{s}_0(X(t), t)\, dt. \tag{9.5}$$

The coefficients to be found, i.e., $a_{ij}(x)$ are buried within the construction of $X(t)$ according to

$$dX(t) = \sum \sigma_{ij}(x(t))\, dB_j(t) + n(X)\, dL$$

where $n(x)$ is the inner normal vector field at ∂D, and $L(t)$ is the local time of $X(t)$ at ∂D. (Recall that the last term in Skorohod's equation describes the kick the diffusing particle gets back into D as it gets to ∂D.)

To give the problem a simpler appearance, assume that $a_{ij}(x) = c^2(x)\delta_{ij} = c_0^2\delta_{ij}/n^2(x)$, where $n(x)$ is the refractive index of the medium. Denote $\frac{1}{2}(\alpha/c_0)^2$ by β, $n^2(x)$ by $q(x)$ and rewrite (1.3) as

$$\left(\beta q(x) - \frac{1}{2}\Delta\right)\tilde{u}(\alpha, x) = \frac{q(x)\tilde{s}_0(\alpha, x)}{2c_0^2},$$

where, of course, c_0 is assumed known. Now, instead of (1.5), we have

$$\tilde{u}(\alpha, x) = E^x \int_0^\infty \frac{e^{-\beta A(t)} q(x(t))\tilde{s}_0(\alpha, X(t))\, dt}{2c_0^2}, \tag{9.6}$$

where now $X(t) = (B_1(t), B_2(t), |B_3(t)|)$, and $B(t) = (B(t), B_2(t), B_3(t))$ denotes the standard three-dimensional imposed motion, and $A(t) = \int_0^t q(X(s))\, ds$.

Again, the problem consists of finding $q(x)$, for $x \in D$, from the knowledge of $\tilde{u}(\alpha, x)$ for $x \in \partial D$, $\alpha > 0$. Even though (1.6) looks simpler than (1.5) as far as the parametric dependence goes, the problem is still quite complicated.

That is so, regardless of the fact that usually $q(x)$ happens to be well known at the support of $\tilde{s}_0(\alpha, x)$.

Homework: What would $f_e(x)$ be if $u(t, x) = \delta(vt - qx)$?

9.1.2. Equivalent source problem

To recover the inhomogeneities in the medium, that is, to use (9.6) for finding $q(x)$ in D from the measurements of $u(t,x)$ on the boundary of ∂D is quite a complicated problem. See Section 9.2 below for the description of some attempts at this problem.

We present now a different approach, which we call the *equivalent source problem* and which may be a mathematical version of what geophysicists call the exploding reflectors technique.

Assume that we are given surface data $u(t,x)$ for $r > 0$, $x \in \partial D$, and we want to find $s_e(t,x)$ for $t > 0$, $x \in D$ such that the solutions to

$$\frac{\partial^2 v}{\partial t^2} = c_e^2 \Delta v + S_e(t,x), \quad t > 0,\ x \in D \tag{9.7}$$

with homogeneous initial conditions and a reflecting boundary condition

$$\frac{\partial v}{\partial x}(t,x) = 0, \quad t > 0,\ x \in \partial D \tag{9.8}$$

has the boundary values $v(t,x) = u(t,x)$. To simply further, we assume that $D = \{x \in \mathbb{R}^3 \mid x_3 > 0\}$ and $\partial D = \{x \in \mathbb{R}^3 \mid x_3 = 0\}$.

In (9.7), we have introduced c_e as an equivalent constant propagation speed which is determined from experimental considerations.

We also allow for $S_e(t,x)$ to incorporate the initial conditions, i.e., $S_e(t,x) = s_e(t,x) + \delta(t) f(x) + \delta(t) g(x)$ where the first term describes the true sources and the other two describe impulsive sources, equivalent to initial conditions.

This problem could be restated as one of three alternatives: Find either the initial displacement $f(x)$ of the medium, or the initial velocity distribution $g(x)$ in the medium, or the true source distribution $s_e(t,x)$, such that the wave field $v(t,x)$ they produce according to (9.7)–(9.8) is such that $u(t,x) = v(t,x)$ for $t > 0$, $x \in \partial D$.

We know that the solution to (9.7)–(9.8) can be written as

$$\tilde{v}(\alpha, x) = E^x \int_0^\infty \mathrm{e}^{-\beta t} \frac{\widetilde{S}_e(\alpha, X(t))}{2c_e^2}\, \mathrm{d}t,$$

where $\beta = \frac{1}{2}(\frac{\alpha}{c_e})^2$ and $X(t) = (B_1(t), B_2(t), |B_3(t)|)$ is the imposed motion reflected on ∂D. If we denote by the same symbol $\widetilde{S}_e(\alpha, x)$ the even extension of $\widetilde{S}_e(\alpha, x)$ to all of $\mathbb{R}^3$, we can write $\tilde{v}(\alpha, x)$ as

$$\tilde{v}(\alpha, x) = \frac{1}{2c_e^2} E^x \int_0^\infty \mathrm{e}^{-\beta t} \widetilde{S}_e(\alpha, B(t))\, \mathrm{d}t = \frac{1}{4\pi c_e^2} \int_{\mathbb{R}^3} \frac{\mathrm{e}^{-\frac{\alpha}{c_e}|x-y|}}{|x-y|} \widetilde{S}_e(\alpha, y)\, \mathrm{d}y.$$

Passing to a coordinate centered at x, and then introducing spherical coordinates, we rewrite the last integral as

$$\tilde{v}(\alpha, x) = \int_0^\infty t' \mathrm{e}^{-\alpha t'} \langle \widetilde{S}_e(\alpha, \cdot) \rangle (x, c_e t')\, \mathrm{d}t', \tag{9.9}$$

where we have set $\rho = c_e t'$, ρ being the radial coordinate, and as before

$$\langle \widetilde{S}_e(\alpha, \cdot)\rangle(x, c_e t) = \frac{1}{4\pi}\int \widetilde{S}_e(\alpha, x + ct\xi)\, d\sigma(\xi).$$

Note now that $e^{-\alpha t'}\widetilde{S}_e(\alpha, x)$ is the Laplace transform of $\theta(t - t')\widetilde{S}_e(t - t', x)$, therefore (9.9) is the Laplace transform of

$$\begin{aligned} v(t, x) &= \int_0^\infty t'\theta(t - t')\langle S_e(t - t', \cdot)\rangle(x, c_e t') \\ &= \int_0^t t'\langle S_e(t - t', \cdot)\rangle(x, c_e t')\, dt'. \end{aligned} \tag{9.10}$$

To particularize a bit, we separate the three possible terms in (9.10), that is, we have either true sources, or so an expression like (9.10) is obtained. When the sources correspond to the initial displacement $f(x)$, (9.10) reduces to

$$v(t, x) = \frac{\partial}{\partial t}(t\langle f_e\rangle(x, c_e t)) \tag{9.11}$$

and, when the sources describe initial velocities, we obtain

$$v(t, x) = t\langle g_e\rangle(x, c_e t) \tag{9.12}$$

for all $t > 0$ and $x \in \mathbb{R}^3$. Recall as well that f_e, g_e are assumed to be even under reflections on ∂D. We have already seen how to obtain (9.11) and (9.12) from (4.12) in Chapter 4. Actually (9.11) and (9.12) are quite simply related to each other.

Our problem can be now restated as: Given $u(t, x)$ for $t > 0$, $x \in \partial D$, find $f_e(x)$ (or $g_e(x)$ or $S_e(t, x)$) for $x \in D$ such that (9.11) (or (9.10), or (9.9)) holds.

Actually, it is easier to invert (9.12) and then to use that result to invert (9.11). But before doing that, let us look at an easier question, namely, if the medium was originally homogeneous and isotropic, and $u(t, x)$ describes the propagation of say, an initial velocity distribution $g(x)$, with the speed c_0 would we recover it?

This means, is it possible to find $g_e \neq g_0$ and $c_e \neq c_0$ such that

$$\langle g_0\rangle(x, c_0 t) = \langle g_e\rangle(x, c_e t)$$

for $t > 0$ and $x \in \partial D$?

To see that it is not possible, assume that g_0 and g_e can be recovered from their Fourier transforms and rewrite the last identity as

$$\int e^{ik\cdot x}\hat{g}_0(k)\left(\int e^{ic_0 tk\cdot\xi}\, d\sigma(\xi)\right)\frac{dk}{(1\pi)^3} = \int e^{ik\cdot x}\hat{g}_e(k)\left(\int e^{ic_e tl\cdot\xi}\, d\sigma(\xi)\right)\frac{dk}{(2\pi)^3}$$

and, carrying out the inner integration, we obtain

$$\int e^{ik\cdot x}\hat{g}_0(k)\frac{\sin(|k|c_0 t)}{|k|c_0}\, dk = \int e^{ik\cdot x}\hat{g}_e\frac{\sin(|k|c_e t)}{|k|c_e}\, dk,$$

and, from the uniqueness of Fourier transforms, we have

$$\hat{g}_0(k)\frac{\sin(|k|c_0t)}{c_0}=\hat{g}_e(k)\frac{\sin(|k|c_et)}{c_e}$$

for all $t>0$ and all $k\in\mathbb{R}^3$. Differentiating with respect to t and letting $t\downarrow 0$, we obtain $\hat{g}_0(k)=\hat{g}_e(k)$, and from that we obtain $c_0=c_e$.

9.1.3. *Solving the integral geometry problem*

In this section, we present a solution to the integral geometry problem consisting in finding a function $g_e(x)$ in $\mathbb{R}^3$ (which we assume even under reflections on the plane $\{x_3=0\}$), from the knowledge of its averages over spheres of radius c_et at all points on the plane.

Again, since we invoke Fourier transforms, functions are assumed to decay rapidly enough at infinity. Let us begin by rewriting (9.19) as

$$\begin{aligned}u(t,x)&=\frac{1}{4\pi c_e}\int_{\mathbb{R}^3}\mathrm{e}^{-ik\cdot x}\frac{\hat{g}_e(k)}{(2\pi)^3}\left(\int_{|\xi|=1}c_et\mathrm{e}^{-ic_etk\cdot\xi}\,\mathrm{d}\sigma(\xi)\right)\mathrm{d}k\\&=\int\mathrm{e}^{ik\cdot x}\frac{\hat{g}_e(k)}{(2\pi)^3}\frac{\sin(c_e|k|t)}{c_e|k|}\,\mathrm{d}k.\end{aligned}$$

Set now $k=(\bar{k},k_3)$ and take a (two-dimensional) Fourier transform on both sides to obtain

$$\begin{aligned}\hat{u}(t,p)&=\int_{\mathbb{R}}\mathrm{e}^{ip\cdot x}u(t,x)\,\mathrm{d}x=\int_{-\infty}^{\infty}\frac{\hat{g}(p,k_3)}{2\pi}\frac{\sin(c_e\sqrt{p^2+k_3^2})}{c_e\sqrt{p^2+k_3^2}}\,\mathrm{d}k_3\\&=\frac{1}{\pi}\int_0^{\infty}\hat{g}_e(p,\xi)\frac{\sin(c_et\sqrt{p^2+\xi^2})}{c_e\sqrt{p^2+\xi^2}}\,\mathrm{d}\zeta,\end{aligned}$$

where, in the last step, we invoked the reflection symmetry of g_e and changed from k_3 to ζ.

Now differentiate both sides with respect to t to obtain:

$$\frac{\partial\hat{u}}{\partial t}(t,\rho)=\frac{1}{\pi}\int_0^{\infty}\hat{g}(\rho,\zeta)\cos\bigl(c_et\sqrt{p^2+\zeta^2}\bigr)\,\mathrm{d}\zeta.$$

One more change of variables will bring the right-hand side above into the form of a standard cosine transform. Set $\psi=c_e\sqrt{p^2+\zeta^2}$ and

$$\hat{h}(p,\zeta)=\theta(\psi-p)\psi\hat{g}\bigl(p,\sqrt{(\psi/c)^2-p^2}\bigr)/c^2\sqrt{(\psi/c)^2-p^2}$$

to obtain

$$\frac{\partial\hat{u}}{\partial t}(t,p)=\frac{1}{\pi}\int_0^{\infty}\hat{h}(p,\psi)\cos(t\psi)\,\mathrm{d}\psi$$

and, inverting the cosine transform, we obtain

$$\hat{h}(p,\psi) = \frac{1}{2}\int_0^\infty \left(\frac{\partial \hat{u}}{\partial t}\right)(t,p)\cos(t\psi)\,\mathrm{d}t$$

and now, undoing the changes of variables, we have

$$\hat{g}_e(p,\zeta) = \frac{c\zeta}{1\sqrt{p^2+\zeta^2}}\int_0^\infty \left(\frac{\partial \hat{u}}{\partial t}\right)(t,p)\cos\left(c_e t\sqrt{p^2+\zeta}\right)\mathrm{d}t. \tag{9.13}$$

Thus, given the boundary data, first compute the cosine transform indicated in (9.13) and then invert the Fourier transform to obtain $g_e(x)$.

To take care of (9.11), we leave it for the reader to verify that it suffices to apply what we have just done, to find $f_e(x)$ in $\mathbb{R}^3$, symmetric under reflections on ∂D such that

$$u_1(t,x) = \int_0^t u(s,x)\,\mathrm{d}s = t\langle f_e\rangle(x,c_e t)$$

which, according to (9.13), is

$$\hat{f}(p,\zeta) = \frac{c_e\zeta}{2\sqrt{p^2+\zeta}}\int_0^\infty \hat{u}(t,p)\cos\left(c_e t\sqrt{p^2+\zeta^2}\right)\mathrm{d}t. \tag{9.14}$$

These computations can be collected under

PROPOSITION 9.1. *Assume that $u(t,x)$ decays rapidly enough, both in t and $x \in \partial D$. Then the function $g_e(x)$ exists in $\mathbb{R}^3$ invariant under reflections on ∂D, which satisfies* (9.12) *and such a function is given by* (9.13).

COROLLARY 9.1. *Let $u(t,x)$ be such that $u,(t,x) = \int_0^t u(s,x)\,\mathrm{d}s$ decays fast enough in $t > 0$ and $x \in \partial D$. Then a function $f_e(x)$ exists on $\mathbb{R}^3$, invariant under reflections on ∂D, which satisfies* (9.11) *and such a function is given by* (9.14).

To invert (9.10) is less trivial. Let us carry out a few steps. First, write $s(t,x) = \int \mathrm{e}^{-ik\cdot x}\hat{s}(t,x)\frac{\mathrm{d}k}{(2\pi)^3}$, insert it in (9.10), take a (two-dimensional) Fourier transform with respect to $x \in \partial D$, and integrate over the unit sphere to obtain

$$\hat{u}(t,p) = \frac{1}{\pi}\int_0^t \mathrm{d}t' \int_0^\infty s(t',p,\zeta)\frac{\sin[c_e(t-t')\sqrt{p^2+\zeta^2}]}{c_e\sqrt{p^2+\zeta^2}}\,\mathrm{d}\zeta,$$

and, differentiating with respect to t both sides, we obtain

$$\left(\frac{\partial \hat{u}}{\partial t}\right)(t,p) = \frac{1}{\pi}\int_0^t \mathrm{d}t' \int_0^\infty s_e(t',p,k_3)\cos\left[c_e(t-t')\sqrt{p^2+\zeta^2}\right]\mathrm{d}\zeta.$$

If we now take a Laplace transform with respect to t, we obtain

$$\tilde{u}(\alpha, p) = \frac{1}{\pi} \frac{\hat{\tilde{s}}_e(\alpha, p\zeta)\,\mathrm{d}\zeta}{(\alpha^2 + c_e^2(p^2+\zeta^2))} = \frac{1}{\pi}\int_{-\infty}^{\infty} \frac{\hat{\tilde{s}}_e(\alpha, p\zeta)\,\mathrm{d}\zeta}{(\alpha^2 + c_e^2(p^2+\zeta^2))},$$

and the problem consists of solving, for $\hat{\tilde{s}}_e(\alpha, p, \zeta)$, and then inverting the Fourier and Laplace transforms.

9.1.4. Some very simple examples

The following examples are as simple as you can get. The two first ones could perhaps be related to what geophysicists call exploding mirrors.

Example 1: Assume that $u(t,x) = h(t)$ on ∂H, i.e., $u(t,x)$ is independent of x. Suppose we want to reproduce $h(t)$ by an initial velocity distribution $g_e(x)$ in the underground. Then, according to (9.13),

$$\hat{g}_e(p,\zeta) = \frac{c\zeta}{2\sqrt{p^2+\zeta}} \int_0^{\infty} 2\pi\delta(p)h' \cos\bigl(c_e t\sqrt{p^2+\zeta^2}\bigr)\,\mathrm{d}t.$$

$\delta(p)$ comes in via $\hat{u}(t,p) = h(t)(2\pi\delta(p))$ and therefore we can clearly set $\hat{g}_e(p,\zeta) = 2\pi\delta(p)\hat{g}_e(\zeta)$ to obtain

$$\hat{g}_e(\delta) = \frac{c_e}{2}\int_0^{\infty} h'(t)\cos(c_e t\zeta)\,\mathrm{d}t = \frac{1}{2}\int_0^{\infty} h'\left(\frac{t}{c_e}\right)\cos(t\zeta)\,\mathrm{d}t$$

and therefore

$$g_e(z) = h'\left(\frac{z}{c_e}\right). \tag{9.15}$$

Example 2: The same as in Example 1, but this time we want to have an initial displacement instead of an initial distribution of velocities responsible for the signal on the surface.

Applying (9.14), we are rapidly led to

$$f_e(z) = h\left(\frac{z}{c_e}\right). \tag{9.16}$$

In both cases, the procedure translates the time dependence observed on the surface into a depth dependence underground.

Example 3: Assume that at the surface we observe the following superposition of plane waves: $u(t,x) = \sin(wt)\cos(q\cdot x)$, and that we want to find $g_e(x)$ such that (9.12) holds.

This time $\hat{u}(t, p) = \pi \sin(wt)\{\delta(p+q) + \delta(p-q)\}$, therefore

$$\begin{aligned}
\hat{h}(p, \zeta) &= \frac{\pi w c_e \zeta}{2\sqrt{p^2 + \zeta^2}} \{\delta(p+q) + \delta(p-q)\} \int_0^\infty \cos(wt) \cos\left[c_e t \sqrt{p^2 + \delta^2}\right] \mathrm{d}t \\
&= \frac{w c_e \delta}{\sqrt{p^2 + \delta}} \{\delta(p+q) + \delta(p-q)\} \\
&\quad \times \left\{\delta\left(w - c_e\sqrt{p^2 + \zeta}\right) + \delta\left(w + c_e\sqrt{p^2 + \zeta^2}\right)\right\}.
\end{aligned}$$

Now we invert the Fourier-cosine transform to obtain $g(\xi, z)$ with $\xi = (x_1, x_2)$

$$\begin{aligned}
g_e(\xi, z) &= \frac{1}{(2\pi)^2} \int_{\mathbb{R}^2} \int_0^\infty \mathrm{e}^{-ip\cdot\xi} \cos(z\zeta) \hat{g}(p, \zeta)\, \mathrm{d}p\, \mathrm{d}\zeta \\
&= \frac{1}{(2\pi)^2} \cos(q\cdot\xi) \int_{-\infty}^\infty \cos(\zeta z) \\
&\quad \times \left\{\delta\left(w - c\sqrt{p^2 + \zeta^2}\right) + \delta\left(w + c\sqrt{q^2 + \zeta}\right)\right\} \frac{c_e \zeta\, \mathrm{d}\zeta}{\sqrt{q^2 + \zeta^2}} \\
&= \frac{w\theta\left(\frac{w}{c_e} - |q|\right)}{(2\pi)^2} \cos(q\cdot\xi) \cos\left(z\sqrt{\left(\frac{w}{c}\right) - q^2}\right).
\end{aligned}$$

Here we see that, unless our choice of the equivalent propagation speed c_e is such that $w = |q| c_e$, an "exploding reflector" cannot be found. The spatial dependence along the surface stays the same. The time dependence is appropriately transformed into depth dependence.

Example 4: Suppose now that the observed wave field $u(t, x)$ is produced by an interface located at $z = L$ in a stratified medium with two different propagation velocities: c_1 for $0 < z < L$ and c_2 for $L < z < \infty$.

To find the solution to the Laplace transformed version of (9.1) that is, of (9.3)–(9.4), is easy if we do some solution matching across boundaries. The solution can be described as follows.

For $0 < z < L$

$$\tilde{u}(\alpha, p, z) = \left\{\frac{Q}{2w_1}\left(\mathrm{e}^{-w_1|z-z_0|} + \mathrm{e}^{-w_1|z+z_0|}\right) + 2R\mathrm{e}^{-w_1 L} \cosh wz\right\},$$

where $w_1 = ((\frac{\alpha}{c_1})^2 + p^2)^{1/2} Q$ denotes the intensity of the sources and R is the reflection coefficient at the interface at $z = L$. For $L < z < \infty$, we have

$$\tilde{u}(\alpha, p, z) = \frac{T}{2w_2} \mathrm{e}^{-w_2(z-L)},$$

where $w_2 = ((\frac{\alpha}{c_2})^2 + p^2)^{1/2}$. And, to be more explicit,

$$\tilde{u}(\alpha, p, z) = \int_{\mathbb{R}} e^{ip\cdot\bar{x}} \left(\int_0^\infty e^{-\alpha t} u(t, \bar{x}, z)\, dt \right) d\bar{x} \quad \text{with } \bar{x} = (x_1, x_2).$$

The boundary matching process yields

$$R = \frac{Q\gamma(\alpha, p)e^{-2w_1 L}}{1 - \gamma(\alpha, p)e^{-2w_1 L}},$$

$$T = \left(\frac{2w_2}{w_1 + w_2}\right) \frac{e^{-w_1 L}}{1 - \gamma(\alpha, p)e^{-w_1 L}}.$$

Note that as $p \to 0$ $\frac{2w_1}{w_1+w_2} \to \frac{2c_1c_2}{c_1+2_2}$ and $\gamma(\alpha, p) \to \gamma = \frac{c_2-c_1}{c_2+c_1}$.

Suppose the "pulse" is produced by a homogeneous plane distribution of sources at $z_0 = 0$ and the reflected field $(\tilde{u}(\alpha, p, z=0) - \frac{Q}{2w_1} \equiv \delta u(\alpha, p, z=0))$ is observed.
That is, we observe

$$\delta u(\alpha, x) = \gamma Q e^{-\frac{2\alpha}{c_1} L} \sum_{n=0}^{\infty} \gamma^n e^{-\frac{\alpha L n}{c_1}}$$

or

$$\delta u(t, x) = Q \sum_{n=0}^{\infty} \gamma^{n+1} \delta\left(t - 2(N+1)\left(\frac{L}{c_1}\right)\right).$$

According to Example 1 or 2 this corresponds to the initial distribution of velocities of displacements at planes, parallel to the surface (and the interface), but distributed at depths $z_n = 2(n+1)L\frac{c_e}{c_1}$. Observe that the positioning (in this simple) example is "correct" only when the equivalent velocity is the same as the true velocity in the first interface.

9.2. Probabilistic approach to a discrete inverse problem: Non-uniqueness questions in discretized models

This section is concerned with the discrete version of the problem of finding $q(x)$ for $x \in D$, from the knowledge of $\tilde{u}(\alpha, x)$ for $\alpha > 0, x \in \partial D$; where $\tilde{u}(\alpha, x)$ satisfies

$$\left(\beta q(x) - \frac{1}{2}\Delta\right)\tilde{u}(\alpha, x) = \widetilde{S}(\alpha, x)$$

and the symbols have the same meaning as in Section 1.

In order to solve this problem numerically, the domain D must be discretized and appropriate boundary conditions must be imposed.

We denote by Q a discretization of the Laplacian as explained in Chapter 8, and denote by $X(t)$ a Markov chain moving in $\mathbb{N}^3$, having Q as an infinitesimal generator. Denote by D the finite sublattice $[0, N] \times [0, N] \times [-N, N]$, and let T be the first hitting time of the boundary ∂D of D by $X(t)$. For this to make sense, we assume that $X(t)$ jumps among the nearest neighbors only. Also assume that $\widetilde{S}(\alpha, x)$ and $q(x)$ are symmetric with respect to the plane $[0, N] \times [0, N] \times \{0\}$. Our discretized equation is

$$\left(\frac{\alpha^2}{2}q(i) - Q\right)\tilde{u}(\alpha, i) = \widetilde{S}(\alpha, i) \tag{9.17}$$

supplemented with the boundary condition

$$\tilde{u}(\alpha, i) = 0, \quad \alpha > 0, i \in \partial D.$$

The solution to (9.17) can be represented probabilistically by

$$\tilde{u}(\alpha, i) = E^x\left\{\int_0^T e^{-\beta A(t)}\widetilde{S}(\alpha, x(T))\,dt\right\}, \tag{9.18}$$

where, again, $A(t) = \int_0^t q(X(s))\,ds$. The problem now is to find $q(i)$ for $i \in [0, N]^3$ from the knowledge of $\tilde{u}(\alpha, i)$, $i \in D_0 = \{(i_1, i_2, 0) : 0 \leqslant i_1, i_2 \leqslant N\}$. For $N = 100$, for example, when you want to find q to the depth of 10 km with a resolution no less than 100 m, you need to solve a nonlinear system (9.18) of equations with 10^6 unknowns and $10^4 \times$ (#α values) data points.

We shall see in the next section that for fixed $\widetilde{S}(\alpha, x)$, the mapping $q \to \tilde{u}|_{\partial D}$ is continuous. What we shall do here is to exploit the discreteness of the situation to examine the non-uniqueness of the reconstruction problem. Let us rewrite (9.18) a bit.

Put $L^j(t) = \int_0^t I_{\{j\}}(X(s))\,ds$ for the total time spent by $x(t)$ at $j \in D$. If we put $q \cdot L(t) = \sum q(j)L^j(t)$, then (17) becomes

$$\tilde{u}(\alpha, i) = E^x\left\{\int_0^T e^{-\beta q\cdot L(t)}\widetilde{S}(\alpha, X(t))\,dt\right\} \tag{9.19}$$

for $\alpha > 0, i \in D_0$. Recall that $\widetilde{S}(\alpha, x(t)) = \tilde{s}(\alpha, X(t))q(X(t))/2c_e^2$ and usually $q(j)$ is known at the points $\operatorname{supp} \tilde{s}(\alpha, \cdot) = \{j : \tilde{s}(\alpha, j) \neq 0\}$, and we shall not worry about those values of $q(j)$. It is easy to see that, for $\alpha > 0$, the right-hand side of (9.19) is as differentiable as you want as a function of $q(j)$'s. If we denote the index set of the data points by collective i, we can write (9.19) as

$$\Phi_i(q) = d_i, \quad i = 1, \ldots, k, \tag{9.20}$$

where k is the number of data points and Φ_i denotes the right-hand side of (9.19) as the function of $q = (q_1, \ldots, q_\mu)$, $\mu = 2N^3$.

Since each q_j satisfies $q_{min} \leqslant q(j) \leqslant q_{max}$, we have $\Phi_i = (q_{min}, q_{max})^M \to \mathbb{R}^k$ is a differentiable function, and assume that $M > k$, thus

$$S = \{q \in (q_{min}, q_{max})^M : \Phi_i(q) = d_i, i = 1, \ldots, k\}$$

is an $(M-k)$-manifold, if not empty.

The procedure to find a point $q_0 \in S$ could be to minimize

$$\sum_{i=1}^{k} (d_i - \Phi_i(q))^2$$

over $(q_{min}, q_{max})^M$. Not a simple problem, even in the one-dimensional case.

Assume that for some $d^0 \in \mathbb{R}^k$ the manifold

$$S_0^0 = \{q : \Pi(q) \doteqdot d^0\}$$

is not empty. Then, the implicit function theorem tells us that it is an $(M-k)$-manifold. We are going to provide an explicit way to move around it, starting from any point q^0 on it.

At any $q \in S_0^0$ define $n_1(q) = \nabla\Phi_i / \|\nabla\Phi_i(q)\|$ to be k vector fields, normal to S^0. Now, let $t_i(q)$ be $(M-k)$ mutually orthogonal vectors, each perpendicular to all $n_i(q)$. That is, $t_i(q)$ generate the tangent space at $g \in S_0^0$. Let $t_{ij}(q)$ denote the j-th component of $t_i(q)$, and define the metric

$$g_{ij}(q) = \sum_{\ell=1}^{M-k} t_{\ell i}(q) t_{\ell j}(q), \quad q \leqslant i, j \leqslant \mu.$$

One could think of $q_{ij}(q)$ as the components of the matrix that projects $\mathbb{R}^M$ on the tangent space to S_0^0 at q. Denote by $g^{ij}(q)$ the matrix inverse to $g_{ij}(q)$ and the Hamiltonian function

$$H(p, q) = \frac{1}{2} \sum_{i,j=1}^{M} p_i p_j g^{ij}(q)$$

and solve the Hamiltonian set of equations

$$\dot{q}_j = \partial H / \partial p_j, \qquad \dot{p}_j = -\partial H / \partial q_j, \quad 1 \leqslant j \leqslant M$$

with the initial condition $q(0) = q_0 \in S_0^0$, $\rho(0) = \sum t_j(q^0)(t_j(q^0), \eta)$ where η is any (you pick it a piacere) vector in $\mathbb{R}^M$. This system has solutions for t in some interval $(-\delta, \delta)$, and along the solution curve $(q(t), \rho(t))$ it is clear that $\Phi(q(t)) = d^0$. This follows from the fact that $\dot{\Phi}_i(q(t)) = \|\nabla\Phi_i\|(q(t))(n_i(q(t)), q^0(t))$, the velocity being tangent to S_0^0. Thus $\dot{\Phi}_i(q(t)) = 0$ for all i, and therefore $\Phi_i(q(0)) = d_i^0$.

9.3. Dependence of boundary data on propagation velocity

Again, we refer to the set up of Section 1; and we are going to verify that the surface data $\tilde{u}(\alpha, x)$, $\alpha > 0$, $x \in \partial D$, depend continuously on $q(x)$ for fixed $\tilde{s}(\alpha, x)$. Actually, a similar but easier argument would yield the continuous dependence on the (Laplace transformed) source term $\tilde{s}(\alpha, x)$.

Recall that we are considering both the velocity profile of rather associated $q(x)$ well as the source term $\tilde{s}(\alpha, x)$ to be defined on $\mathbb{R}^3$ and symmetric with respect to reflections about the surface $\partial D = \{x \mid x_3 = 0\}$.

As we saw in Section 1, the solution to (9.3) subject to (9.4) is given by (9.6), which becomes

$$\tilde{u}(\alpha, x) = E^x \int_0^\infty e^{-\beta A(t)} \tilde{s}(\alpha, B(t))\, dt, \tag{9.21}$$

where use was made of the symmetry of $q(x)$ and $\tilde{s}(\alpha, x)$ to replace the reflected imposed motion $(B_1(t), B_2(t), |B_3(t)|)$ by the standard imposed motion $B(t)$.

Let us denote by $q_i(x)$ the quantities $(c_0/c_i(x))^2$ and $(c_0/c_2(x))^2$, and by $\tilde{u}_1(\alpha, x)^2$, $\tilde{u}_2(\alpha, x)$ the corresponding surface data they generate. Recall that $q_{\min} \leqslant q_i(x) \leqslant q_{\max}$.

PROPOSITION 9.2. *Let $q_1(x)$ and $q_2(x)$ be such that $\delta q = \|q_1 - q_2\| < q_{\min}$. Let $s(t, x)$ be such that $|S|_\infty := \int_0^\infty \|s(t, 0)\|_\infty\, dt < \infty$. Then*

$$\|\tilde{u}_1(\alpha, 0) - \tilde{u}_2(\alpha, 0)\| < K \delta q$$

for some appropriate constant K.

PROOF. Note that

$$\begin{aligned}
&|\tilde{u}_1(\alpha, x) - u_2(\alpha, x)| \\
&\quad \leqslant E^x \left\{ \int_0^\infty \left| e^{-\beta A_1(t)} - e^{-\beta A_2(t)} \right| |q_1(B(t))| |\tilde{s}(\alpha, B(t))|\, dt / 2c_0^2 \right\} \\
&\quad + E^x \int_0^\infty e^{-\beta A_2(t)} |q_1(B(t)) - q_2(B(t))| |\tilde{s}(\alpha, B(t))|\, dt / 2c_0^2 \\
&\quad \leqslant \frac{|s|}{2c_0^2} \left\{ q_{\max} \int_0^\infty e^{-\beta q_{\min} t} \left(e^{\beta \delta q} - 1 \right) + \delta q \int_0^\infty e^{\beta q_{\min} t} \right\} \\
&\quad = \frac{|s|_\infty}{2c_0^2} \frac{q_{\max}}{q_{\min}} \left\{ \frac{q_2}{q_{\min} - \delta q} + 1 \right\} \delta q = K \delta q.
\end{aligned}$$

□

Comments: This comes out so easy because of the explicit representation of the solution in terms of path integrals. Even though our estimate is valid for all x, we compare the values of x at which the wave field is observed.

9.4. The Born approximation

Assume that c_0 is the maximum velocity in the range of $c(x)$ and write $q(x) = 1 + a(x)$. Then $0 \leqslant a(x) \leqslant a_\mu = (\frac{c_2}{c_1})^2 - 1$, with $c_1 \leqslant c(x) \leqslant c_2$. Put $c(t) = \int_0^t a(B(s))\,\mathrm{d}s$ and rewrite (9.21) as

$$\tilde{u}(\alpha, x) = E^x \int_0^\infty \mathrm{e}^{-\beta C(t)} \tilde{s}(\alpha, B(t))\,\mathrm{d}t.$$

Note that

$$\mathrm{e}^{-\beta C(t)} = 1 - \beta \int_0^t a(B(s))\mathrm{e}^{-\beta C(s)}\,\mathrm{d}s$$

with the previous representation for $\tilde{u}(\alpha, x)$ becomes

$$\tilde{u}(\alpha, x) = \tilde{u}_0(\alpha, x) - \beta E^x \int_0^\infty \mathrm{e}^{-\beta t} \left(\int_0^t a(B(s))\mathrm{e}^{-\beta C(s)}\,\mathrm{d}s \right) \tilde{s}(\alpha, B(t))\,\mathrm{d}t,$$

where $\tilde{u}_0(\alpha, x)$ stands for

$$\tilde{u}_0(\alpha, x) = E^x \int_0^\infty \mathrm{e}^{-\beta t} \tilde{s}(\alpha, B(t))\,\mathrm{d}t. \tag{9.22}$$

Note that $\tilde{u}_0(\alpha, x)$ is known as long as $q(x)$ is known on the support of $\tilde{s}(\alpha, x)$. Exchanging the integration over s with the integration over t, the second term in the previous identity becomes

$$\begin{aligned} & E^x \left\{ \int_0^\infty \mathrm{e}^{-\beta C(s)} a(B(s)) \left(\int_s^\infty \mathrm{e}^{-\beta t} \widetilde{S}(\alpha, B(t))\,\mathrm{d}t \right) \mathrm{d}s \right\} \\ &= E^x \left\{ \int_0^\infty \mathrm{e}^{-\beta A(s)} a(B(s)) \left(\int_0^\infty \mathrm{e}^{-\beta t} \tilde{s}(\alpha, B(t)\,\mathrm{d}t) \cdot \theta_s \right) \mathrm{d}t \right\} \\ &= E^x \left\{ \int_0^\infty \mathrm{e}^{-\beta A(s)} a(B(s)) \tilde{u}_0(\alpha, B(s))\,\mathrm{d}s \right\} =: \tilde{w}(\alpha, x). \end{aligned}$$

To obtain the last step, we invoked the standard Markov property of the imposed motion $B(t)$. The previous to the last consists of a simple change of variables plus the definition of a shift operator and the fact that $A(t) = t + C(t)$.

A glance at (9.6) and at $\tilde{w}(\alpha, x)$ should convince us that it satisfies

$$\left(\beta a(x) - \frac{1}{2}\Delta \right) \tilde{w}(\alpha, x) = a(x) \tilde{u}_0(\alpha, x)$$

plus the free boundary condition $\partial \tilde{w}(\alpha, x)/\partial n = 0$ for $\alpha > 0$ and $x \in \partial D$. To sum up

$$\tilde{u}(\alpha, x) = \tilde{u}_0(\alpha, x) - \beta E^x \left\{ \int_0^\infty \mathrm{e}^{-\beta A(t)} a(B(t)) \tilde{u}_0(\alpha, B(t))\,\mathrm{d}t \right\}. \tag{9.23}$$

Note that, if we put $\tilde{u}(\alpha, x) = R^{\beta}\widetilde{S}(\alpha, x)$, $\tilde{u}_0(\alpha, x) = R_0^{\beta}\widetilde{S}(\alpha, x)$, then (9.23) is just $R^{\beta} = R_0^{\beta} - \beta R^{\beta} a R_0^{\beta}$ as expected. Again, using $A(t) = t + C(t)$, (9.18) becomes

$$\begin{aligned}\tilde{u}(\alpha, x) = \tilde{u}_0(\alpha, x) - \beta E^x\left\{\int_0^{\infty} \mathrm{e}^{-\beta t} a(B(t))\tilde{u}_0(\alpha, B(t))\,\mathrm{d}t\right\}\\ -\beta E^x\left\{\int_0^{\infty} \mathrm{e}^{-\beta t}\big(1 - \mathrm{e}^{-\beta C(t)}\big) a(B(t))\tilde{u}_0(\alpha, B(t))\,\mathrm{d}t\right\}.\end{aligned} \tag{9.24}$$

The Born approximation consists of dropping the last term in (9.24) and regarding the resulting equation as an integral equation in $a(x)$. There are a couple of estimates of the size of the last term or its decay rate.

PROPOSITION 9.3. *Let $f :\to \mathbb{R}$ be some bounded, measurable function. Assume that for $\rho > 0$, $\sup_{x\in D}\langle a\rangle(\alpha, \rho) \leqslant A\rho^{-(b+2)}$ with $0 < b < 1$ and, as usual,*

$$\langle a\rangle(x, \rho) = \frac{1}{4\pi}\int_{|\xi|=1} a(x + \rho\xi)\,\mathrm{d}s\sigma(\xi).$$

Then

$$\left|E^x\int_0^{\infty} \mathrm{e}^{-\beta t}\big(1 - \mathrm{e}^{-\beta C(t)}\big) a(B(t)) f(B(t))\right| \leqslant KA\gamma,$$

where $\gamma = (\sqrt{a_\mu + 1} - 1) = (c_2 - c_1)/c_1$, K being same generic constant.

PROOF. Start computing

$$\begin{aligned}&\left|E^x\int_0^{\infty} \mathrm{e}^{-\beta t}\big(1 - \mathrm{e}^{-\beta c(t)}\big) a(B(t)) f(B(t))\,\mathrm{d}t\right|\\ &\quad\leqslant \|f\|_\infty E^x\int_0^{\infty} \mathrm{e}^{-\beta t}\big(1 - \mathrm{e}^{\beta a_\mu t}\big) a(B(t))\,\mathrm{d}t\\ &\quad= \|f\|_\infty \int\left(\frac{\mathrm{e}^{-\sqrt{2\beta}|x-y|} - \mathrm{e}^{-\sqrt{2\beta(a_\mu+1)}|x-y|}}{2\pi|x-y|}\right) a(y)\,\mathrm{d}y,\end{aligned}$$

where we have made use once again of

$$\int\frac{\mathrm{e}^{-\beta t}\mathrm{e}^{-x^2/2t}}{(2\pi t)^{3/2}}\,\mathrm{d}t = \frac{\mathrm{e}^{-\sqrt{2\beta}|x-y|}}{2\pi|x-y|}.$$

Now, switch to spherical coordinates centered at x to replace the last term of the chain by

$$\begin{aligned}&K\int_0^{\infty}\big(\mathrm{e}^{-\sqrt{2\beta}\rho} - \mathrm{e}^{\sqrt{-2\beta(a_\mu+1)}}\rho\big)\rho\langle a\rangle(x, \rho)\\ &\quad= K\gamma\sqrt{2\beta}\int_0^{\infty}\mathrm{e}^{-\sqrt{2\beta}\rho}\left(\int_0^{\rho}\mathrm{e}^{-\gamma\sqrt{2\beta}s}\,\mathrm{d}s\right)\rho\langle a\rangle(x, \rho)\,\mathrm{d}\rho\end{aligned}$$

$$= K\sqrt{2\beta}\gamma \int_0^\infty e^{-\gamma\sqrt{2\beta}s}\left(\int_s^\infty e^{-\sqrt{2\beta}\rho}\rho\langle a\rangle(x,\rho)\,d\rho\right)ds$$

$$\leqslant K\sqrt{2\beta}\gamma A \int_0^\infty \frac{e^{-2\beta}(\gamma+1)s}{s^b}$$

$$\leqslant KA\sqrt{2\beta}\gamma[\sqrt{2\beta}(\gamma+1)]^{b-1}\int_0^\infty \frac{e^{-s}\,ds}{s^b}.$$

The integral is finite for $0 < b < 1$. Also, $\sqrt{2\beta}(\gamma+1) = (\frac{\alpha}{c_1})$, which, inserted above, yields the desired conclusion. □

Comment: To apply this to estimate the last term in (9.24), note that the boundedness of $\widetilde{S}(\alpha, x)$ rapidly yields boundedness of $\tilde{u}_0(\alpha, x)$.

A standard assumption in this game is that $a(x)$ is of bounded support, contained in a ball of radius R. In this case, a better estimate, i.e., in terms of $\|a\|_\infty$ can be obtained.

PROPOSITION 9.4. *The rest being as in Proposition* 9.3, *assume now that $a(x)$ vanishes off the ball of radius R and is bounded by δ inside. Then*

$$\left|E^x \int_0^\infty e^{-\beta t}\left(1 - e^{-\beta c(t)}\right)a(B(t))f(b(t))\,dt\right| \leqslant \frac{K\delta}{\alpha}(c_2 - c_1)$$

for some constant k.

Comment: A very simple computation yields the estimate $\|f\|_\infty\delta/\beta$. Try it.

PROOF. Proceeding as above, we achieve

$$K\int_0^\infty \left(e^{-\sqrt{2\beta}\rho} - e^{-\sqrt{2\beta(a_\mu+1)}\rho}\rho\right)\langle a\rangle(x,\rho)\,d\rho$$

$$= K\gamma\sqrt{2\beta}\int_0^R e^{-\sqrt{2\rho}\rho}\left(\int_0^\rho e^{\gamma\sqrt{2\beta}s}\,ds\right)\rho\langle a\rangle(x,\rho)\,d\rho$$

$$= K\gamma\sqrt{2\beta}\int_0^R e^{-\gamma\sqrt{2\beta}s}\left(\int_0^R e^{-\sqrt{2\beta}\rho}\rho\langle a\rangle(x,\rho)\,d\rho\right)ds$$

$$\leqslant RK\gamma\sqrt{2\beta}\delta\int_0^R e^{-\gamma\sqrt{2\beta}s}\left(\int_s^R e^{-\sqrt{2\beta}\rho}\,d\rho\right)ds$$

$$\leqslant \frac{RK\delta\gamma}{\sqrt{2\beta}(\gamma+1)} = \frac{k}{\alpha}\delta(c_2 - c_1).$$

Surprisingly, within the same setup a lower bound in terms of the moment of inertia of $\langle a\rangle(x,\rho)$ can be obtained. □

PROPOSITION 9.5. *With the some notation and under the same assumptions of Proposition 9.4, we have*

$$\left|E^x \int_0^\infty \mathrm{e}^{-\beta t}\big(1-\mathrm{e}^{-\beta c(t)}\big)a(B(s))f(B(s))\,\mathrm{d}s\right| \geqslant K\mathrm{e}^{-R\alpha/c_1}\int_0^R \rho^2\langle a\rangle(x,\rho)\,\mathrm{d}\rho.$$

PROOF. Start from the second link of the chain in the proof of 9.4 to obtain

$$\begin{aligned}
&K\gamma\sqrt{2\beta}\int_0^R \mathrm{e}^{-\rho\sqrt{2\beta}}\left(\int_0^\rho \mathrm{e}^{-\gamma\sqrt{2\beta}s}\,\mathrm{d}s\right)\rho\langle a\rangle(x,\rho)\,\mathrm{d}s\\
&\quad\geqslant K\gamma\sqrt{2\beta}\int_0^R \mathrm{e}^{-\rho\sqrt{2\beta}(\gamma+1)}\rho^2\langle a\rangle(x,\rho)\,\mathrm{d}\rho\\
&\quad\geqslant K\gamma\sqrt{2\beta}\mathrm{e}^{-R\alpha/c_1}\int_0^R \rho^2\langle a\rangle(x,\rho)\,\mathrm{d}\rho.
\end{aligned}$$

Again, it is the starting point that makes this so simple.

To finish with this line of inquiry, we prove that $\tilde{u}(\alpha,x)$ cannot be the same as its Born approximation unless $a\equiv 0$. □

PROPOSITION 9.6. *Assume that our sources are such that* $\tilde{s}(\alpha,x)>0$. *Then* $\tilde{u}(\alpha,x)=\tilde{u}_0(\alpha,x)-\beta E^x\int_0^\infty \mathrm{e}^{-\beta t}a(B(t))\tilde{u}_0(\alpha,B(t))\,\mathrm{d}t$ *if and only if* $a\equiv 0$ *a.c.*

Comment: The positivity condition on $\widetilde{S}(\alpha,x)$ is to insure that $\tilde{u}_0(\alpha,x)>0$. Usually sources are such that $\tilde{u}_0(\alpha,x)$ is a positive sum of terms like $\mathrm{e}^{-(\alpha/c)|x-y|}/|x-y|$.

PROOF. If $a\equiv 0$, there is not much to verify. Conversely, assume that $\tilde{u}(\alpha,x)=\tilde{u}_0(\alpha,x)-E^x\int_0^\infty \mathrm{e}^{\beta t}a(B(t))\tilde{u}_0(\alpha,B(t))\,\mathrm{d}t$. Therefore

$$E^x\int_0^\infty \mathrm{e}^{-\beta t}\big(1-\mathrm{e}^{-\beta c(t)}\big)a(B(t))\tilde{u}_0(\alpha,B(t))\,\mathrm{d}t=0.$$

Since everything under the integral signs is positive and $\tilde{u}_0(\alpha,x)>0$, we conclude $a(x)=0$, for $B(t)$ sweeps the whole of $\mathbb{R}^3$. □

9.4.1. Inversion of the Born approximation

This subsection is designated to the issue of finding $a(x)$ starting from the Born approximation.

Assume we are given $\tilde{u}(\alpha,x)$ and we know how to compute $\tilde{u}_0(\alpha,x)$ for $\alpha>0$ and $x\in\partial D$. We want to use

$$c_0^2\delta(\alpha,x)=\frac{1}{2}E^x\int_0^\infty \mathrm{e}^{-\beta t}a(B(t))\tilde{u}_0(\alpha,B(t))\,\mathrm{d}t, \tag{9.25}$$

where $\delta(\alpha, x) = (\tilde{u}_0(\alpha, x) - \tilde{u}(\alpha))/\alpha^2$, to solve for $a(x)$. Again, set

$$\langle a\tilde{u}_0(\alpha, 0)\rangle(x, \rho) = \frac{1}{4\pi}\int_{|\xi|=1} a(x+\rho\xi)\tilde{u}_0(\alpha, x+\rho\xi)\,\mathrm{d}\sigma(\xi).$$

With this (9.25) becomes

$$\begin{aligned} c_0^2\delta(\alpha, x) &= \int \frac{\mathrm{e}^{-\frac{\alpha}{c_0}|x-y|}}{4\pi(x-y)} a(y)\tilde{u}_0(\alpha, y)\,\mathrm{d}y \\ &= \int_0^\infty \rho \mathrm{e}^{-\frac{\alpha}{c_0}\rho}\langle a\tilde{u}_0(\alpha, 0)\rangle(x, \rho)\,\mathrm{d}\rho \end{aligned}$$

and setting $\rho/c_0 = s$

$$\delta(\alpha, x) = \int_0^\infty s\mathrm{e}^{-\alpha,s}\langle a\tilde{u}_0(x, \cdot)\rangle(x, c_0 s)\,\mathrm{d}s. \tag{9.26}$$

Now, note that

$$\mathrm{e}^{-\alpha,s}\tilde{u}_0(\alpha, x) = \int_0^\infty \mathrm{e}^{-\alpha t}\theta(t-s)u_0(t-x, x)\,\mathrm{d}t$$

which can be inserted in (9.26) and, having the integrals over s and t exchanged, we obtain

$$\delta(\alpha, x) = \int_0^\infty \mathrm{e}^{-\alpha t}\left(\int_0^t s\langle au_0(t-s, \cdot)\rangle(x, c_0 s)\,\mathrm{d}s\right).$$

Next, denoting by $\delta(t, x)$ the Laplace transform of $\delta(\alpha, x)$, we have

$$\delta(t, x) = \int_0^t s\langle au_0(t-s, \cdot)\rangle(x, c_0 s)\,\mathrm{d}s \tag{9.27}$$

which looks pretty close to (9.10). To simplify further, we consider a situation ten of described in the literature, not readily deduced from the above.

Assume that the source term in (9.1) is of the type $s(t, x) = Q\delta(t)\delta(x - x_0)$. Now, to get to the analogues of (9.5) or (9.6), we have to proceed differently.

If we denote by $\rho_0(t, x, x_0)$ the transition density of the imposed motion reflected on ∂D, then it satisfies

$$\frac{\partial\rho}{\partial t}(t, x, x_0) = \frac{1}{2}\Delta\rho_0(t, x, x_0)$$

with the initial condition $\rho_0(t, x, x_0) \to \delta(x - x_0)$ as $t \downarrow 0$, and the boundary condition $\frac{\partial\rho_0}{\partial\eta} = 0$ for $t > 0$, $x \in \partial D$. In terms of $\rho_0(t, x, x_0)$ one defines (with the notation adapted to the problems we deal with)

$$\tilde{u}_0(\alpha, x, x_0) = \int_0^\infty \mathrm{e}^{-\beta t}\rho_0(t, x, x_0)\,\mathrm{d}t, \tag{9.28}$$

where, recall, $\beta = \frac{1}{2}(\frac{\alpha}{c_0})^2$. Clearly

$$\left(\beta - \frac{1}{2}\Delta\right)\tilde{u}_0(\alpha, x, x_0) = \delta(x - x_0). \tag{9.29}$$

To obtain the solution to

$$\left(\beta q(x) - \frac{1}{2}\Delta\right)\tilde{u}(\alpha, x, x_0) = \delta(x - x_0), \tag{9.30}$$

one could iterate either of

$$\tilde{u}(\alpha, x, x_0) = \tilde{u}_0(\alpha, x, x_0) - \beta \int \tilde{u}_0(\alpha, x, y)a(y)\tilde{u}(\alpha, y, x_0)\, \mathrm{d}y \tag{9.31}$$

or

$$\tilde{u}_0(\alpha, x, x_0) = \tilde{u}(\alpha, x, x_0) - \beta \int \tilde{u}(\alpha, x, y)q(y)\tilde{u}_0(\alpha, y, x_0)\, \mathrm{d}y \tag{9.32}$$

which is just a rewritten version of (9.26). The last term of (9.32) is just

$$E^x \int_0^\infty \mathrm{e}^{-\beta t} a(x(t))\tilde{u}(\alpha, B(t)x_0)\, \mathrm{d}t.$$

It is not difficult to see, but not entirely trivial that $\rho_0(t, x, x')$ is symmetric in x, x_1, and so are $\tilde{u}_0(\alpha, x, x')$ and $\tilde{u}(\alpha, x, x')$.

Anyway, the Born approximation is obtained iterating either of (9.31) or (9.32) and dropping the remainder. Also, observe that, if one changes the right-hand side of (9.30), say to $\frac{2q(x_))}{c_0^2}\delta(x - x_0)$, one has to multiply $\tilde{u}(\alpha, x, x_0)$ there by $2q(x_0)/c_0^2$ to obtain the correct solution.

To obtain the analogue of (9.27), rewrite (9.31) or (9.32) iterated once as

$$c_0^2\delta(\alpha, x, x_0) = \frac{1}{2}E^x \int_0^\infty \mathrm{e}^{-\beta t} a(B(t))\tilde{u}_0(\alpha, B(t), x_0)\, \mathrm{d}t,$$

where implicitly we have gone back to the flat Earth model and we are extending all functions to $\mathbb{R}^3$ by reflection in $\partial D = \{z_3 = 0\}$. Also, $\delta(\alpha, x, x_0) = (\tilde{u}_0(\alpha, x, x_0) - \tilde{u}(\alpha, x, x_0))/\alpha^2$.

Now things are easy, for the solution to (9.31) is just $\mathrm{e}^{-\frac{\alpha}{c_0}|x - x_0|}/(2\pi|x - x_0|)$, and the problem thus becomes to find $a(y)$ in

$$c_0^2\delta(\alpha, x, x_0) = \frac{1}{2(2\pi)2} \int \frac{\mathrm{e}^{\frac{-\alpha}{c_0}|x-y|}\mathrm{e}^{-\frac{\alpha}{c_0}|x_0-y|}}{|x - y||x_0 - y|} a(y)\, \mathrm{d}y$$

and a further simplification is brought about when we let $x \to x_0$, which physically means to observe the wave field at the point where the source is located.

Let us for, the time being, assume that the limit

$$\lim_{x \to x_0} c_0^2 \delta(\alpha, x, x_0) = c_0^2 d(\alpha, x_0) \tag{9.33}$$

exists (and is necessarily a Laplace transform). Then our problem becomes: find $a(y)$ such that

$$c_0^2 d(\alpha, x_0) = \frac{1}{2\pi} \int \mathrm{e}^{-\alpha\rho} \langle a \rangle (x_0, 2c_0\rho)\, \mathrm{d}\rho.$$

If we denote by $d(t, x_0)$ the inverse Laplace transform of $d(\alpha, x_0)$, then our problem becomes to find $a(y)$ such that

$$\langle a \rangle (x_0, 2c_0 t) = 2\pi c_0 d(t, x_0)$$

which, after another change of variables, becomes

$$\langle a \rangle (x_0, z) = 2\pi c_0 d\left(\frac{z}{2c_0}, x_0\right)$$

for $x_0 \in \partial D$ and $z \in \mathbb{R}$ (or $[0, \infty)$ to be precise). We considered such problems in Section 1.

To verify that singularities cause no problem in (9.33), one just iterates (9.31) or (9.32) to obtain

$$\tilde{u}_0(\alpha, x, x_0) - \tilde{u}(\alpha, x, x_0) = \sum_{n \geqslant 1}^{\infty} I_n(x, x_0),$$

where each $I_n(x, x_0)$ is an expression like

$$\int \frac{\mathrm{e}^{-\alpha|x-y|}}{|x-y|} F_n(y, -, z) \frac{\mathrm{e}^{-\gamma|z-x_0|}}{|z-x_0|}\, \mathrm{d}y\, \mathrm{d}z$$

and $F_n(y, -, z)$ may consist of a bunch of similar integrals. After verifying the convergence of the series for $x \neq x_0$, one sets $x = x_0$ to obtain

$$\int \frac{\mathrm{e}^{-\gamma|x_0-y|}}{|x_0-y|} F_n(y, -, z) \frac{\mathrm{e}^{-\gamma|z-x|}}{|z-x_0|}\, \mathrm{d}y\, \mathrm{d}z$$

and, passing to spherical coordinates centered at x_0, the last integral becomes

$$\int \eta \mathrm{e}^{-\gamma\eta} F_n(\eta\hat{y}, -, \zeta\hat{z}) \zeta \mathrm{e}^{\gamma\zeta}\, \mathrm{d}\eta\, \mathrm{d}\sigma(\hat{y})\, \mathrm{d}\zeta\, \mathrm{d}\sigma(\hat{z})$$

which is non-singular. In other words, the singularity of $\tilde{u}(\alpha, x, x_0)$ is already contained in $\tilde{u}_0(\alpha, x, x_0)$ and $\delta(\alpha, x, x_0)$ contains no singularity.

9.5. Scattering by a bounded object

Consider a domain D_0, which could be either $\mathbb{R}^3$, a half space, or some unbounded domain. Inside $\overline{D}_0$ we place some sources described by $s(t, x)$ and let ∂D_0 denote the boundary of our domain. Assume it smooth enough.

Let us denote by $u_0(t, x)$ the wave produce in D_0 by $s(t, x)$, i.e., $u_0(t, x)$ satisfies

$$\partial_{tt}^2 u_0 = c^2 \Delta u_0 + s$$

with $u_0(0, x) = 0$, $u_t(0, x) = 0$ and $\frac{\partial u_0}{\partial \eta}(t, x) = 0$ for $t > 0$, $x \in \partial D_0$. This last condition will be present when ∂D_0 is not empty, of course. Also, c denotes the velocity of propagation of the waves inside D_0, and we assume it to be constant. We have seen repeatedly that, if $X(t)$ denotes the imposed motion in $\overline{D}_0$, reflected upon reaching ∂D_0, then $\tilde{u}_0(\alpha, x)$ can be written as

$$\tilde{u}_0(\alpha, x) = E^x \left\{ \int_0^\infty e^{-\beta t} \tilde{s}(\alpha, x(t))\, dt \right\} / 2c^2$$

with $\beta = \frac{1}{2}(\frac{\alpha}{c})^2$.

Consider now some bounded domain $D \subset D_0$, and assume that it is soft, i.e., when illuminated with $u_0(t, x)$, the resulting field $u(t, x)$ inside D_0 will vanish on ∂D. We assume that the boundary ∂D is regular, i.e., if $T = \inf\{t > 0 : X(t) \notin D_0 - \overline{D}\}$, then $P^x(T = 0) = 1$ for all $x \in \partial D$.

The resulting wave field $u(t, x)$ inside $D_0 - \overline{D}$ satisfies

$$\frac{\partial^2}{\partial t^2} u(t, x) = c^2 \Delta u(t, x) + s(t, x), \quad t > 0, x \in D_0 - \overline{D}$$

with homogeneous, initial conditions plus the boundary conditions

$$u(t, x) = 0, \quad x \in \partial D, \qquad \frac{\partial u}{\partial n}(t, x) = 0, \quad x \in \partial D_0.$$

Again, the last condition will be absent when $D_0 = \mathbb{R}^3$ and $X(t) = B(t)$, in this case.

If we invoke the superposition principle and write $u(t, x) = u_0(t, x) + v(t, x)$, then $v(t, \alpha)$ is compelled to satisfy

$$\frac{\partial^2 v}{\partial t^2}(t, x) = c^2 \Delta v(t, x), \quad t > 0, x \in D_0 - \overline{D}$$

with homogeneous initial conditions plus the boundary conditions

$$v(t, x) = -u_0(t, x) \quad \text{for } x \in \partial D \quad \text{and} \quad \frac{\partial v}{\partial n}(t, x) = 0 \quad \text{for } x \in \partial D_0,$$

the last condition being void as $D_0 = \mathbb{R}^3$. If we take Laplace transforms, the last problem becomes

$$\left(\beta + \frac{1}{2}\Delta\right)\tilde{v}(\alpha, x) = 0 \tag{9.34}$$

for $\alpha > 0$, $x \in D_0 - \overline{D}$; su 'plemented with

$$\tilde{v}(\alpha, x) = -\tilde{u}(\alpha, x \quad \in \partial D, \qquad \frac{\partial k}{\partial n}(\alpha, x) = 0, \quad x \in \partial D_0. \tag{9.35}$$

LEMMA 9.1. *Assume* *is regular and that the solution to* (9.34) *exists as a function in* $C^2(D - \overline{D}) \cap C^1(C$ *Then, for* $\alpha > 0, x \in D - \overline{D}$,

$$\tilde{v}(\alpha, x) = -E^x\{e \quad , X(T))\},$$

where $T = \inf\{t > 0 : X \quad D\}$.

Comments: We do not explicitly integrate over the set $\{T < \infty\}$, because the exponential factor causes the integrand to vanish over $\{T = \infty\}$. Also, continuity of the trajectories of $X(t)$ permits us to equate t hitting time of $\overline{D}$ with that of ∂D, for trajectories starting in $D_0 - \overline{D}$.

PROOF. Consider the function $e^{\beta t}\tilde{v}(\alpha, X(t))$. According to Ito's formula in Chapter 2, for α

$$\begin{aligned} d\left(e^{-\beta t}\tilde{v}(\alpha, x( ))\right) &= \left(-\beta + \frac{1}{2}\Delta\right)e^{-\beta t}\tilde{v}(\alpha, X(t))\,dt + \nabla\tilde{v}(\alpha, X(t)) \cdot dX_t \\ &= \nabla\tilde{v}(\alpha, X(t)) \cdot dB(t) + \nabla\tilde{v}(\alpha, X(t)) \cdot n(X(t))\,dt_t \\ &= \nabla\tilde{v}(\alpha, X(t) \cdot dB(t)). \end{aligned}$$

We have made use of the fact that $\tilde{v}(\alpha, X)$ satisfies (9.34), $dx_t = dB_t + n(X(t))\,dL_t$, and that $\partial\tilde{v}/\partial n = \nabla\tilde{v} \cdot n$, which combined with the fact that L_t grows only when $X(t) \in \partial D_0$, yield the desired identity. Now, since the last term defines a martingale, it will drop when taking expectations. To ensure proper application of the optional sampling theorem, we consider the sequence of bounded stopping times $T \wedge n$ to obtain

$$\tilde{v}(\alpha, x) = E^x\left[e^{-\beta(T \wedge n)}\tilde{v}(\alpha, X(T \wedge n))\right]$$

which is a kind of averaging property for $\tilde{v}(\alpha, x)$. Now, as $n \to \infty$, $T \wedge n \uparrow T$, $\tilde{v}(\alpha, X(T \wedge n)) \to \tilde{v}(\alpha, X(T)) = -\tilde{u}_0(\alpha, X(t))$ on $\{T < \infty\}$, thus continuity yields

$$\tilde{v}(\alpha, x) = -E^x\left[e^{-\beta T}\tilde{u}_0(\alpha, X(T))\right]. \qquad \square \tag{9.36}$$

Comment: When $D_0 = \mathbb{R}^3$, $X(T)$ is replaced by $B(T)$, and (9.36) satisfies the Sommerfeld radiation condition!

Exercise: We leave for the reader to verify that

$$\tilde{u}(\alpha, x) = \frac{2}{2c^2} E^x \left\{ \int_0^T e^{-\beta t} \tilde{s}(\alpha, X(t))\, dt \right\} \tag{9.37}$$

solves $(\beta - \frac{1}{2}\Delta)\tilde{u} = \tilde{s}(\alpha, x)/2c^2$ with $\tilde{u}(\alpha, x) = 0$ on ∂D and $\frac{\partial u}{\partial n}(\alpha, x) = 0$ on ∂D. To make it fun, start from (9.36) plus $\tilde{u}(\alpha, x) = \tilde{u}_0(\alpha, x) - \tilde{v}(\alpha, x)$.

Also note that (9.37) allows us to say that $\tilde{u}(\alpha, x) = 0$ for x in D as well. Just because of the definition of T.

Let us append the subscript D to $\tilde{u}(\alpha, x)$ to explicitly show the obstacle we are illuminating. Thus $\tilde{u}_D(\alpha, x)$ is given by (9.37). Now let D_1 and D_2 be two open domains. It is simple to verify that

$$|u_{D_1}(x) - u_{D_2}(x)| = \frac{1}{2c^2} \left| E^x \int_{T_1 \wedge T_2}^{T_1 \vee T_2} e^{-\beta t} \tilde{s}(\alpha, x(t))\, dt \right|. \tag{9.38}$$

PROPOSITION 9.7. *Let $s(t, x)$ be a positive function and s_0 a closed subset of $\overline{D}_0$ such that $s_m I_{S_0}(x) \leqslant \tilde{s}(\alpha, x) \leqslant s_M I_{S_0}(x)$ for all $\alpha > 0$, $x \in D_0 - \overline{D}$, for appropriate $s_m < s_M$. Then*

$$\begin{aligned} \frac{s_m}{2c_0^2} E^x \left[\int_{T_1 \wedge T_2}^{T_1 \vee T_2} e^{-\beta t} I_{S_0}(x_t)\, dt \right] &\leqslant |u_{D_1}(x) - u_{D_2}(x)| \\ &\leqslant \frac{s_M}{2c^2} E^x \left[\int_{T_1 \wedge T_2}^{T_1 \upsilon T_2} e^{-\beta t} I_{S_0}(x_t)\, dt \right]. \end{aligned} \tag{9.39}$$

PROOF. Quite obvious starting from (9.38). □

Comments: Note that $\{T_1 \wedge T_2 > T_1 \wedge T_2\} = \{T_1 > T_2\} \cap \{T_2 > T_1\}$.

Since $T_1 \wedge T_2 = T(\overline{D} \cup \overline{D}_2)$ the hitting time of $\overline{D}_1 \cup \overline{D}_2$, and since $\delta_{12} = \int_{T_1 \wedge T_2}^{T_1 \upsilon T_2} I_{S_0}(X(t))\, dt$ is the time spent by $X(t)$ at the support S_0 of $\tilde{s}(\alpha, x)$ between successive hits of D_1 and D_2, or D_2 and D_1, we see that, if $\delta_{12} = 0$ with probability 1, then $u_{D_1}(x) = u_{D_2}(x)$. The reciprocal is true as well.

Also, any form of convergence of $D_1 \to D_2$ such that $T_1 \vee T_2 \to T_1 \wedge T_2$ would yield continuity of the mapping $D \to u_D$. (For a fixed source, or a class of sources.)

An open problem of interest for uniqueness question may go as follows. Assume that for a subset of $\overline{D}_0 - \overline{D_1 \cup D_2}$, we know that $u_{D_1}(x) = u_{D_2}(x)$, and S_0 is fat enough, that is, $P^x(X(t) \in S_0) > 0$ for all x and all $t > 0$, can we conclude that $P^x(T_1 > T_2$ or $T_2 > T_1) = 0$? And hence $D_1 = D_2$?

To consider a setup closer to standard practice, let $D_0 = \mathbb{R}^3$. Now (9.36) becomes

$$\tilde{v}(\alpha, x) = -E^x\big[\tilde{u}(\alpha, B(T))e^{-\beta T}\big] \tag{9.40}$$

and the analogue of (9.37) is now

$$\tilde{u}(\alpha, x) = \frac{1}{2c^2} E^x \left\{ \int_0^T e^{-\beta t} \tilde{s}(\alpha, B(t))\, dt \right\}. \tag{9.41}$$

The comments made above apply as well.

Assume that $\tilde{s}(\alpha, x)$ can be recovered from its Fourier transform according to

$$\tilde{s}(\alpha, x) = \int e^{ikx} \hat{s}(\alpha, k) \frac{dk}{(2\pi)^3}.$$

Then $\hat{u}_0(x) = \hat{u}_0(\alpha, k, x) = \frac{e^{ikx}}{\alpha^2 + (kc)^2}$ satisfies

$$\left(\beta - \frac{1}{2}\Delta\right)\hat{u}_0(x) = \frac{1}{2c^2} e^{ik\cdot x}$$

and $\tilde{u}_0(\alpha, x) = \int \hat{s}(\alpha, k)\hat{u}_0(x, \alpha, k)\frac{dk}{(2\pi)^3}$ is the Laplace transform of the wave produced by the source to illuminate our object. The corresponding $\tilde{v}(\alpha, x, k)$ this time is

$$\tilde{v}(\alpha, k, \alpha) = -\frac{1}{\alpha^2 + (kc)^2} E^x \left[e^{-\beta T} e^{ik\cdot B(T)}\right]. \tag{9.42}$$

If we denote by $\hat{u}(\alpha, k, x)$ the solution to

$$\left(\beta - \frac{1}{2}\Delta\right)\hat{u}(\alpha, k, x) = \frac{e^{ik\cdot x}}{2c^2},$$

then, as above,

$$\hat{u}(\alpha, k, x) = \frac{1}{2c^2} E^x \left\{ \int_0^T e^{-\beta t} e^{ik\cdot B(t)}\, dt \right\} \tag{9.43}$$

and if we know $\delta(\alpha, k, x) = \hat{u}_0(\alpha, k, x) - \hat{u}(\alpha, k, x)$ as a function of k, we can, in principle, invert the Fourier transform in (92) and obtain the support of the hitting distribution of $B(T)$, i.e., we can obtain ∂D.

The study of the mapping $D \to U_D()$, using (9.42)/(9.43), is still an open business.

9.5.1. *Point sources*

Let us consider another case which is rather common, just because of the simplification it brings in. A point source is usually described mathematically by a distribution like $s(t, x) = f(t)\delta(x - x_0)$ for some $x_0 = D_0$. We consider $D_0 = \mathbb{R}^3$, but the extension to $D_0 = \{x \in \mathbb{R}^3 : x_3 \geqslant 0\}$ is rather simple.

We have repeatedly seen that $u_0(\alpha, x, x_0) = \frac{e^{-\frac{\alpha}{c}|x-x_0|}}{2\pi|x-x_0|}\widetilde{F}(\alpha)$ satisfies

$$\left(\beta - \frac{1}{2}\Delta\right)\tilde{u}_0(\alpha, x, x_0) = \tilde{f}(\alpha)\delta(x - x_0),$$

where $F(\alpha) = \tilde{f}(\alpha)/2c^2$. The analogue of Lemma 1 and (9.35)–(9.36) being now

LEMMA 9.2. *Assume that there exists a $\tilde{v}(\alpha, x, x_0)$ in $C^2(D_0 - \overline{D}) \cap C^1(D_0 - \overline{D})$ such that*

$$\left(\alpha - \frac{1}{2}\Delta\right)\tilde{v}(\alpha, x, x_0) = 0$$

for $\alpha > 0$, $x \in D_0 - D$, $\tilde{v}(\alpha, x, x_0) = -\tilde{u}_0(\alpha, x, x_0)$, $x \in \partial D$. Then

$$\tilde{v}(\alpha, x, x_0) = E^x\left[e^{\beta T}\tilde{u}(\alpha, B(T)x_0)\right], \tag{9.44}$$

where $T = \inf\{t > 0 : B(t) \in \partial D\}$.

PROOF. For the reader. □

Comments: Note, again, that due to the presence of the exponential, we do not worry about what happens on $\{T = \infty\}$. Also, we saw (where?) that (9.44) satisfies the Sommerfeld radiation condition, which makes $\tilde{v}(\alpha, x, x_0)$ unique.

Let us put $\tilde{u}(\alpha, x, x_0) = F(\alpha)\tilde{w}(x, x_0) + \tilde{v}(\alpha, x, x_0)$, therefore $\tilde{w}(\alpha, x, x_0)$ satisfies

$$\left(\beta - \frac{1}{2}\Delta\right)\tilde{w}(\alpha, x, x_0) = \delta(x - x_0) \tag{9.45}$$

with the boundary condition $\tilde{w}(\alpha, x, x_0) = 0$ for $\alpha > 0$, $x \in \partial D$.

If we set $\delta(\alpha, x, x_0) = \frac{e^{-\frac{\alpha}{c}|x,x_0|}}{2\pi|x,x_0|} - \tilde{w}(\alpha, x, x_0)$, then

$$\delta(\alpha, x, x_0) = \frac{1}{2\pi}E^x\left[\frac{e^{-\beta T}e^{-\frac{\alpha}{c}|B(T)-x_0|}}{|B(T) - x_0|}\right]. \tag{9.46}$$

If we denote by $\rho(x, s, y)$ the density of the point probability distribution of $(T, B(T))$ with respect to P^x, then (42) looks like

$$\delta(\alpha, x, x_0) = \int_0^\infty ds \int_D e^{-\beta s}\frac{e^{-\frac{\alpha}{c}|y-x_0|}}{2\pi|x - x_0|}\rho(x, s, y)\, d\sigma(y).$$

The inverse problem now reduces to finding the support of $\int_0^\infty e^{-\beta s}\rho(x, s, y)\, ds$ from the knowledge of $\delta(\alpha, x, x_0)$. Even though this may be possible in theory, in practice it may be hard to get.

But we can console ourselves with some estimates.

Denote by $d(D, x_0)$ and $d(D, x_0) + \ell$ the distances from x_0 to D and to the point in $\overline{D}$ further away from x. Then

$$\frac{\mathrm{e}^{-\frac{\alpha}{c_0}(d(x_0,D)+\ell)}}{2\pi d(D, x_0) + \ell} E^x\big[\mathrm{e}^{-\beta T}\big] \leqslant \delta(\alpha, x, x_0) \leqslant \frac{\mathrm{e}^{\frac{\alpha}{c_0}d(x_0,D)}}{2\pi d(D, x_0)} E^x\big[\mathrm{e}^{-\beta T}\big].$$

If $\ell/d(D, x_0)$ is small compared to 1, then

$$\frac{\mathrm{e}^{-\frac{\alpha}{c}d(D,x_0)}}{2\pi d(D, x_0)}\left(1 - \frac{\ell}{d(D, x_0)}\right) E^x\big[\mathrm{e}^{-\beta T}\big] \leqslant \delta(\alpha, x, x_0) \leqslant \frac{\mathrm{e}^{-\frac{\alpha}{c}d(x_0,D)}}{2\pi d(x_0, D)} E^x\big[\mathrm{e}^{-\beta T}\big]$$

from which some interesting conclusions can be drawn:

(i) when D is "flat" and happens to be contained in a spherical cap centered at x_0, clearly

$$\delta(\alpha, x, x_0) = E^x\big[\mathrm{e}^{-\beta t_1}\big]\frac{\mathrm{e}^{-\frac{\alpha}{c_0}} d(x_0, D)}{2\pi d(x_0, D)}.$$

(ii) If the imposed motion does not "see" $\overline{D}$, i.e., as $T = \infty$, then $\delta(\alpha, x, x_0) = 0$. Thus, the only type of nonuniqueness in using (9.42) to solve for $\operatorname{supp} \partial D$ may come from the addition of sets to D which are missed by the paths of $B(t)$.

(iii) If our object is bounded and the source is far away from it, then

$$\delta(\alpha, x, x_0) \sim E^x\big[\mathrm{e}^{-\beta T}\big]\frac{\mathrm{e}^{-\frac{\alpha}{c}d(x_0,D)}}{2\pi d(D, x_0)}$$

which provides a probabilistic interpretation for the scattering amplitude plus the fact that details about the surface influence the hitting time of D only.

We close this section with a couple of applications of (9.46). First the basic probabilistic version of a potential theoretic construct.

Wee say that A is a polar set if $P^x(T_A < \infty) = 0$ for any $x \in \mathbb{R}^3$. That is, A is polar if almost all the trajectories of $B(t)$ miss it.

Also that, for any two Borel sets D_1 and D_2, $T_{D_1 \cup D_2} = T_{D_1} \wedge T_{D_2}$, and if D_2 happens to be a polar set, $T_{D_1 \cup D_2} = T_{D_2}$. With this it is clear that the following lemma is true.

LEMMA 9.3. *Let D_1 and D_2 differ by a polar set, then they produce the same scattered field, given by* (9.46).

Comment: Since points and segments are polar for the three-dimensional imposed motion, countably many holes and/or hairs a surface ∂D cannot be detected by an acoustic wave.

A bit of stochastic calculus allows us to complete the following uniqueness result.

LEMMA 9.4. *Let D_1 and D_2 be two bounded domains, with all points of ∂D_i regular for D_i, $i = 1, 2$. Then, if they produce the same scattered field, they differ by a polar set.*

PROOF. If they produce the save scattered field, then

$$\begin{aligned} 0 &= E^x\left[e^{-\beta T_1}\tilde{u}_0(\alpha, B(T_1)x_0)\right] - E^x\left[e^{-\beta T_2}\tilde{u}_0(\alpha_1 B(T_c), x_0)\right] \\ &= E^x\left\{\int_{T_1\wedge T^2}^{T_1\vee T_2} e^{-\beta t}\tilde{u}_0(\alpha, B(t), x_0)\,dt\right\}. \end{aligned}$$

Since the integrand is positive, it means that $T_1 \wedge T_2 = T_1 \vee T_2$ or $T_1 = T_2$ a.s., which, in turn implies that $B(t_1) = B(T_2)$ a.s. Since the support of the distribution of $B(T_i)$ differs from ∂B_i most a polar set, we are through. □

Comment: But if both ∂D_2, are smooth enough, then they are coincidental. This comes about because $P^x(B(t_i) \in d\sigma(y))$ has a strictly positive density, which means that $B(T_i)$ sweeps ∂D_i!

References

Standard references in this field are:

1. Romanov, V. G.: *Inverse Problems of Mathematical Physics*, VNU Science Press, Utrecht, 1987.
2. Romanov, V. G.: *Integral Geometry and Inverse Problems*, Springer-Verlag, Berlin 1974.
3. Bleistein, N., Cohen, J. and Stockwell, J.: *Mathematics of Multidimensional Seismic Imaging, Migration and Inversion*, Springer-Verlag, New York, 2000.
4. Claerbout, J. F.: *Imaging the Earth's Interior*, Blackwell Scientific, 1985.
5. Ramm, A. G.: *Scattering by Obstacles*, Kluwer Acad. Publ., Dordrecht, 1986.

Section 4 is pretty much modeled on

6. Ramm, A. G.: Is the Born approximation good for solving the inverse problem when the potential is small, *J. Math. Anal. Appl.* **146** (1990), 480–485.

A line of work going well beyond the issues treated in Section 5 can be traced back from

7. Colton, D.: Inverse scattering problem for time harmonic waves, *SIAM Rev.* **26** (1984), 323–350.

The following reference is relevant to Section 1

8. Fawcett, J.: Inversion of n-dimensional spherical averages, *SIAM J. Appl. Math.* **45** (1985), 336–341.

The basic stoff for Section 9 appeared in

9. Gzyl, H.: Probabilistic remarks in the theory of scattering by soft obstacles, *Studies in Appl. Math.* **109** (2000), 379–389.

SUBJECT INDEX

MIX
Papier aus verantwortungsvollen Quellen
Paper from responsible sources
FSC® C105338

If you have any concerns about our products,
you can contact us on
ProductSafety@springernature.com

In case Publisher is established outside the EU,
the EU authorized representative is:
Springer Nature Customer Service Center GmbH
Europaplatz 3, 69115 Heidelberg, Germany

Printed by Libri Plureos GmbH
in Hamburg, Germany